Oriental Adventures of the French Air Force 1914–1918

David Méchin

Dedication

The author wishes to dedicate this book to Benoît Henriet, great-nephew of the pilot Paul Andrillon, and who for ten years accompanied the author in the archive rooms of the fort of Vincennes to dissect the fifty or so boxes of the Orient Army, allowing the writing of this work. The author associates with these thanks his father Jean-Claude Henriet who helped with the research.

The author would also like to warmly thank Boris Ciglic, Serbian aeronautical historian (https://wingsofserbia.com) for their numerous exchanges of documents and photographs which have greatly enriched this book.

Finally, among his many contacts among First World War aviation enthusiasts, the author would like to thank the following people for their specific assistance in collecting data on Oriental aviation: MM. Michel Bénichou, Albin Denis, Robert Feuilloy, Lucien Morareau, Alexis Rocher (France), Bernard de Broglio, Colin Owers (Australia), Ian Burns (Canada), Alexis Mehtidis, Paschalis Palavouzis (Greece), and Paolo Varriale (Italy).

The author gives his special thanks to Mr. Paschalis Palavouzis and Aaron Weaver, who helped to proofread this book.

Acknowledgements

For our aviation books in print please see our website at: **www.aeronautbooks.com**

Thanks to Mike O'Neal for the cover painting.

For questions or to help with photographs please contact Aeronaut Books at **jherris@me.com**

ISBN: 978-1-953201-55-3

Design and layout: Jack Herris
Cover design: Aaron Weaver
Digital photo editing: Jack Herris

Aeronaut Books

Books for Enthusiasts by Enthusiasts
www.aeronautbooks.com

Table of Contents

Chapter 1: Precursors: French Aviation Maritime in Montenegro (September–October 1914)

Above: Colorized postcard representing the mouths of Cattaro, views of Mount Lovcen, from the place where the French Navy tried to install cannons to aim at Austro-Hungarian ships. (Author's collection)

1.1 – Against the Austro-Hungarian Fleet

At the dawn of the Great War, Montenegro was a small kingdom of 435,000 inhabitants located on the shores of the Adriatic Sea north of Albania, covered by mountains and poorly equipped with communication routes. On its northwestern border, its powerful neighbor, the Austro-Hungarian Empire, had extended its sovereignty over the mouths of Cattaro (or Kotor), a magnificent mountainous fjord offering shelter to ships of great tonnage, thus making it a natural naval base of the first order for the Austrian navy.

The assassination of Duke Franz Ferdinand (1863–1914) and his wife resulted in the declaration of war by Austria-Hungary against Serbia, which was followed by all other declarations of war that set Europe ablaze. Montenegro was also caught up in the turmoil, because its pro-Serbian public opinion pushed the King to declare war on Austria-Hungary.

The small Montenegrin army was basically composed of poorly-equipped peasant militia with little artillery and no aviation assets. Grouped in four columns, the main part of the forces was massed in the north of the country and was going to join forces with the Serbian army for the purpose of putting the Austro-Hungarian army in difficulty. But this front did not have the primary priority for the Habsburg Empire, which had concentrated most of its forces in Bucovina (Bukowina) facing the Russian giant. The main concern for the Austro-Hungarians was to secure their southern border and more specifically their naval base at Cattaro. For this, the powerful guns of its fleet were supposed to be enough, but also their small air force would work to disrupt the enemy supply...

From the beginning of the international crisis, on July 20, 1914, the Austrian Navy sent three of its major ships,

Above & Facing Page: Bucolic image of the Voisin seaplane n° V.6 of LV Janvier moored at Antivari. The appearances are deceptive: the wind of the Adriatic, the Bura, proves to be formidable and will quickly get the better of the frail French aircraft, which was which will find itself embedded on the roof of a house... Nothing being wasted among French sailors, the half destroyed aircraft recovered, brought back to France France and repaired, ending its career as a land bomber at Dunkirk in 1915! (Collection Boris Ciglic)

the "*Radetzky*", "*Erzherzog Franz Ferdinand*" and "*Zrinyi*" as reinforcements to Cattaro. Each one carried on deck a Löhner seaplane, coded E 18, E 20, and E 21 with their respective pilots, the Fregattenleutnant (Frglt.) Konstantin Maglic, the Linienschiffsleutnant (Lschlt.) Wenzel Wosecek who commanded the detachment, and Fregattenleutnant Gottfried Banfield, who would become the most famous pilot of the Austrian Navy. The three aircraft were landed at the Teodo arsenal on July 24 and were transferred on July 29 to Kumbor on the North Shore of the Mouths, sheltered from possible enemy artillery fire; from there they flew to familiarize themselves with the area, usually taking an artillery officer as an observer.

War was declared on August 8, 1914, and the Austrian navy quickly carried out a total naval blockade of Montenegro. The three Austrian seaplanes were used to guide the fire of the navy's guns towards the heights where the Montenegrins were. On August 15, they tried their hand at tactical support when the E 18 piloted by Frglt. Konstantin Maglic and taking Oberleutnant (Oblt.) Pieler as observer, dropped two bombs on the enemy positions in Krstac, on the heights of Cattaro. It was the first war action of Austrian naval aviation.

But the Allies would not remain idle. France, which by a naval agreement with England had concentrated most of its fleet in the Mediterranean to face Austria, responded immediately by sending a squadron to the aid of its ally, with the intention of taking on the Austrian navy... On August 16, the French ships were in sight of the coast of Montenegro and a deluge of artillery rained down on the S.M.S. *Zenta*, an Austrian destroyer cruising off the coast of Antivari (today Bar in Montenegro). All the Austrian fast patrol boats quickly returned to the mouth of Cattaro to take shelter, and were thus caught in a trap, although the Austrian Navy's submarines and fast ships did not hesitate to make sorties... For Admiral Anton Haus (1851–1917), commander of the Austrian navy, it would be useless as well as suicidal to attempt a sortie against the French ships: his fleet was of greater service to remain a potential threat to the enemy, who were far from their bases: the nearest supply point was located in Malta... The opposing fleets battled for securing their positions and the confrontation reached a stalemate. and the French navy contribution was thus limited to helping the Montenegrin troops as best it could by landing heavy artillery pieces and by offering the services of their aviation.

1.2 - The Voisin in Antivari (6 September – 28 September 1914)

On the morning of September 6 at 8:30 am, the hydroplane carrier "*Foudre*" launched the Voisin V.7 piloted by Lieutenant Fournié off Antivari. The sea was agitated by a light swell which did not reach 1m of hollow, but which was very short. The aircraft was started up at full power and tried to take off, but hit hard in the swell, broke its front float, and ended up capsizing: it could not be repaired by local means. The V.6 piloted by LV Janvier was luckier and managed to reach Antivari after a 1h10 flight. Janvier was joined by the

destroyer *Faulx*, which landed two mechanics and some spare parts during the day.

The Voisin was hoisted onto the beach, and the three men were left to their own devices, as the *Faulx* joined the French fleet offshore... The Austrians greeted its arrival in their own way by bombing the villa of the King of Montenegro with the Löhner E 18 piloted by Frglt. Konstantin Maglic, taking along an observer. Antivari was only a small fishing port without any military installation. There was a jetty to protect it from the waves and, on the quay, a small crane served by personnel of an Italian company, which Janvier decided not to use because of its weak wing, which would make any seaplane hoisting risky. He then had his aircraft hoisted by hand onto the sandy beach. On September 7, he conducted a short 25-minute flight in search of floating mines.

The next day, the Löhner E 18 dropped a few bombs on the port warehouses, without much success, but the civilian population was beginning to worry about this aerial ballet. Thunderstorms and rains forbade any new flights until September 13, when Janvier could take off for a short test flight, his aircraft being very tired from the bad weather that accumulated during two one-month stays on the *Foudre*... The varnish of the wings was flaking and the rudder was warped. On September 15, he could not even take off! It was only the next day that his aircraft, properly revised, managed to rise and allowed him to fly over Cattaro where the Austro-Hungarian fleet was located. He counted the ships lined up there and returned to Antivari after an hour and a half flight. He crossed the route of the E 20 seaplane piloted by Frglt. Gottfried Banfield and carrying Lschlt. Erich Edler von Kunsti as an observer, returning from a bombing mission on Sutomore. The two Austrians saw the enemy Voisin seaplane, but did not engage in combat, delaying the first aeronaval confrontation in history...

On September 18, 180 French sailors disembarked with six pieces of heavy artillery, which the men practically climbed to Mount Lovcen (1759m). On September 19, while the French and Austrian navies were bombing land positions, Janvier left for another hour and a half flight to Cattaro. Mechanical problems combined with bad weather tied the Voisin back to its beach, from which it did not leave until the 26th for a short 20-minute test flight. A strong wind came up, forcing Janvier to moor his aircraft to four piles planted in the sand.

Two days later, the aircraft was still in this position, its engine dismantled by the mechanics to revise it. After a lunch break in town, the small French detachment returned to the beach to discover the Voisin embedded on the roof of a nearby building... The aircraft had the tail twisted and the wings torn off, it was irreparable by local means and once the wind had died down all the parts were recovered that can be. This was the end of the first French detachment that returned to the port of Antivari at the first passage of an allied ship...

During this period, the wind also blew at the Austrians who, on September 14, saw two of their canvas hangars fly away, damaging the E 18 seaplane. Lschlt. Wenzel Woseceka left Kumbor for Pola on the E 21 and was replaced by Lschlt. Hugo Ockermüller arrived from Pola on the E 17, he took command of the air detachment after a short interim by Frglt. Banfield, who was himself recalled to Pola on his E 20 and replaced by the Lohner E 33 piloted by Lschlt. Glauko Prebanda. This small air force would be used intensively. On October 2, the E 18 piloted by Lschlt. Hugo Ockermüller and taking along Frglt. Observer Heinrich Fontaine von Felsenbrunnen bombed Budua, then mounted by the crew Lschlt. Prebanda and Oblt. Pieler, conducted a reconnaissance on Mount Lovcen which the guns landed by the French are beginning to climb... To try to disrupt these deliveries of material to the Montenegrins, the Austrians organized the next day a "mass" bombardment with three aircraft on the station of Antivari, targeting the only

Voisin "13,50 m" n° V.6 of Lieutenant de vaisseau Émile Janvier, Antivari (Monténégro) September 1914. (David Méchin)

locomotive that was there, without any great results other than to panic the civilian population.

During the following days, on October 4 and 9, the E 33 led by the crew Lschlt. Prebanda and Frglt. Fontaine von Felsenbrunn dropped bombs on the port of Antivari. On 15 October, the French guns were operational on Mount Lovcen and started to fire at the mouths of Cattaro, but the counter-battery fire of the Austrian ships led them to cease fire. The next day, Austrian seaplanes also targeted a radio station.

1.3 - The Nieuports in action (October 17–November 23, 1914)

But the French Navy was not left behind, as it landed two new Nieuport seaplanes at Antivari on 17 October 1914. The leader of the small naval air detachment was Lieutenant de vaisseau Cintré, who had a pilot's licence, but as he had not been trained to fly the Nieuports he could only serve as an observer. The two pilots of the seaplanes coded N 6 and N 14 were LV Destrem and the sailor (soon to be quartermaster) Levasseur. As soon as he arrived in Antivari, Cintré reported that his small detachment was welcomed "with gunfire" by the local population, frightened by the Austrian bombing... who greeted the arrival of the French detachment by carrying

Below: One of the first opponents of the French, Lohner E-18 flown by Frglt. Konstantin Maglic. (Coll. Paolo Varriale)

Above & Below: The Lohner E 20 being put to water. This plane, flown by Frglt. Gottfried Banfield, met French Voisin V 6 in flight on September 16, 1914, but did not engage combat. (Collection Paolo Varriale)

Above: Nieuport seaplane being put to water from the French seaplane carrier *Foudre*. (Author's collection)

out no less than two raids in the course of the day, with in particular the E 33 flown by the Lschlt. Prebanda and Frglt. Fontaine von Felsenbrunn which also attacked the French cruiser "*Waldeck Rousseau*" with bombs.

Cintré had to react and the next day, as soon as an Austrian seaplane was sighted, he took off armed with a pistol as a passenger on the Nieuport N 14 piloted by sailor Levasseur. They chased the enemy, which turned out to be the Löhner E 18, flown by the crew of Lschlt. Prebanda and Fl.Gast. Brenner. Cautious, Prebanda turned around and returned to Kumbor along the coast, pursued by the Nieuport which was travelling at more than 100 km/h! Unfortunately, Cintré and Levasseur heard their engine making disturbing noises eight times. They abandoned the pursuit in Budua and returned to Antivari, where they found that their mad pursuit had cracked a cylinder and broken a piston. The Nieuport N 14 was immobilized for some time and Cintré decided that evening to evacuate Antivari and settle down further east, on the lake of Scutari (currently Shkodër or Skadar), near the village of Virpazar, which was served by the railroad.

The small detachment began its move on 21 October by taking the only railroad that snaked through the mountains; the N 6 of Destrem reached its new base by air. The Austrians received a significant reinforcement that same day: the battleship *Radetzky* broke the blockade and reached Cattaro, taking on its deck the coded Löhner E 34 with its pilot, Lschlt. Franz Mikuleczky. It was the ship's 305 mm guns that were most useful, as they permanently silenced the French battery on Mount Lovcen, forcing the naval gunners to evacuate the position.

On October 23, 1914, Destrem attempted a reconnaissance over Cattaro alone on his Nieuport, which left the waters of Lake Scutari with difficulty. It gained enough height to pass the mountain, but a mechanical problem forced it to land on the lake, controls locked. The Nieuport rolled over and was destroyed, but the engine was unharmed. The French detachment on the lake of Scutari was left without planes! Cintré had only to wait for the train that would bring the N14... In the meantime, he telegraphed his superiors to request the dispatch of other Nieuports stored in Malta.

During this time, Antivari experienced a "massive" bombardment by the entire Austrian air force: the E 33 (Lschlt. Ockermüller/ Hptm. Dvořák), the E 34 (Lschlt. Mikuleczky/ Frglt. Fontaine von Felsenbrunn), and the E 18 (Lschlt. Prebanda/ Oblt. Pieler) dropped 10 bombs on the station, without much result, since the next day a detachment of French sailors brought the Nieuport N14 (which had a

Above: After the failure of the Voisin seaplanes, the Navy disembarked a new contingent of airmen at Antivari to put two Nieuport seaplanes on line. In this picture, the sailors led by LV Destrem (bearded, in the middle) are ready to embark on the Antivari railway to reach the lake of Scutari where the Nieuports should operate. (ARDHAN collection)
Below: In navy blue uniforms appearing dark on this black and white photo, LV Cintré (left) and LV de l'Escaille (right) with Montenegrin officers. (ARDHAN collection)

Nieuport seaplane N 14 with 80 Le Rhône engine, flown by Matelot Levasseur, Antivari (Montenegro), October 18, 1914. (David Méchin)

cracked cylinder) from Antivari to Virpazar. The seaplane was then repaired and a test was carried out by the sailor Levasseur taking Cintré as an observer: it was impossible to take off with two people on this fresh water lake...

The Austrians made new sorties in the following days, attacking the Antivari train with three aircraft on November 1 and 2, 1914. On November 3, the reinforcements requested by Cintré left Malta: the "*Foudre*" embarked the Nieuport section of LV de l'Escaille, including EV Barthélémy de Saizieu (observer) and LV Delage (pilot), to bring them to Scutari. The arrival of the French detachment in Antivari was greeted by the Austrian air force, which bombed the city by night on November 10, as well as the French fleet by day the following day, leading moreover the same day a raid with three aircraft on the small port of Dulcigno (today Ulcinj in Montenegro)!

On November 15, while Antivari suffered a new raid, the LV de l'Escaille signed a report suggesting the return of French aircraft. Virpazar was indeed unsuitable for a war service, and Antivari was vulnerable to raids by the Austrian air force. But above all, the usefulness of an aeronautical service remains to be demonstrated on this front whose main objective, the port of Cattaro, is perfectly observable by the troops on the ground from the surrounding heights...

Following the report of the Escaille, the order to evacuate Montenegro was issued shortly afterwards, on October 23, putting an end to the first "naval aviation" intervention in the history of the French Navy. The Nieuports of the "*Foudre*",

Above: All of the Austro-Hungarian air force – 3 seaplanes – shown on this photo. (Collection Paolo Varriale)

requested by the British Admiralty, were then directed towards the Suez Canal, where they were successfully used to monitor the Turkish troops. If the Austrian naval aviation can be considered as having won this engagement, it is necessary to measure this statement by recalling that the planes of 1914 were only fragile machines with limited offensive capacities, subject to the vagaries of the weather... It was a bad gust of wind which would become the ally of the French navy, by destroying on December 15, 1914 the seaplanes E 33, E 34, and E 37 as well as their hangar!

Always supplied by the French navy, which remained off its coast, Montenegro resisted (without aviation) the Austrians until January 1916 when the Serbian defeat sealed its loss. The mouths of Cattaro remained until the end of the war a major Austrian base for its submarines and seaplanes, and were only occupied by the French Navy after the armistice.

Chapter 2: Nieuport Hydroplanes in Port-Saïd (December 1914–April 1916)

Above: Lieutenant Gustave Delage (right), designer of the Nieuport monoplane seaplanes, pictured with Lieutenant De L'Escaille, head of the Port Said Maritime Aviation Center where Delage served as a pilot at the end of 1914, before returning to France in January 1915 and taking over the general management of the Nieuport company. (Coll Geraud Guillerez)

While the Nieuports of the French navy flew in Montenegro, the Ottoman Empire joined Germany and Austro-Hungary in the war. With an important consequence: Egypt under British rule was threatened by the Turkish troops present in Palestine, and particularly the Suez Canal which was only protected by the Sinai Peninsula, 200 km wide and entirely deserted. Admiral Carden, head of the British squadron in the Eastern Mediterranean, then asked Admiral Boué de Lapeyrère, the Chief of Staff of the French Navy, for airborne resources to monitor the approaches to the Suez Canal. The latter, somewhat disappointed by the poor results of his aircraft in Montenegro, was easily convinced, especially since their use in Egypt was precisely a proposal that the LV de l'Escaille had made to him in a report dated November 24, 1914. Gathered in Malta were the Nieuports N7, N11, N12, and N13 from Bizerte plus the N14 from Montenegro. They were convoyed by La *Foudre* to Port-Saïd on 30 November 1914, under the orders of Lieutenant de vaisseau de l'Escaille and with Lieutenant de vaisseau Destrem and Quartier Maître Hervé Grall as pilots, as well as two reservists specializing in the Nieuport, seaman Julien Levasseur and Gustave Delage, the director of the Nieuport company, who had been recalled to the armed forces with his rank of lieutenant de vaisseau. In addition, the detachment included two lieutenants de vaisseaux, Louis Barthélémy de Saizieu and Alfred Cintré, both licensed civilian pilots but not qualified on seaplanes, who served initially as observers.

The squadron settled on a quay in Port Said on December 2 and was soon ready to work under the orders of the British Admiralty. On December 5th the first flight was made by Destrem with Cintré as observer who flew 56 km east along the Mediterranean coast towards El-Arish; they saw the canvas of their wings pierced by the fire of Turkish horsemen. The British Admiralty requested that reconnaissance be carried out on the cities of El-Arish (on the Mediterranean coast), on the Gulf of Aqaba (Red Sea) and on the city of Beersheba, in Palestine, 40 km from the shore, which is the obligatory rear base of any army wishing to cross the Sinai. But the Nieuports did not have the performance to carry out such distant missions, as a British observer, Lt Kenneth Williams, testifies: "*The maximum speed was about 85 km/h at sea level, with a ceiling of about 1500m. The climb to 1000m, with three hours of fuel, the pilot, an observer, a rifle, and forty rounds of ammunition, plus a 10kg bomb or a camera, took us over 30 minutes. We were not allowed to venture inland at an altitude of less than 1,000 meters; therefore our range was greatly handicapped by this low rate of climb.*" De l'Escaille officially complained in a report about the poor quality of the 80 hp Le Rhone engines that equipped all the aircraft of his initial contingent (except for the N13, which had a more reliable 80 hp Clerget) and that he attributed to the fact that they had been produced for the Turks... To remedy the poor range, the British Admiralty decided to put the Nieuports on ships and launch them near the target from where they would take off in the open sea. Kenneth Williams: "*When we were launched, which in itself was a slow and difficult job handling winches and other lifting equipment, the business of starting the engine began. The starting handle was located in the observer's station, which was quite small and made the task of manipulating it quite difficult for anyone larger than a small boy. One had to struggle violently for a good five minutes to start the engine, which coughed, backfired, or caught fire. In this case, you had to rush out of the machine, get under the shrouds of the floats, and plug the air intake pipe with anything – most often your flight helmet. When the engine finally started, a game of ping-pong with the*

Above: In Port-Saïd, the Nieuport X H N.14 undergoing a stowage test on the cargo ship *La Lorraine* of the Compagnie Générale des Transports in April 1915. A veteran of the Montenegro expedition, N.14 was convoyed to Port Saïd on the *Foudre* on 30 November 1914. It ended up in the desert on 10 October 1915 near Beersheba, its crew being captured. (Coll ARDHAN via Robert Feuilloy)

Above: The Nieuport serial N 14 found empty by a British ship on 28-01-1915. The crew, QM Le Gall and Lt Partridge, had to abandon it due to an engine problem, and came back to the British lines in Suez on foot, but were shot down by mistake by the Gurkhas.

ocean began, and with each leap the observer's head would hit the cabin masts... An 80 hp engine did not give much power to lift an aircraft at once. Therefore, if there were no waves, we sometimes took off, but not always, and if there were waves the takeoff never happened: we ended up as if surrounded by a buzz of angry bees, covered with foam."

Two cruisers of the Royal Navy, the *Doris* and the *Minerva*, left on a mission in the Mediterranean and in the Red Sea respectively, each carrying a Nieuport which was difficult to accommodate on their decks, which were

Above: View of the quays of Port-Saïd, photographed in July 1914. Six months later, the Nieuport seaplanes moved into the quay seen in the left background, behind the large office building of the Suez Canal Company. (Léon Busy)

Below: The Suez Canal, a strategic route that the British troops helped by the French Nieuports worked to defend from the threat of the Turkish army. (Photo Paul Casteinau)

Above: One of the Indian sailors from HMS Hardinge, one of the British ships using Nieuports on board. Their crews were very cosmopolitan. (Paul Casteinau)

Left: Lieutenant de vaisseau Marcel Destrem (1884–1923), one of the very few pre-war French navy pilots, shown here after the war with the rank of capitaine de corvette. (Coll. ARDHAN via Robert Feuilloy)

Above: The lines of defense against Turkish troops in Sinai. On January 27, 1915, QM Jean-Marie Le Gall and Lt G.N.B. Partridge, his British observer, had to return on foot after having to abandon their Nieuport and were mistakenly shot down by the Gurkas posted in the trenches. (Paul Casteinau)

Above: Hangars of CAM Port Saïd in 1915. (Coll ARDHAN via Robert Feuilloy)

Right: The type X H N11 had a long career in the Port Said escadrille, visiting the coasts of Palestine, Lebanon, and Turkish Cilicia. It sank off the Sinai on December 4, 1915, its crew (QM Grall and 2nd Lt H.M.C. Ledger) managing to swim to shore. The wreckage was recovered and used as a spare parts store. (Coll ARDHAN via Robert Feuilloy)

Above: The N13 crashed in the south of Palestine on December 31, 1914, 30 km from the Gulf of Aqaba where its ship, *Minerva*, was waiting. Its heroic crew, composed of QM Grall and Cpt Stirling, managed to escape the Turkish patrols on foot in the desert. (Coll ARDHAN via Robert Feuilloy)

not adapted to this task. The reconnaissance carried out on December 10, 1914 over Beersheba by LV Delage with a British observer, Captain James H. Herbert, failed due to engine trouble. A new mission was finally successful on December 14 on Beersheba by Destrem and Herbert after a new trip from the *Doris*, where they observed only 80 tents and no particular garrison. Other missions were conducted on the coast, until December 19 when the Nieuport damaged the canvas of its wings, unable to be repaired on the ship which continued its mission by attacking the coasts of Palestine, Lebanon, and Syria. Meanwhile, the *Minerva* made two trips to the Gulf of Aqaba where each time its Nieuport failed to reach the Jordanian city of Ma'an where the railroad to Mecca runs. The flight made on the fourth trip of the *Minerva*, carried out at dawn on December 31, 1914 by the Nieuport N13, gave a lot of cold sweat to its crew composed of QM Hervé Grall and Cpt Walter Françis Stirling, who took off unarmed to save weight. They followed due north the course of the Wadi Araba, a dry river that connects the Gulf of Aqaba to the Dead Sea, not observing any large detachments of troops. Having reached the limits of their range after seeing the Dead Sea, they turned back and the engine failed 30 km from Aqaba. The Nieuport crashed into the desert and Stirling was left with his head in the sand almost unharmed, while Grall was left with broken ribs. The two men walked south, but Grall could not keep up... They agreed to split up so that Stirling would have a chance to call for help. The latter managed to avoid a Turkish patrol launched in pursuit by burying himself around the only palm tree in the landscape, not being spotted by the soldiers stopping there for a rest! Arriving exhausted on the Gulf of Aqaba, he managed to be picked up by his ship, which was also the case for Grall the following evening, after having spent a night in the desert. The valiant pilot was awarded the French *médaille militaire* (Military Medal) and the Distinguished Service Medal by the British.

As the British warships were unsuitable to carry Nieuports on their decks, the Admiralty fitted out two German freighters captured in the canal at the time of the declaration of war, the *Aenne Rickmers* and *Rabenfels*, which were renamed *Anne* and *Raven II*. Each of them was manned by an unlikely crew composed of British Empire officers,

N.16.

Above: At the end of June 1915, LV Cintré (left) and 2nd Lt H.G. Hillas, a British observer, made leaflet drops in Nieuport N.18 from the Hardinge transport on the Arabian coast of the Red Sea, flying over Jeddah on June 30th. The picture was taken five days later. (Coll ARDHAN via Robert Feuilloy). (Coll ARDHAN via Robert Feuilloy)

Facing Page Above: Nieuport N.16 ready to go for a mission. (Coll ARDHAN via Robert Feuilloy)

Facing Page, Below: View of same N.16 recovered after an accident. The aircraft will fly again... We note a different inscription "N16" compared to the previous picture, testimony that Nieuport seaplanes were regularly overhauled and re-covered by the mechanics of Port-Saïd. (Coll ARDHAN via Robert Feuilloy)

Nieuport hydro NB2 Type X Clerget 100 hp

Above: Photo taken from a magazine of the time, but very interesting because it shows one of the Nieuports of the Port Said squadron (probably the N.17) used on land by the detachment sent to the Dardanelles in April 1915. (Coll Ian Burns)

Greek sailors, and French pilots and mechanics. These two ships, which could carry two Nieuports on operations (plus others in crates), were from then on the designated mobile bases for the seaplanes of the Port Said squadron. The first mission was carried out from January 17 to 22, 1915 by the *Anne* along the Sinai coast, at the very moment when the Turkish army attempted an offensive towards the Suez Canal. Their advance was spotted by the French Nieuports, but they suffered their first loss when QM Jean-Marie Le Gall's N.14, with Lt G.N.B. Partridge as observer, was reported missing in the afternoon of January 27th after taking off from Port Said for an observation mission along the Mediterranean shore. A fast patrol boat was immediately dispatched and at nightfall found their Nieuport stranded on a beach, empty, which they brought back to Port Said. The two men had left on foot to reach the British lines... Which they did indeed reach at night, but to be shot down by mistake by Indian Army Gurkas on duty. From 31 January to 3 February 1915, the Turkish attack took place, which the Nieuports contributed to stop, by revealing its progression. The enemy was dispersed by the artillery of the Franco-British fleet, occasionally regulated by seaplanes which also attacked by dropping darts (small spindle-shaped metal projectiles dropped in whole boxes) and bombs. The Suez Canal was never worried again...

The flights of the Port-Saïd squadron, which was to become administratively a Maritime Aviation Center (*Centre d'Aviation Maritime* - CAM), continued for more than two years, mainly from the ships *Anne* and *Raven II*. The equipment situation improved somewhat with the arrival of 80 hp Clerget-powered Nieuports: the N15 to N17 delivered in January 1915, then the N18 to N20 the following March. The N21 to N23, delivered in November 1915, were equipped with an even more powerful Clerget 100 hp engine, as were the last three aircraft (NB1 to NB3) in March 1916, which were also converted to standard controls with foot rudder and stick warp.

These aircraft were mainly concentrated on the coast of Sinai and Palestine, and especially on the city of Beersheba, whose strategic importance was reinforced when the Turks, with the help of the Germans, connected it to their rail network at the end of 1915. The Nieuports managed to fly over it fourteen times, but also had eight aborted missions when trying to approach it, either because of sand storms or mechanical problems. Two crews were lost during these missions: on October 10, 1915, the N14 of SM Georges Trouillet with Lt Sir Robert Paul as observer, crashed in the desert after returning from the mission and its crew was captured. Then, on December 24, 1915, the N17 of LV Louis Barthélémy de Saizieu and 2nd Lt H.M.C. Ledger, hit by ground fire at Beersheba, had to land on the ground. While the pilot attempted to fire the aircraft, Ledger fired at the Bedouins who had come to capture them and was killed in the engagement, and de Saizieu was captured.

Less frequently, Nieuports were sent to more distant shores, from the coasts of Lebanon and Syria to Turkish Cilicia – the cities of Adana, Alexandrette, and Tarsin were flown on two trips by the *Raven II* and the *Anne* in June and August 1915. In early June 1915, LV Cintré, trained as a pilot in Port Said, flew over the southern points along the Red Sea coast on board the troop carrier *Hardinge* to launch leaflets inciting the Arab population to rise up against the Turks, as far as the city of Jeddah on the N18 on June 31st 1915.

Above: Nothing was lost in maritime aviation: after the dissolution of the Brindisi squadron (Italy) in January 1916, the NB 1 and NB 2 served on the deck of the cargo ship *Anne* in the Red Sea, as early as February 1916. NB 1 sank on March 13th; NB 2 was brought to Argostoli in June after the dissolution of the Port Said CAM. (Coll. Colin Owers)

Among the distant expeditions of the Nieuport CAM from Port-Saïd, one is particularly noteworthy, as its aircraft took part in the Battle of the Gallipoli. On March 3rd, the *Anne*, then on a mission along the Palestinian coast with the N11 and N17 piloted by Destrem and Grall, received the order to leave for Smyrna where it was damaged by a Turkish torpedo and had to take refuge on the island of Mudros for repairs, disembarking its aircraft which were in no condition to fly. The *Raven II* with on board the N14, N15, and N16 and as pilots the LV de l'Escaille, de Saizieu and the QM Trouillet, also left on a mission to Smyrna where the N16 made a reconnaissance on March 26. At Mudros, it received orders to return to Port Said but, before leaving, it landed the Nieuport N14 and N17 on April 2, 1915 with De Saizieu and Trouillet as pilots, two observers, and some mechanics. Used as ground aircraft thanks to the "kit" allowing the installation of a Nieuport IV landing gear, of which the Navy had purchased seven, they were a very valuable reinforcement at the beginning of the Dardanelles landing, where the Royal Navy Air Service, among its jumble of aircraft, had only five really operational ones. The Nieuports carried out observation missions of the Turkish positions before the landing, which took place on 25 April, and that same day they supported the French troops who made a diversionary landing on the Asian side of the Dardanelles at Kum Kale, before re-embarking. When Captain Césari's MF 98 T squadron settled on the island of Mudros on 9 May 1915, the presence of the Nieuports was no longer indispensable and they were reembarked towards Port Said in a British cargo ship at the end of the month.

In late 1915, as the British Air Force in Egypt was growing in power and had efficient aircraft, the French Nieuport presence became less and less necessary. The CAM of Port-Saïd was then disbanded in April 1916. Its surviving aircraft (N15, N16, N18 to N21, N23, NB2, and NB3) were recovered and transported on May 14th 1916 by the *Campina* seaplane carrier to Argostoli, on the island of Cephalonia, where a new airfield was opened to monitor the approaches to the Corinth Canal. Flights started immediately from the ship even before the shore installations were completed. Gradually replaced from October 1916 by 100hp FBA seaplanes, the last of the robust Nieuport monoplanes were retired in May 1917, ending an operational career of more than two and a half years, which was exceptional for WWI aircraft.

Above & Facing Page: Maps of the eastern Mediterranean and Red sea. The areas flown by the Nieuports are shown in red.

Nieuport XH "N11" flew from December 1914 to January 1916. It was equipped with a 80 hp Rhône engine.

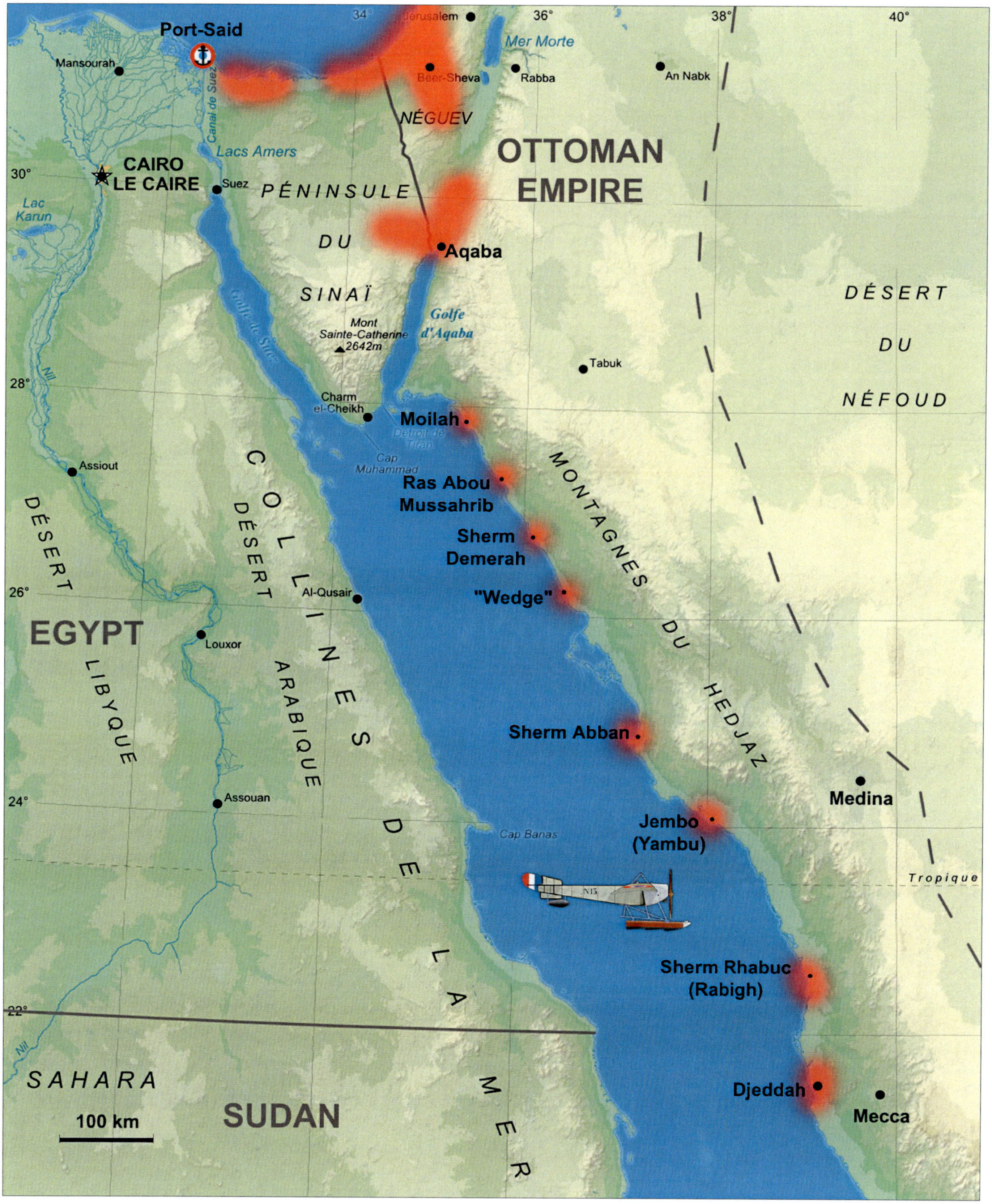
Port-Said
Mansourah
Canal de Suez
Jerusalem
Mer Morte
Beer-Sheva
Rabba
An Nabk
NÉGUEV
Lacs Amers
CAIRO
LE CAIRE
Suez
OTTOMAN
EMPIRE
PÉNINSULE
DU
SINAÏ
Lac Karun
Aqaba
Golfe de Suez
Mont Sainte-Catherine 2642m
Golfe d'Aqaba
Tabuk
DÉSERT
DU
NÉFOUD
Nil
Charm el-Cheikh
Moilah
Détroit de Tiran
Cap Muhammad
Assiout
Ras Abou Mussahrib
Sherm Demerah
"Wedge"
MONTAGNES DU HEDJAZ
COLLINES
DÉSERT ARABIQUE
DÉSERT LIBYQUE
Al-Qusair
EGYPT
Louxor
Sherm Abban
Medina
Assouan
Jembo (Yambu)
Cap Banas
DE LA MER
Tropique
N15
Sherm Rhabuc (Rabigh)
SAHARA
Djeddah
Mecca
100 km
SUDAN
34°
36°
38°
40°
30°
28°
26°
24°
22°

Nieuport N 13 was lost during a reconnaissance flight to the Dead Sea, crashing near Yotvéta after engine problems. The heroïc crew, QM Hervé Le Gall and British observer, Captain Walter Francis Sterling, managed to return by foot to their supporting ship *Minerva* stationed in the Aqaba Gulf despite Turkish patrols.

Nieuport XH "N16" with 80 hp Clerget engine had a long career of 16 months of flights in Port-Saïd, and at least one overhaul as shown in the two profiles above & below.

Nieuport XH "N 17" used as a landplane in Gallipoli during the months of March and April 1915.

Nieuport XH "N 18" with 80 hp Clerget engine flew from March 1915 to April 1916 in Port-Saïd.

Nieuport XH "NB1" with 100 hp Clerget engine began its career in Brindisi, then came in March 1916 to Port-Saïd and finished its career in Argostoli.

Chapter 3: MF 98 T in Gallipoli (January 1916)

Above: British landing in Dardanelles.

3.1 - The Gallipoli Campaign

The Gallipoli campaign, or Dardanelles expedition for the French, was decided by Winston Churchill, then First Lord of the Admiralty. It was supposed to conquer the city of Constantinople and force the Sultan to take his country out of the war. A Franco-British fleet was assembled for this purpose in early 1915 and went on the attack on February 19, trying to destroy the Turkish forts guarding the entrance of the Straits by the power of their guns. Seaplanes of the Royal Navy Air Service abroad seaplane carrier *Ark Royal* demonstrated their inefficiency while trying to make artillery spotting. The Turkish forts were still operational after the firing campaign ... The Franco-British fleet decided, despite all, to force the Straits on March 18, 1915. They suffered a bloody defeat by meeting naval mines which sunk British battleships *Irresistible* and *Ocean* and the French *Bouvet*, and also damaged the British *Inflexible* and *Agramemnon* and the French *Gaulois* and *Suffren*. The British high command understood the need to use land forces to take the straits, and the need to have reliable aviation.

The UK would then send on the island of Bozcaada a squadron of landplanes, the 3rd Squadron RNAS, under the command of Squadron Commander Charles Rumney Samson, who moved on 23 March 1915 to a former vineyard west of the island. But the majority of his 22 planes, of different models, were obsolescent and only 5 could be considered operational. Two French Nieuport hydroplanes from the French navy, transformed into landplanes, were sent from Port-Saïd as reinforcements until the month of May (see last chapter). But the main French air unit was escadrille MF 98 T, formed in Lyon on February 1915. It moved to Marseilles on March 24, and sailed for Mudros Island on March 28 on cargo ship *Ganges* but the boat was diverted to Alexandria, where pilots and mechanics camped three weeks near the pyramids, with all the soldiers of the British Empire in garrison for the defense of Egypt. As on April 29, MF 98 T arrived by boat on the island of Tenedos, the Dardanelles landing has already begun for four days.

While French troops made a diversion landing on the Asian side of the Straits (Kum Kale), the main attack in the Gallipoli Peninsula, was led by the troops of the British Empire (including many Australians and New Zealanders), with very heavy losses. After five days of fierce fighting, the British established a very short bridgehead with no real depth so that no place was safe from enemy fire: every man, from the simple private to the highest general, could be suddenly volatilized by a Turkish artillery shell.

Escadrille MF 98 T began during the battle to land its equipment in Tenedos on May 6, 1915, and settled in the west of the island, near the camp of the British RNAS squadron. The word "landing" deserves attention, because the hundred French aviators and mechanics must manage without any port infrastructure to unload their horses and

Above: The British ship *River Clyde* during the landing in Cape Helles – Photo taken by the French observer François Masnou from escadrille MF 98 T.

Above: Maurice Farman MF 11 from escadrille MF 98 T in Tenedos Airfield, September 1915 – Photo René Germain.

Left: Capitaine Antoine Césari, commanding officer of escadrille MF 98 T, stands near a Maurice Farman MF 11 in Tenedos. Note his medal, the légion d'honneur, won in 1914 in France after bombing a Zeppelin on the ground.

Facing Page, Above: French airfield in Tenedos in late 1915. A Maurice Farman MF 11 and a Morane L can be seen on the ground.

Facing Page, Below: Maurice Farman MF 11 flying back from Gallipoli to Tenedos in late 1915.

their boxes of equipment, then climb the latter with the help of the first to the airfield ... Hangars Bessonneau are mounted to house the equipment. Men will live in tents and empty crates of their aircraft. The first French Farman takes the air on May 11 and the entire escadrille would soon take part in the fighting on the side of British aircraft. Samson and Césari became friends and the entente cordiale works on the Greek island, they also exchange – unofficially – airplanes as they receive reinforcements. Both air units will use a secondary landing field built on a beach of the peninsula at Cape Helles, extremely dangerous because under the scope of the enemy artillery. Very soon, the allied planes see land forces digging into a costly war of position, similar to the one taking place on the French front, but in a very small area, with incredible concentrations of troops. The supply lines become crucial on both sides and airplanes will play a role to harass them.

3.2 - Franco-Britsh Air Superiority and War Against Turkish Supply Lines

On May 16, a French archive tells us the initial use of all the 7 Farmans (one was lost at sea in an accident) of Escadrille MF 98 T: two of them made observation flights and artillery spotting based in the dangerous landing field of Cape Helles to immediately communicate the results of their flight to the troops. Two Farmans in Tenedos conduct a photographic work, two more are reserved for long-distance reconnaissance flights and one for bombardment.

This activity would soon take a major part to harass Turkish supply columns that have only poor roads for their convoys. The French Farmans, like their British comrades,

Above: The Turkish city of Canakkale taken in picture by French observer François Masnou from a Maurice Farman MF 11.

Above: A Voisin LAS bomber from the RNAS. British planes in Gallipoli were wearing French cockades.

Facing Page, Below: The city of Tenedos.

soon take with them a bomb during each mission that they drop on anything that moves. If the men of Samson are well equipped with projectiles, French aviators have to improvise their bombs by transforming naval artillery shells on which they welded fins and adapt a primer delayed. Samson recalls, "*Cesari was not a man to back down. He left himself one day to bomb Soghan Deré, carrying a bomb equipped with a primer delayed. The release lever was operated by an observer, that he would take in Helles landing field. On his flight from Tenedos to Helles, he suddenly realized that the primer was pressed. He knew that the only thing to do was to drop the bomb and that it would explode in a few seconds. From his cockpit he could not reach the lever of the observer. He woke up to reach the rear of the nacelle. But if you release the controls, the Farman does not fly exactly alone ... The bomb fell 70 meters down. Then he landed in Helles, loaded a bomb of the same type and left for his bombing in Soghan Deré ...*"

On 21 June 1915, the squadron MF 98 T was very active in the offensive of French troops in the ravine Kerevès Deré, deploying a permanent detachment of two aircraft in flight throughout the day, for reconnaissance and bombardment. For this action the unit would receive a mention in dispatches although this offensive resulted in another failure on the ground. On 23 June 1915, a ship unloaded seven new airplanes for MF 98 T, four Farmans with 80 hp engines and three Morane parasols to be used as fighters. A Morane would be permanently made ready to use on the dangerous airfield of Cape Helles to try to intercept enemy aircraft on their way home.

Skirmishes would multiply with the German-Turkish aircraft, which were currently not armed and defended themselves with carbines. But the Entente force, multiplying reconnaissance and bombings in the area of the Straits, would bring to the enemy a decisive blow on July 5, 1915 by destroying its air force on the ground. Anything that flies and can carry bombs, or 14 French aircraft from MF 98 T and eight British aircraft, takeoff at 7 PM to the aerodrome of Canakkale on the Asian side, previously identified by Samson in the month of April. While a rain of bombs fell on the Turks, a lucky drop from a British pilot fell right on the main hangar, destroying the 3 Albatros B.I parked there and composing all of Germano-Turkish aviation! Deliveries from Germany allowed the Turkish air force to be quickly reconstituted on 13 July 1915, but it had to settle on the Gatala airfield, further away on the European side of the Straits, with 4 Rumpler B.I. still unarmed, but equipped with a powerful 150 hp engine that allowed them to keep most of the Franco-British aircraft behind them. The latter,

Above: Charles Rumney Samson, the British commanding officer of RNAS in Gallipoli, near his Nieuport 10 in September 1915.

in the meantime, were having a great time. For the 14th of July 1915, the French experimented with incendiary bombs built with 4 petrol cans around a 10 kg shell. They attacked with several objectives: Cannakale, Maidos (which received 2.8 tons of bombs in 3 trips of the squadron), the Turkish HQ of Soghan Deré... On the evening of July 25, a new visit of the French planes on Canakkale created a huge fireworks by setting fire to the gasoline depot. The airfield would be regularly bombed, but without consequences for the Turks who left only a dummy plane there!

Also reinforced with two Gotha seaplanes, the Germano-Turkish air force became bolder to the point of bombing the Franco-British on their base of Tenedos. But the air superiority of the Entente was real and the events of the 31st of July 1915 are an illustration. During the night, two Rumpler B.I and two Gotha seaplanes dropped five bombs and 500 metal darts on Tenedos. It was at 4:45 am and dawn was beginning to break; British Flight Lieutenant Marix was in fact about to leave on a mission and immediately took off towards an enemy seaplane that he saw. The latter dropped its bombs at sea and ran to the Asian coast, pursued also by French Corporal Barnay's Morane parasol of MF 98 T. The Gotha landed in the direction of Canakkale, Marix managed to cut it off and shoot at it at short range. The Gotha landed on a beach in front of the Soghan Deré. Seven planes of MF 98 T came to bomb its landing during the day! In the evening, a Rumpler ventured out to sea again and was intercepted by Cesari, alone on board his Farman, who fired three shots from his rifle and pursued him in flight to his airfield, assisted by Corporal Barnay's Morane, which was guarding the runway at Cape Hellès.

Faced with the stalemate of his troops on the peninsula, Sir Hamilton, the British Chief of Staff, decided to land further north, in the Bay of Sulva, in order to bypass the Turkish lines and reach the Strait. The landing operation took place on August 6, 1915 at two points in the Sulva Bay. The troops landing near the Salt Lake met almost no opposition. But the incompetence of the British leaders meant that their troops remained inactive in their landing zone for too long instead of rushing inland – time used by the Turkish General Mustapha Kemal to send reserve troops to take up positions and settle in impregnable trenches.

For MF 98 T, the day of August 6, 1915 was marked by the disappearance of Farman n°459. Leaving late in the afternoon, Sergeant Dumas took Captain Aubert, an observer, inland on the Asian side in the Ezine region, due east of the island of Tenedos. Their Farman was shot at and an awful noise of scrap metal told them that the engine had

Above: Fokker Eindecker wearing Turkish markings, piloted by German ace Hans-Joachim Buddecke.

Above: Gymnastic teacher in civilian life, Sergeant Charles Denti became a Morane L pilot in escadrille MF 98 T in late 1915.

Above: Escadrille MF 98 M in the airfield of Mytilène in late January 1916.

been hit – it ceased to work while they were at an altitude of 1800 meters and well inland. Dumas recounts his adventure: "*The captain asks me what I am going to do. We had to sail upwind to get back and the coast was 6 kilometers away. I think I'm done for and I go down to the sea losing as little height as possible. It is understood that we will try to reach the beach or the sea and that we will advise. At first it was not going well, but when I reached 1,200 meters, I was in a favorable wind current. In a few seconds I feel that it changes; we gain speed. I flatten out and the plane goes towards the sea, which we soon reach. There are still nearly 8 kilometers to go before touching down; I lengthen the descent by the most improbable dead leaf that I have ever allowed myself; the aircraft drags itself like a greasy paper in the wind; I pass the lighthouse... Finally, that's it and we are saved; I land on the water in sight of Tenedos and we patiently wait for help. We were seen and, a quarter of an hour later, a trawler picked us up, barely anchored. We have come back from far away! Nearly 15 kilometers in the air, half of which at less than 1,200 meters, and under the intensive fire of the Turks, furious to see a prey that they had thought they were holding escape. I was happy to shake my wet boots on the deck and I saw my poor old cuckoo clock sink without much regret, as two hundred hours of flight had not demolished it.*" Three days later, a Morane was shot down by French flak, fortunately without damage, and Césari had to give the infantry units a drawing of the aircraft so that they would avoid shooting at anything that flew. On 17 August 1915, the seaplanes of the RNAS were to make news by sinking a Turkish supply ship in the Dardanelles Strait. The next day, MF 98 T took over and bombed the entire Ak Bashi Liman landing stage, with Sergeant Lecompte sinking a transport with a well-placed 155 mm shell.

On August 25, MF 98 T deplored its first loss among the flying personnel: a Farman having just taken off from Tenedos was caught in a whirlwind and crashed into the sea, killing its crew composed of S/Lt Saint André (pilot flying in the place of the observer) and Sergeant Dumas that luck did not save a second time. During this critical period, MF 98 T carried out numerous bombing raids in groups, including on the enemy airfield of Gatala, without obtaining too many results because the enemy had learned to disperse and camouflage its aircraft.

MF 98 T then received a new reinforcement of aircraft, including a Farman 130 hp (MF XI bis) with which Césari envisaged bombing Constantinople, but finally limited himself to a raid on Enos in Thrace, carried out by Adjudant Lecompte, a sporting feat for which he had to justify himself, as this town was not one of the objectives set for his squadron. The Entente forces now had more than forty aircraft and an apparent complete mastery of the sky. However, this supremacy was disputed by Major Serno's Turkish air force, which received in September 1915, in addition to new seaplanes, two-seater Albatros C.I armed

Above: Future ace Gabriel Thomas and gunner Charles Weisshaupt (in the plane) and their Morane L damaged after a bad landing in Mytilène in early 1916.

Below: Pilots from escadrille MF 98 M in Mytilène, early 1916. In the center of the first ranks, sits Capitaine Antoine Césari, commanding officer. In the extreme right sits Sergeant Gabriel Thomas, future 7-victory-ace. Standing on 2nd rank, 2nd from left, is Sergeant Charles Denti.

Above: Maurice Farman MF 11 and Morane Parasol from escadrille MF 98 M landed in Long Island, during their commando raid in Smyrna on February 7, 1916.

with a machine gun on a turret ring. This was an effective defense against the Franco-British aircraft, many of which were equipped with individual weapons. More importantly, with these new deliveries, Serno was able to order daily raids on Entente positions, which caused some casualties and undermined the morale of the soldiers stuck in their trenches dominated by those of the Turks. The German-Turkish airmen went so far as to provoke the British by bombing the island of Imbros, causing a few casualties, including one airman – an attempt at retaliation on the part of Césari on the Gatala airfield came to nothing because of strong gusts of wind. With the month of October, the weather conditions deteriorated and the Dardanelles were gradually hit by a vigorous cold. The soldiers in the trenches were the first to suffer, but the MF 98 T squadron, somewhat forgotten by the quartermaster, also suffered. Poorly supplied, the faces of the airmen became emaciated. The worn-out uniforms were patched, for lack of anything better, by the rags of the mechanics who used straw to work on the engines instead.

3.3. - Withdrawal

Faced with a military impasse, the British government decided to evacuate its troops from the Dardanelles. On 7 December 1915, the order was given to evacuate the Sulva sector. The Commonwealth troops retreated discreetly and in good order, while in the sky the wind of war was turning in favor of the German-Turks. They now had eight operational land aircraft, six of which were armed (Albatros C.I), not counting the seaplanes. As they were now receiving new aircraft by rail, they would redouble their activity.

On November 30, 1915, one of their Albatros C.I (registration AK.1) flown by a Turkish crew (pilot Mülâzim-i Evvel (lieutenant) Ali Riza, observer Mülâzim-i Sani (second lieutenant) Ibrahim Ohran) claimed an aerial victory over a French aircraft which it atacked over the bay of Sulva and whose tank exploded, the aircraft crashing on the hill of Intepe near Seddülbahir. No loss was recorded that day at MF 98 T – it could have been a British aircraft, or a French aircraft returned safely to its base. The French airmen also claimed a victory on an unspecified date at the end of 1915, awarded to Sergeant Charles Denti, a French reinforcement pilot who arrived at Tenedos on 15 November 1915 and was assigned to fly Morane fighters. On an unspecified date, he was on patrol with his machine gunner, Corporal René Weisshaupt, and encountered a Turkish aircraft. Constantly harassed by them on their airfield, they decided to attack it. Denti skillfully positioned himself and Weissaupt fired a burst from his Saint-Etienne machine gun. The surprised Turkish fled in a dive, and the Morane's mission continued

Right: The village on the island of Uzunada (Long Island) where, according to French pilot Garsonnin, all the Greek inhabitants were slaughtered a few days before by the Turks.

without further incident to the point that Denti concluded his report with a laconic "nothing to report". But from the ground, the air battle was followed enthusiastically by British sailors on their ship – they rushed to send a message of congratulations to the MF 98 T squadron for this downed Turkish aircraft. The letter arrived on Césari's desk, and he was astonished to discover it: he shouted at the pilot who had shot down a Turkish aircraft without even mentioning it in his report! No loss was recorded that day among the Turks, but the victory would nevertheless be duly recognized to Denti and Weisshaupt on September 4, 1916, by a citation at the army order, while they both served on the Orient Front in Macedonia.

After the evacuation of Sulva, the British government ordered the evacuation of the entire peninsula on December 28, 1915. The withdrawal operation was the only success of the campaign, even if, as Churchill would say some 25 years later, a war is not won by evacuations: the withdrawal of the soldiers was carried out at night with the utmost discretion under the noses of the Turks who did not launch any offensive. In the air, on the other hand, the situation was completely changed: three formidable Fokker Eindeckers were put into service by the Turks in December 1915. Their results would be spectacular, they obtained aerial victories as soon as the day after their arrival when Leutnant Buddecke shoots down on January 6, 1916 Farman MF XI n°942 piloted by Lieutenant Jules Lecompte, who kills himself by crashing in the Turkish lines near Gatala. On 9 January 1916, the evacuation actually took place and the Turks discovered empty trenches.

3.4 - A Detachment in Mytilene

MF 98 T left the Dardanelles theater to join the city of Salonica, just like the French troops. But its war against the Turks did not stop there, because a detachment of the squadron (named MF 98 M) led by Captain Césari was formed and ordered to reach Mytilene on the island of Lesbos on January 13th 1916. The pilots flew to the island in their Farman aircraft; the mechanics followed by boat. They were given a fortnight's rest, which was not too much for the pilots, who were able to recharge their batteries in the villa provided by the French consul – they also took the opportunity to refurbish their outfits... Once again operational on January 29, 1916, the French airmen prepared a raid on the city of Smyrna, where the Entente command wished to organize a mock landing in order to force the Turks to fix their troops there unnecessarily. That represented a round trip flight of more than 200 kilometers, at the extreme limit of the possibilities of Farman MF XI 80hp. It was thus envisaged to make it possible to fly while carrying the maximum of payload that Farmans would refuel in secrecy behind the enemy "lines", on the "Long Island", one of the islands of the gulf of Smyrna whose inhabitants are all Greeks. The Farmans took off for the adventure on the afternoon of February 7, 1916. They landed on the "long island" at 5 p.m. and found it deserted: the Greek inhabitants had been massacred by the Turks, and the pilots made the macabre discovery of corpses rotting in the village's Orthodox church. The only inhabitant of the island was a mangy donkey which came to sniff the planes and which the French aviators did not manage to keep away... They have anyway another concern: the English ship which should supply them was not there! Césari decided to wait for it, and the French spent a scary night with their musket in their fist – a Turkish patrol would have easily shot them. The next day the supply ship finally showed up – a delay in the transmission of orders – and allowed the Farmans to prepare for the raid on Smyrna that they could see on the horizon. Césari was able to take off at the head of his men on the objective, throwing 155 mm shells on the port and proclamations on the city announcing the arrival of the Franco-British troops. The six Farmans were fired upon by the flak installed on Mount Pagus. Sergeant Garsonnin testified: "*The view was enchanting. The gulf drew its elegant curve, bordered to the north by the white hills where the ancient Phocaea was located, whose sailors colonized Marseille. The city detached its white houses in amphitheater on the dark green of the cypresses and was crowned by the ancient Genoese citadel. We had at our feet the Aimable, the crown of Ionia, the pearl of the East. Then we returned to the Long Island, and, after having refueled, we took again the way of Mytilene. The diversion was finished.*"

MF 98 M stayed for some time in Lesbos, then received the order on February 27, 1916 to fly to Salonica. The airmen left for their last flight over the Aegean Sea on February 29, all equipped with an inner tube and a bicycle pump to serve as an improvised buoy. This precaution was not useless, because the Farman of MdL Fourcade, victim of mechanical problems, had to land on the island of Strati where he broke his plane and was saved two days later by a British ship. Captain Césari was given a forced bath 15 kilometers from Lesbos, where he was quickly rescued by a ship. The five other Farmans, which were guided by the summit of Mount Athos, reached Salonica without any problems, and the mechanics reached it on 6 March. The Dardanelles campaign was then over; each pilot of the first detachment had flown more than a thousand hours, for a total of one killed in action, two by accident, and 11 aircraft lost, eight of which following their forced landing due to mechanical problems. Whereas Captain Césari was repatriated to France, MF 98 M was dissolved after having lasted a few months. MF 98 T would join the aviation of the Orient Front in Salonica by being assigned to the Serbian army.

Maurice Farman MF 11 n°458, escadrille MF 98 T, Tenedos, 1915.

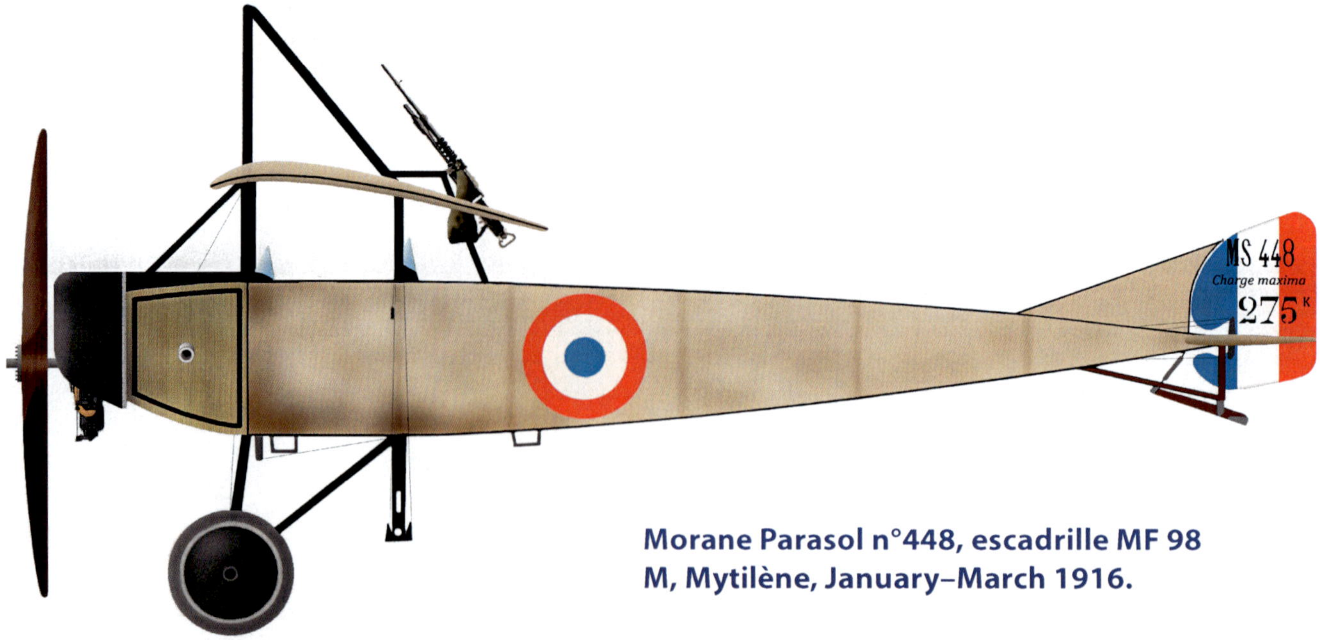

Morane Parasol n°448, escadrille MF 98 M, Mytilène, January–March 1916.

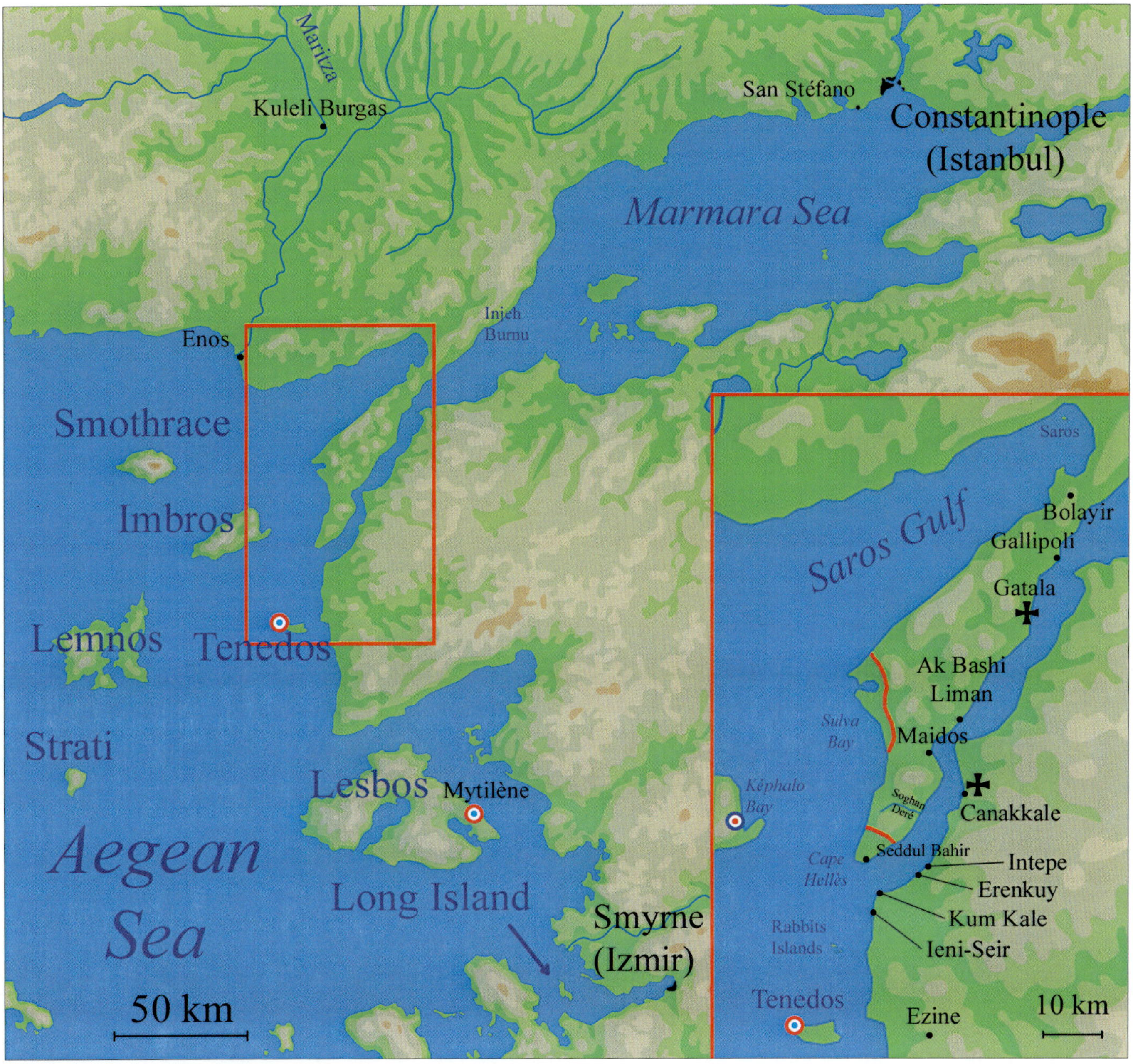

Above: Map of the operating area.

Chapter 4: Adventures of Escadrille MF 99 S in Serbia

Above: Officers of escadrille MF 99 S after their arrival in Serbia, in late March 1915. Left to right, standing : Cne François Tulasne (observer), Cne Jules de Baillardel de Lareinty-Tholosan (pilot), Cne Richard Mavet (observer), médecin aide Major Dominique Dumas (medical officer), Cne André Mortureux (observer), Cne Louis Paulhan (pilot). Sitting: Cne Jean Dufaur de Rochefort (observer), Cdt Roger Vitrat (pilot and commanding officer), Lt Robert Martinet (pilot). (Collection Boris Ciglic)

4.1 - A Squadron for the Serbs

On June 28, 1914, a Serbian nationalist named Gavrilo Princip fired several shots at the heir to the Austro-Hungarian throne, Archduke Franz Ferdinand of Habsburg, who succumbed to his injuries. The consequences of what history will remember as the Sarajevo bombing are well known: the beginning of the First World War, with the declaration of war by the Austro-Hungarian Empire on Serbia on July 28, 1914, which led to the declaration of war by all the European powers through the game of alliances. A few weeks later, the North of France and the plains of Poland were transformed into a huge battlefield. This completely overshadowed the fate of Serbia, which at the time had 4 million inhabitants, no maritime outlet, and which only the Danube and its tributaries separated from its powerful Austro-Hungarian neighbor. The capital, Belgrade, was only a few rowing strokes away from the Habsburg Empire!

During the night of the declaration of war, the Austrian river gunboats (or monitors) *Temes*, *Bodrog*, and *Szamos* fired the first cannon shots of the war by shelling the fortifications of the Serbian capital, while the troops of its 3rd army crossed the Sava river and penetrated the territory from the west. The small Serbian army defended itself valiantly and was indirectly helped by the Russian military victories in Bukovina, which forced the Austro-Hungarians to assign their troops to this sector as a priority. The Serbian soldiers took advantage of this and pushed the Austrians back to the border... The Serbian David thus defeated the Austrian Goliath, but the situation remained worrying because the Serbian army lacked everything, even rifle cartridges! Its allies, very busy with the war on their own soil, could only send a little technical help. Thus the Russians provided military equipment and engineering specialists by ships sailing up the Danube through Romania. France, for its part, sent three 140 mm cannons served by a small contingent of sailors commanded by Lieutenant de Vaisseau Edouard

Picot. The guns were unloaded in Salonica, a Greek port in which Serbia, thanks to a treaty signed in 1912, had the exclusive use of certain landing docks as well as the freedom to use the Macedonian railroad that linked to Belgrade. This passage through neutral Greece was to be the main supply route for Serbia, but it was necessary to take care to spare the Greeks, whose king's sympathies lay with the central empires. The French sailors had to travel in civilian clothes and the story goes that Picot was disguised as a mufti... Installed on November 21, 1914 in pillboxes near Belgrade, they kept the Austrian monitors in check as they had to move away from the capital.

In December 1914 a new Austrian offensive was launched and managed to occupy Belgrade on December 2. But the Serbian army counter-attacked once again with energy and pushed back the enemy, liberating their capital on the 14th. This new Serbian military victory was likely to be the last one: the country was bloodless, the army lacked ammunition, and a terrible typhus epidemic was spreading. Calls for help from foreign powers became more urgent and were heard this time with the arrival of several medical missions. The French government decided to send additional military aid

Right: Commandant Roger Vitrat, commanding officer of escadrille MF 99 S and future commander of Serbian aviation in Salonica, during years 1916–1917. This courageous man died in 1944 while fighting for the Resistance. (Collection Boris Ciglic)

Below: Maurice Farman MF XI n°194 during spring 1915. This plane was usually flown by Adjudant Marcel Thirouin who scored a victory at its controls on June 9, 1915, with Sergeant-Major Magnoni as his gunner. (Collection Boris Ciglic)

Above: A bomb is loaded abroad Maurice Farman MF XI n°197 in Banjica airfield, during spring 1915. This plane was usually flown by Adjudant-Chef Joseph Seret. (Collection Boris Ciglic)

Above: Capitaine François Tulasne (observer) seems lost in his thoughts in front of Maurice Farman MF XI n°452. Note the reinforced carriage. (Collection Tulasne)

Right: During one of the first flights of the unit, Capitaine Paulhan, sitting in front of his Maurice Farman MF XI n°193, carries his observer, Capitaine André Mortureux. (Collection Passot-Seguin)

in the form of a squadron to relieve the Serbian air force, which was only composed of a single Blériot XI and an observation balloon...

The French general staff then gathered aircraft and personnel in Lyon for the formation of a new squadron of six Maurice Farman MF XI which was numbered MF 99 S (Serbia). Its command was entrusted to Captain Roger Vitrat, an active soldier who had already distinguished himself for his reconnaissance missions in the MF 25 and MF 21 squadrons. Under his command were seven pilots, Lieutenants Louis Paulhan, Robert Martinet, Jules de Lareinty-Tholozan as well as Adjudant Marcel Thirouin and Sergeants Alcide Peté, Léonce Selaquet, and Joseph Seret. Louis Paulhan was a famous aviation pilot (pilot's license n°10 in October 1909!), known to the general public for several years for his participation in air shows and air races during the Belle Epoque. The squadron included four observation officers (Lieutenants Jean Dufaur De Rochefort, François Tulasne, Richard Mavet, and André Mortureux), as well as a doctor, the Assistant-Major Dominique Dumas, who would not be out of place due to the health situation in the country. Mechanics, administrative staff, and vehicle drivers completed the staff – among these simple soldiers, we note the presence of 2nd class Maurice Arnoux, future ace with 8 victories killed in action aboard his Morane 406 in 1940.

4.2 - Trip to Belgrade

Three weeks were spent in Lyon-Bron to assemble the squadron, a period during which most of the men were able to taste the joys of the regional capital after several months spent at the front. The officers were able to sew on new stripes: all were temporarily promoted to higher rank for this assignment in a foreign country. The small troop of 120 men dressed in horizon blue embarked in Marseille on March 3, 1915 on the liner *Mossoul*. After a stopover in Malta, where they learned the news of the Franco-British landing in the Dardanelles, the ship reached Piraeus on 9 March, where the French military attaché made a point of ensuring that the airmen were in civilian disguise, in order to spare the Greeks' sensibilities. Some officers took advantage of the stopover to visit the Parthenon... The *Mossoul* left for Salonica and the group of French "tourists" landed there, probably not deceiving the German consul who, standing on the quays, observed this landing with a suspicious eye.

The departure for Serbia was by train on March 14. The suffocating smell of camphor and phenic acid in the

Maurice Farman MF XI n°193 of Capitaine Louis Paulhan, Banjica (Serbia), Spring 1915. In October this plane carried Serbian cockades. (David Méchin)

Left: Under the carriage of the Maurice Farman, the city of Belgrade. (Collection Passot-Seguin)

compartments reminded the French of the typhus epidemic that was raging towards their destination. After a long journey up the dreary Vardar valley, the train arrived the next day in Niš, the country's second largest city and seat of the Serbian government, which had evacuated Belgrade. As the officers crossed the city to get to the embassy where they were invited, they discovered streets covered with snow and mud under a grey sky, crossed by ox carts. They passed

Above: Capitaine François Tulasne, Miss Zineida Ivanov (daughter of the Russian commander of the Danube Flotilla), Capitaine Jean Dufaur de Rochefort, and an unknown Montenegrin officer in May 1915 in the village of Prahovo where escadrille MF 99 S sent a detachment. (Collection Boris Ciglic).

ragged Serbian soldiers in convalescence and Austrian prisoners in uniform, circulating without any surveillance and recognizable by their raised kepi. Black flags of mourning hung from the windows of the houses where typhus had taken its toll... Some children, mistaking the horizon blue uniform for that of the Austrians, called them "Chwabe" which quickly turned into "Hurrah Francsousti!" when they discovered their mistake. The French ambassador, Mr. Boppe, gave them the best welcome and all were impressed by the stature of the military attaché, Colonel Fournié, an energetic man who insisted strongly on the hygiene instructions that Doctor Dumas should apply.

The trip to Belgrade continued by train, which Captain Tulasne described as being "*in shabby second-class carriages that were fortunately disinfected beforehand.*" In their first sleep, the Frenchmen heard, as in a dream, the accents of the Marseillaise, then the sound of the doors opening and bringing them back to reality. Tulasne, the first to get off the train, recounts: "*I was immediately surrounded by 15 Serbian officers. Mutual presentation. They are cavalry officers of the independent division who want to celebrate our passage to Palanka and offer us a wine of honor and supper; the most generous wines, the most varied meats that we are served in abundance. When we got back into our wagon, we found it cluttered with food for the men and for us.*"

The detachment arrived in Belgrade on the third day of the trip, March 16, 1915, where they were welcomed by Serbian officials and by Lieutenant de vaisseau Picot, who had come with his sailors. They met their liaison officer, Captain Préditch, a highly educated academic who was also a Belgrade city councilor. While waiting for the arrival of the equipment which will follow in the next rail convoy, the squadron benefited from a few days of rest. The Serbian intendance lodged the officers in a hotel, then in a villa, and the men in a barracks of the capital. Belgrade, some of whose buildings were destroyed by Austrian artillery, was partly emptied of its population, but the presence of many soldiers gave it a bit of life. The doctors of the French military mission arrived in large numbers and the quartermaster's office kept them wearing the red pants (madder) of 1914: During the following days, the French airmen visited the artillery casemates around the city, which contained the naval guns of the Lieutenant de vaisseau Picot, who in fact instituted a balance of terror by forcing the Austrians to stop their bombardments of Belgrade, for fear of retaliation. On the other hand, they saw an Austrian plane flying over the military installations with impunity, without the slightest piece of flak going into action! Taking advantage of this period of technical unemployment, Captain Mortureux went to the fortress-town of Kragujevac, headquarters of

Above: The French National day, July 14, 1915, sees a ceremony on Banjica airfield where decorations are given to pilots and observers. Left to right: X, Cne André Mortureux, Serbian officer, Cdt Roger Vitrat, Serbian officer, Cne Richard Mavet, and Cne Louis Paulhan. (Collection Boris Ciglic)

Left: On July 15, 1915, Capitaine Louis Paulhan conducts in his plane (MF 193) Colonel Fournié, the French military attaché, to the Serbian headquarters in Kragujevac. (Collection Passot-Seguin)

the Serbian HQ, to request the creation of anti-aircraft artillery pieces, for which he tinkered with the sights and instructed the servants. Two batteries were thus assembled in Belgrade as well as in other points of the front and obtained some results during the campaign. The other officers took advantage of this forced vacation to do some sightseeing and improve their lives. A lady living near their their villa put all her silverware at their disposal. Speaking perfect French, she invited the airmen to have tea at her house, asked to interview the famous airman Paulhan, and pressed the others

Above: French officers in Serbia pose with Serbian officers on July 15, 1915, just a few days after a contingent of reinforcements arrived in escadrille MF 99 S – and Commandant Robert Dangelzer, standing on the far right. On the far left in white uniform is Capitaine Louis Paulhan and standing on the center is Commandant Roger Vitrat. All officers in dark uniforms are French navy officers serving in the artillery detachment protecting the city of Belgrade against Austrian monitors. (Collection Passot-Seguin)

with questions... She impressed the officers with her great culture and some found her quite charming. Fortunately, Lieutenant Picot warned the airmen in time not to fall into the trap of this formidable adventuress, who had already blackmailed the British naval attaché in Belgrade by claiming to be pregnant by him!

In the meantime, the convoy of equipment finally arrived and the planes were unloaded in crates at the station of Palanka, south of Belgrade, where it was decided to install the squadron's park. While the mechanics were busy reassembling the aircraft, a detachment under the command of Captains Mavet, de Rochefort and Tulasne remained in Belgrade with the aim of finding a location close to the capital that would serve as an operating base for the squadron. They chose Banjica ("Banitza", according to the French pronunciation in the documents), a village 10 km to the south, with several villas belonging to the country's wealthy elite, as well as a maneuvering field that would serve perfectly as a landing strip. The detachment camouflaged its vehicles in a small wood, set up the first tents, and requisitioned the villas as well as a barracks that would be used to house the troops. Tulasne wrote in his diary: "*The squadron arrived from Palanka by air on Sunday morning, April 18, at 8 o'clock. The weather being very uncertain, we did not expect them. The tractors and all the equipment came loaded on a special train but they were unloaded only at 6 o'clock in the evening. The roads leading from the terminal station in Belgrade to the Banjica field were so impassable that two tractors left there were derailed by oxen the next day. The following days were used to install the hangars and the personnel.*"

As soon as he arrived, Dr. Dumas had the buildings disinfected by bleaching them with lime, burning sulfur, and cleaning the floors with petroleum. One of the villas was used as a command post with a radio and a photo lab. A long table set up under the shades that was lined with a tent will serve as an open-air dining room. The mechanics set up the

Above & Facing Page: Spectacular photos showing in action MF 193 of Capitaine Louis Paulhan and his gunner, probably his mechanic, Corporal André Lestradet. (Collection Passot-Seguin)

canvas hangars housing the aircraft, and always resourceful, recovered the boxes that contained the aircraft to make barracks. A village was born, and the mechanics soon erected a grey piglet as a mascot, which they named *Baptiste* and decorated with a green ribbon.

4.3 - First Flights and Time for Medals

Despite a gust of wind that destroyed Farman No. 196, the squadron was quickly operational and the first war flights took place on April 23, 1915, when three Farmans took off: the Serbian general staff requested daily reconnaissance of the front line. The squadron divided the task between three aircraft: Captain de Rochefort, piloted by Martinet, observed Syrmia, Captain Mortureux, piloted by Adjudant-Chef Thirouin, spotted Bosnia while Captain Paulhan led Captain Tulasne over the Hungarian Banat. Tulasne noted in his notebook: "*God, it's so quiet over this country! Not a cannon shot, not a rifle shot that I know of, because no trace of bullets in the canvas.*" The population of Belgrade was enthusiastic about the sight of the aircraft with the tricolored cockades, and from Sunday, April 25, an atmosphere of an air show of the good old days reached the Banjica airfield, where a crowd of civilians and Serbian and foreign military personnel

rushed in. Senior Serbian officers asked to take a first flight... A British military doctor, very chic, was driven to the airfield in a squadron car, and, a great prince, gave a tip to his French driver, a very young soldier of 18 years. The latter accepted with a blush... It was private 2nd class Rothschild, from a very rich and famous family of French bankers and industrialists!

Aerial reconnaissance was very much appreciated by the Serbian staff, and quickly became the daily task of the squadron when the weather allowed flying. The Serbian artillery having only few shells to fire, artillery-sporting missions would be very rare. On the other hand, the aerial bombardment would be immediately practiced on the enemy positions and in particular on the river gunboats, the main enemies of the Serbian defenders of Belgrade who ask that they be attacked. On the evening of 29 April 1915, five Farmans took off and bombed the monitors from an altitude of 2,000 meters. None of the bombs hit, but some military buildings were hit and, more importantly, the population of Belgrade was able to see the action from their windows, which boosted their morale. The night that fell was clear enough, and Captain de Lareinty-Tholozan, a specialist in night flights, set off again for the objective. Captain Mortureux, jealous of the other officers, all decorated with the Croix de Guerre, only wanted to see some action and took the observer's place. He would be well served in strong sensations... While the Farman was approaching the monitors, a flak projector sweeps the night. The two men flew over the target, and Mortureux removed the safety from the firing pin and threw the bomb overboard by hand. No

Above: In August 1915, the plane of Adjudant Thirouin (center) and Captaine Tulasne had to make a forced landing in Sabac. Serbian peasants gather around the plane and Serbian women filled it with flowers. (Collection Boris Ciglic)

sound of explosion was heard... They soon discovered to their horror that the bomb had caught a landing gear turnbuckle with one of its fins, and was hanging upside down! Landing would mean certain death. After getting rid of his three other bombs, Mortureux succeeded in a feat of acrobatic flying by kneeling on the lower wing, and, leaning dangerously forward towards the void with his arms outstretched, managed to detach the bomb which fell into a field. Needless to say, he did not have a parachute like all the other airmen of the time. He did not steal his *croix de guerre*!

The flights continued and were similar, with the squadron carrying out, as soon as the weather permitted, a daily reconnaissance of the three sectors of the front, in addition to occasional bombing raids, such as the Banaviste airfield, which was attacked on 1 May 1915. At the request of the Serbs, who wished to provide air cover for the river port of Prahovo, two Farmans piloted by Paulhan and Seret flew to this town downstream from the Danube, where equipment from Russia was unloaded. The detachment returned on June 9, but new pilots left later for the same task: Commander Vitrat did not appreciate this "big gap" imposed on his squadron and asked for reinforcements in Paris.

However, not all these flights were carried out in a sky devoid of any opposition. The Austrians had two Fliegerkompagnie, Flik 15 in Novi Sad and Flik 9 in Bavaniste, from which their aircraft decorated with red and white stripes left for reconnaissance and bombing raids. As early as April 26, three days after the first flights, a Farman was launched in pursuit of an Austrian... A primitive lookout network was set up and an aircraft was always kept on alert to go on interception, but the results were long in coming, as the Farman's climbing speed was so mediocre. The first encounter occurred on April 27, when Adjudant-Chef Thirouin encountered an enemy aircraft and approached it from less than 100 meters. But his gunner, Corporal Janio, was so surprised that when he was about to fire he realized that he had forgotten to cock the machine gun! The time to grab the musket, the Austrian woke up and shots were exchanged (one bullet hit the Farman), the enemy aircraft managing to escape. In the meantime, the flak guns installed by Captain Mortureux proved to be more effective in repelling a raid on the Banjica airfield on 9 May (they shot down two enemy aircraft during the campaign). But on June 9, Adjudant Thirouin fulfilled his dreams of revenge. The telephone rang in Banjica: the Serbian headquarters in Kragujevac reported that it was being bombed by an enemy squadron! Vitrat took off three Farmans, which had the time to gain altitude and place themselves in a good position to cut off the Austrians returning to their base. Adjudant Thirouin's Farman MF 194, flying at 2,500 m, was lucky enough to come across one of them as it crossed the Danube near the town of Smederevo. He spiked and allowed his machine-gunner, Sergeant Major Magnoni (who was also the squadron's accountant), to fire a good burst that hit the target and shot down the enemy on the Austrian side. The result of the fight was confirmed by the ground observatories as well as by a photo of the wreckage, taken from the Farman of Paulhan. As for the Austrian archives, they do not record any losses on that date, but they do tell us that on July 7, 1915, an aircraft of Flik 9 engaged in an aerial combat with the French and was hit by its engine, having to make a forced landing in the marshes near Homokbalvanyos (Bavaniste,

Above: Austro-Hungarian Knoller Albatros B.I 23.09 of *Flik* 15, in the airfield of Novi Sad during autumn 1915. (Collection Boris Ciglic)

north of Smederevo on the Austrian side). The place of the crash being very similar to the one claimed by the French, the date of July 7 is undoubtedly an error of transcription of the date of June 9. In any case, the victory was duly certified to the two aviators and raised the enthusiasm of the Serbian officials, first of all the regent Alexander who personally wanted to give the military medal to the heroes of the day.

A few days later, on June 14, two airmen also received decorations that they would not soon forget. Returning from Sabac, from where it was to carry out a reconnaissance, the Farman flown by Adjudant Peté, with Captain Tulasne as observer, was caught in a thunderstorm and had to make an emergency landing. Peté landed in a field full of wild grass, mowing down the vegetation with his landing gear tensioners, one of which broke. Judging that it was dangerous to take off in these conditions, the two aviators went to look for help and found only peasants who fled – they managed to get close by shouting "Fransuzki! None of the Serbs spoke French, Tulasne succeeding with great difficulty in making them understand that he had to be taken to the nearest telephone station to join his squadron. While Peté stayed to watch the airplane, Tulasne left the field strewn with a peasant on a plow horse, in direction of a farm where he is offered food while another horse is saddled, on which he then follows his host for several kilometers in the wilderness to a post office where there is no "telephon" nor fresh horses. While Tulasne thinks about Michel Strogoff's misfortunes (the famous hero of writer Jules Verne), his host invites him to continue the road and he finally finds a military post from where he can call Sabac, not being able to reach Belgrade directly. He manages to tell his adventure to a correspondent who knows three words of French, then goes back to his plane accompanied by a detachment of soldiers who relieve Peté in the guard of the plane while the night falls. The next day, after a night spent in a farmhouse, the airmen were able to repair their turnbuckle with a wire while the Serbian soldiers mowed down a runway for takeoff. The Farman could take off towards Banjica, filled with bouquets of flowers with which the Serbian peasant women had entirely decorated the aircraft. Very touched by this attention, Tulasne

Above: Austro-Hungarian aircraft on Bavaniste airfield during year 1915, the opponents of escadrille MF 99 S, who claimed two of them. (Collection Boris Ciglic)

will note in his diary: "*O small bouquet of Serbian women (...) you are more precious to me than the honors and official decorations.*"

On June 28, a game of cat and mouse took place between the Austrian and French planes, where the former took advantage of the slightest absence of the protection patrols of the latter to attempt an aerial bombardment. They succeeded in launching projectiles on Banjica... But the flak guns transformed by Mortureux came into action and a Serbian servant, Corporal Stanislav Juric, managed to hit an enemy aircraft which had to land urgently in its lines north of Zemun, its crew being composed of the pilot named Foret and the observer, Oberleutant Ludwig Hautzmayer, a future ace with 7 victories. On July 1, 1915, Vitrat telegraphed to Paris that his squadron had made 217 war flights representing 344 flight hours, and launched 170 bombs on the enemy. He had every reason to be satisfied: the requested reinforcements arrived from France on July 13, in the form of new aircraft and three pilots (Second Lieutenant Maurice Maire, Adjudant René Cornemont, Corporal Pierre Bastide) led by Captain (temporary commander) Robert Dangelzer, an active officer and reckless pilot who liked to panic his observer by flying into a flight of migratory birds... While the newcomers were being welcomed, Paulhan intercepted an enemy plane that had come to fly over Serbian territory and fired 100 rounds of ammunition at it from quite a distance, apparently without success. The next day, July 14, was the occasion for Colonel Fournié to award decorations to the pilots who had distinguished themselves – the squadron now had 10 flying aircraft, but was about to lose one pilot, Lieutenant Martinet, who was ill and had to be evacuated to France.

On the morning of July 23, a new aerial bombardment on Belgrade by the Austrians led the Farmans to take off on alert. Captain Paulhan flew towards the enemy, taking on board the soldier André Lestradet, who was none other than his pre-war mechanic when he was a civilian pilot. Their Farman MF 193 crossed paths with an enemy aircraft north of Borča, in a swampy area north of Belgrade. Paulhan approached the enemy from the left and across: "*It was not long. Flanking him with 100 machine-gun bullets at a distance of 80 to 100 m, he saw the plane dive to the ground without direction, falling like a dead leaf from one wing to the other, to come crashing down in the flooded spaces between Temess (Tamich) and the Danube, where the plane turned upside down.*" Paulhan, having nothing to fear from the enemy in these deserted spaces, descended while spinning up to 800 m above his enemy on the ground and to finish him off sent a bomb which fell less than 30 m from the wounded bird, if not already hit to death. Meanwhile, in Banjica, Major Vitrat decided to retaliate against this bombing of Belgrade by personally leading with Thirouin and Dangelzer a retaliatory raid on the airfield of Bavanište. The French aviators had the satisfaction of announcing their exploits to the Belgrade authorities, who were phoning the general staff to complain about "*these Frenchmen that we never see flying*". Tulasne relates the events: "*At that moment Paulhan returned. To stop them in their tracks, they phoned the event*

Above & Two Following Pages: On August 9, 1915, the heir to the throne of Serbia, Aleksandar Karadordević, visited escadrille MF 99 S in Banjica and was offered a flight as passenger of Maurice Farman n°193 piloted by Louis Paulhan. (Collection Passot-Seguin, Boris Ciglic for pages 52–53.)

and so that they could not doubt it, another plane was sent to the scene to photograph the remains of the enemy aircraft. From the rest of Belgrade, several notabilities, among them a captain and the Francophile lawyer Ivanovitch, were able to follow the fight and the fall of the opponent, so that a small demonstration of enthusiasm was organized in the streets cheering the "Franzuski" and counterbalancing the offensive suspicions of a frightened municipality and completely ignorant of the things of the aviation." Paulhan's victory was duly certified and was the subject of a citation. An Austrian document tells us that on July 24, 1915, Oberleutnant Milorad Tarabochia, an observer with Flik 15, was hospitalized. He was wounded in the arm the previous morning in an aerial combat in which his plane, piloted by an unknown pilot, crashed in the swamps. But Oberleutnant Wilhelm Schreiber, an observer, reports in his memoirs that he was personally shot down near Borča with his pilot Oberleutnant Béla von Losonczy, commander of Flik 15 (wounded in the arm by the shot), their Lloyd C I being destroyed by a bomb thrown by the enemy aircraft. The circumstances of the fight are similar to those described by Paulhan, however the fight is dated July 13 instead of July 23... The Austrian archives leave us in the greatest confusion but do not prohibit the hypothesis that Paulhan actually won two aerial victories.

4.4 - Germany Enters the Scene

The summer of 1915 continued but the storm was threatening and the airmen had a front row seat to see the beginnings of it: sooner or later, the enemy was going to

MF 5

MF 193

Above: Another photo of the visit of Aleksandar Karadordević to escadrille MF 99 S on August 9, 1915, (Boris Ciglic)

attack Serbia again. In the Serbian general staff, the far-fetched project of setting fire to the wheat fields of the Hungarian Banat by plane was circulating... More seriously, the opportunity to carry out a preventive attack against Bulgaria is being discussed, whose entry into the war on the side of the central empires would be catastrophic. On August 20, 1915, Adjudant René Cornemont got lost while flying his Farman to the Prahovo detachment and landed, to his great amazement, in Bulgaria where he was interned. He was quickly released after a few days after the intervention of the French ambassador, Mr. de Panafieu. The latter did even better and also obtained, after several other days of negotiation, the return of the aircraft: Cornemont crossed the border again to retrieve it. The local officers, who had all tried it, demanded a "fee" to let it go. The French military attaché paid the baksheesh for him and Cornemont could return to Prahovo with his Farman covered with Bulgarian cockades...

In September, two new Farman aircraft arrived in crates, as well as a final reinforcement pilot, Lieutenant Pilot Milan Štefánik, an officer of Slovak origin and a brilliant scientist renowned for his work in astronomy and meteorology, who

Left: In November 1915 begins the long retreat of escadrille MF 99 S across the mountains of Albania, following the fate of Serbian army. Here, Capitaine de Rochefort stands with Serbian soldiers on a narrow road near the Jankova river. (Collection Boris Ciglic)

Above: On November 2, 1915, what remains of escadrille MF 99 S stands in Raška. (Collection Boris Ciglic)

would lead the Czechoslovak army at the end of the war. But it was also a period when the reconnaissance missions carried out by the squadron confirmed large troop movements. Germany came to the rescue of the Austrians by sending its 11th army, which is assigned to the east of Belgrade, with its air force of six Flieger Abteilungen and especially a powerful artillery. Faced with the imminence of a German attack on this sector, Commander Vitrat, with the agreement of the Serbian General Staff, detached on September 22 three aircraft under the command of Major Dangelzer (Dangelzer, Selaquet, Seret) to Pozarevac, which he reinforced with two others (Maire and Bastide) recalled from Prahovo on September 29. The rest of the squadron left Banjica on September 24 to establish itself in Ralja under the direct orders of Major Vitrat. This evacuation caused one casualty: the pig-mascot Baptiste, too big to be on the trip, who was promptly turned into a ham... The movements of the squadron were judicious, as the German artillery was firing on the south bank of the Danube. The airfield and the town of Pozarevac were subjected to violent aerial bombardment, which the French Farman aircraft tried to intercept, claiming to have encountered twin-engine aircraft armed with two machine guns. One of them, after a difficult aerial battle, seriously damaged the aircraft of Adjudant Seret by cutting off its tailplane; it managed to land on its own ground. Those planes may well be AEG G.II, at this time flying with FFA 1 sent on the Serbian front

For the time being, Vitrat was concerned about avoiding the destruction of his detachment on the ground and moved it to a more discreet location in the South on October 1, 1915. But events were accelerating. The Czar of Bulgaria, strongly solicited by the diplomatic efforts of the central empires, committed himself to leading his country against Serbia and thus taking revenge for his defeat in the second Balkan war of 1913. Faced with the mobilization of Bulgarian troops, the Serbian general staff asked Vitrat to send a detachment to the eastern border; he had the Ralja detachment moved to the town of Niš on 7 October, while the Pozarevac detachment settled on 16 October on the Palanka airfield (which the air park services of the squadron had just evacuated) and ensured the entire surveillance of the northern front on its own. Vitrat went to the fortress town of Kragujevac, where the Serbian headquarters was located, and from where he tried to command his two detachments. The air fleet was evacuated to Kraljevo on 10 October for parts and fuel, while a workshop was set up in Kruševac where mechanics were feverishly assembling the last reserve aircraft.

Above & Above Right: On November 11, 1915 , old Farman MF 193 of Capitaine Paulhan ended its career while landing under a stong wind. Note the Serbian cockades. (Collection Passot-Seguin)

4.5 - The Withdrawal

The stay in Palanka of the Northern detachment of the squadron was short. The German troops, whose artillery had cleared the way, crossed the Danube and forced the Serbs to evacuate their capital once more. The detachment then followed the retreat of the Serbian army from October 19, experiencing several retreats towards the South along the Morava river before joining the park at Kraljevo around October 30.

As for the Eastern detachment, commanded by Captain Paulhan (pilots Lt Štefánik, Adjudant-Chef Thirouin, Adjudants Peté and Cornemont), its operational activity was also short-lived. France, which was not at war with Bulgaria on October 7, wished to avoid any incident and the five Farmans were covered with Serbian roundels... While the Bulgarians attacked on October 15, the detachment was grounded by heavy rain until the 25th of that month. Its aircraft carried out five reconnaissance missions which revealed that the Bulgarians were about to seize Kumanovo, cutting off any retreat towards Salonica, and that they were approaching Niš dangerously... Vitrat then telegraphed Paulhan to evacuate to Prokuplje and from there to Pristina, but the advance of the Bulgarian troops made this route hazardous and he was finally ordered to join the park in Kraljevo by way of Kruševac. While the panic reached the local Serbian authorities, Paulhan did not manage to find a place on a military train and he must resolve to have all the material he cannot transport destroyed. The mechanics evacuated by vehicle, while the five Farmans took off towards their new destination... Except for Lieutenant Štefánik›s MF 499, which broke its propeller at take-off. Adjudant Cornemont's plane saw a spar bend in flight and the pilot had to finish his trip on foot after an emergency landing. Štefánik and his mechanic remained in Niš waiting for a spare part promised by Vitrat. They were afraid that the Bulgarians would take the city any minute... But fortunately, a Farman brought him a spare propeller and after a makeshift repair his plane takes off painfully under the rain to reach Kruševac where his propeller breaks again on landing. But the mechanics of the park, led by Captain de Lareinty-Tholozan, managed to reassemble three aircraft stored in a crate, before receiving the order to join the rest of the squadron on foot (which they would meet again on November 15 in Mitrovica). Štefánik was then able to change planes, as well as two pilots of the Northern Detachment, Adjudant Selaquet and Corporal Bastide. Štefánik and Bastide were the last pilots to leave Krusevac, but a heavy storm was raging and the two pilots were waiting for a lull in the storm as the town was hurriedly crossed by the last Serbian detachments. They finally tried their luck and the planes barely advanced in a headwind. A magneto breakdown forced Bastide to land, he joined his squadron by his own means (on an oxcart...) and found his comrades in Raška. As for Štefánik, the strong gusts of wind broke his spars. He managed to land in a small Serbian town after having razed the roofs of the houses, not far from Bastide's plane. He then tried to repair with his mechanic by taking parts from the other plane, and managed to take off again, but a new damage forced them to stop their flight. They had to burn their Farman and venture on foot through the mountains of Serbia, where they were followed by the Bulgarians and managed to find their comrades in Mitrovica.

Contrary to the fights of 1914, the Serbian forces that had to face the Austrians and the German and Bulgarian armies could not perform miracles. The Serbian general staff decided to evacuate the country but still hoped to be able to reach the Greek border from where Franco-British troops transferred from the Dardanelles arrived. The main part of the squadron stood at Kraljevo on 30 October 1915, reduced to eight aircraft. Despite the congestion of the railroad track, which forced them to abandon a few cars en route, the park services managed to evacuate good stocks of fuel and spare parts, although they lacked emergency aircraft and tents to shelter the Farmans from bad weather. This situation was not to last: it was necessary to evacuate again to the south to Raška to follow the progress of the large Serbian headquarters, which was much more difficult because there was no longer a railroad line but a narrow road that wound along the Morava

Above: Last stand of escadrille MF 99 S in Prizren under the snow on November 17, 1915. (Collection Passot-Seguin)

River... The evacuation had to be carried out by car and the personnel succeeded in evacuating 6,000 liters of fuel, oil, three engines, tools, as well as a few aircraft spare parts. The workshop truck, driven by a particularly skillful driver, even managed to reach Raška where operational service resumed on 2 November. Until November 8, numerous reconnaissance missions were conducted and demonstrate the Austrian advance on Čačak and that of the Bulgarians who had taken the town of Niš. If the planes do not lack anything, food is scarce and Vitrat, intractable, does not want to distribute the cookies and canned food that he was able to find in an abandoned Serbian stock and that he prefers to keep for the future, which looks difficult; the events will prove him right... The Serbian headquarters reached Mitrovica on 9 November, and Vitrat sent his squadron there, but left a detachment of three aircraft in Raška, as well as his precious workshop truck, which helped the Serbian artillery to adapt 75 mm shells to its mountain guns, with orders to evacuate as soon as the three Farmans had to join the rest of the squadron. This order arrived on November 14 and only two Farmans returned, the one belonging to Adjudant Selaquet being destroyed at take-off by a strong eddy which caused it to crash to the ground. Lieutenant Paulhan's usual plane having been destroyed by a storm, there are only six left with fuel that is now running out. Worse, the workshop truck could not drive over a bridge because of its weight and had to be abandoned, as well as many vehicles immobilized on the way by mechanical breakdowns. Vitrat received the order to evacuate again to Prizren on the evening of November 15: it was necessary to act quickly because the only road leading to Prizren was 20 kilometers away from the first Bulgarian troops... He sent Captain Tulasne on the road to Raška, trying to save as much material as possible, in particular the machine tools of the truck workshop, with the order to reach Prizren as soon as possible. This city was reached by the squadron on November 17 under a snowstorm. It was crowded with routed troops and Vitrat had to ask for the reinforcement of Serbian gendarmes to force Albanian

This Page: Most of the personal of escadrille MF 99 S had to evacuate Serbia on foot across the snowy mountains of Albania. (Collection Boris Ciglic)

traders to sell food to his men in exchange for paper money. The final Serbian offensive towards the South having failed, it was necessary to prepare for the retreat towards Albania. Colonel Fournier, head of the French military mission, gathered all the officers of the various detachments and distributed money to organize this crossing. Vitrat and his men added their own money to buy food, mules, fodder, and a small herd of ten sheep which would serve as a reserve of fresh meat! It is planned to reach the city of Scutari (Shkodër in Albanian), an Albanian city on the shores of the lake of the same name recently occupied by the allied troops of Montenegro, from where it will be possible to make contact with the Italian troops of Valona (Vlorë). It is neither more nor less than to consider leading a troop of poorly dressed men on foot for a journey of 100 kilometers through snow-covered mountains, on mule tracks and populated by Albanians at best hostile with their Serbian neighbors.

As for the planes that use the last reserves of fuel, the crossing in flight to Scutari is just as hazardous, given the state of fatigue of the engines and airframes that limit the ceiling. It would be necessary to resolve to fly with the estimate in the foggy valleys under an execrable weather. The five planes succeeded, however, the journey in difficulty, bringing with them a patient each. A sixth aircraft left on November 18 for Valona, in Italian hands, to carry out what seems to be one of the first medical evacuations in the history of aviation. Lieutenant Štefánik, exhausted by his long march from Kruševac to Mitrovica, was unable to move and needed urgent treatment in a hospital worthy of the name. Paulhan, the most experienced aviator, was entrusted with the task of flying to Valona. Carrying a bad map of Albania, still poorly mapped at the beginning of the century, the tired Farman took off painfully in bad weather and sank into the Drin valley, where it was buffeted by powerful air currents that threatened to press it against the rocky walls. One can imagine the anguish of the two aviators lost in a pea soup... Štefánik testifies: "*For fifteen minutes, we could not see anything; however, we knew that some of the mountains we had to cross had an altitude of 1,500 meters and our altimeter, at that moment, indicated 1,200 meters. We were therefore in danger of breaking against the bare rocks of the Albanian mountains. Paulhan gave the pedal, we climbed slowly, slowly... Suddenly, the clouds were torn apart and we found ourselves between two peaks, the wings of the plane barely thirty meters from the trees of the slope. We were in danger of catching them and crashing into*

Above: A detachment left behind, let by Capitaine Tulasne, managed to join Scutari across the mountains of Montenegro. (Collection Tulasne)

the rocks. As soon as we arrived near Alessio, the panorama was marvelous and of a grandiose picturesque. Suddenly, we saw the Adriatic. We were saved, because we had the possibility to land on the shore in case of necessity. Finally, three hours and ten minutes after our departure from Prizren, we touched down in Valona, where the Italian officers celebrated us. The Admiralty, informed of the arrival of two French airmen, one of whom was seriously ill, sent a high-speed destroyer which, in four hours, took us to Brindisi."

For the mechanics and drivers, the journey was much more difficult. The French column left Prizren for the adventure on November 25, 1915 through the Drin valley with all its animals and surviving vehicles, which were abandoned in the river gorges as soon as the road became a mountain path. This one, rather slippery, causes some falls of beasts of burden. After having followed the banks of the Drin for two days, the troops entered the mountains on November 28, heading southwest from the village of Spach. The temperatures became extreme (-18 degrees). A stream, the Mihanya, was transformed into a torrent and Vitrat noted that "*To try to save horses, the men entered the water up to their waists and came out with a shell of ice. The shoes are transformed into ice cubes and the spectacle of this desolate nature is made even more terrible by all the corpses of Serbs who died of hunger and cold that line our route. For the first time, we managed to stay in a "Han" [a traditional hut], but the Albanian owner demanded that the colonel pay twelve gold Napoleons. We have to comply.*" Several horses fell into steep ravines, including the one containing all the archives of the squadron, which were definitely lost. After seven trying days, Scutari was finally in sight, although the food supplies were exhausted. However, the men were not at the end of their troubles as they had to cross the swamps that surrounded it, which were knee deep.

It is finally a ragged troop that reaches Scutari on the morning of December 3, 1915, finding the pilots and the planes that are reduced to four because Adjutant Peté broke his plane on landing. Vitrat had no difficulty in obtaining accommodation and food for his men from the Montenegrin authorities, but several of them had to go to the hospital because their fingers and toes were frozen by the crossing. While the small detachment of Captain Tulasne, lost on November 15 in Mitrovica, reached the city on December 6 after a no less perilous journey through Montenegro, Vitrat tried to organize the flights with local means. The first missions of the four remaining planes consisted in organizing flights to Valona because the telegraph was not working well. He took advantage of the situation to evacuate the sickest men by air. These flights were relatively safe because they followed the shores of the Adriatic. Nevertheless, the planes remained in a very fragile state and Major Dangelzer broke his plane after an emergency landing near Scutari. In the city, which was filling up more and more with refugees from Serbia, Austrian army and navy planes made daily visits and sometimes carried out bombing raids. Vitrat had his Farman planes take off to give chase, although this was a bluff because the machine guns of the planes had been lost in the Albanian mountains... Some of his men

Above: Tired personnel of escadrille MF 99 S arrive in Bari on December 25, 1915, evacuated by the French Navy across the Adriatic. The Serbian adventure ended but most of these men will volunteer to continue the fight with their Serbian brothers in arms in Serbian aviation reformed in Salonica. (Collection Tulasne)

improvised a flak gun at the request of the local authorities. On December 23, 1915, the Lohner L 60 seaplane of the Austrian Navy (pilot Flieger Meist Heinrich Roschitz, observer Linienschiffleutnant der reserve Alois Poljanec – a Slovenian) spotted the installation and bombed it, killing the soldiers Marius Lecaroux and Marcel Mettetal, who would be the only dead of the squadron during the whole campaign. On Christmas Eve, colonel Fournier announced the good news of the evacuation order to Saint Jean de Médua (Shengjinen in Albanian) where ships of the national navy were to transport them to Italy. A final journey on foot was made from December 25 to 28 over muddy roads, while the last two Farmans in flying condition evacuated the most seriously injured. Seven untransportable soldiers had to be left at the hospital in Scutari, as well as four men who were to ensure the functioning of the telegraph station in Podgorica, capital of Montenegro, until the end. The whole squadron embarked for Bari and then for France by train. The men finally ended 1915 in a hospital in the Lyon region.

When the squadron was disbanded, Vitrat wrote a long report on all the activities of his squadron since the beginning of this long retreat, in which he praised the stamina and spirit of his men in these very difficult moments. The Serbian government, grateful, will not spare the decorations to the personnel of the squadron, only constituted troop to have shared the destiny of the Serbian army which will still have in front of it long months of fight in exile before finding its fatherland. The officers of the squadron met again on January 8, 1916 in Paris where they attended the wedding of Captain Mavet. After a long period of rest, those precursors of the Orient front will join the city of Salonica to become the commanding officer of the future Serbian air force, which will be led by Major Vitrat himself.

Empire Austro-Hongrois
0 50 100 km
Novi Sad
Orsova
Danube
Belgrade
Sava
Zemun
Bavaniste
Pojarevatz
Banjica
Roumanie
Sabac
Ralja
Smederevo
Prahovo
Palanka
Kragujevac
Bulgarie
Caçak
Sarajevo
Kraljevo
Kruševac
Niš
Prokuplje
Raška
Serbie
Sofia
Mitrovica
Pristina
Monténégro
Kumanovo
Podgoritza
Drin
Prizren
Spach
Scutari
Antivari
Alessio
St Jean de Médua
Vardar
Salonique
Albanie
Grèce
Valona

Chapter 5: Salonica and the Orient Front

5.1 - Constitution of the Orient Front

When we think of the First World War, we usually think of the trenches that were dug in the north of France and the battlefields of Verdun. This is to forget that clashes also took place on the plains of Russia, in the Italian Alps, in Palestine, Iraq, and Africa, as well as in the north of Greece in what was designated at the time as the Front d'Orient – the Orient Front.

This Orient Front was formed in 1915 with the Dardanelles expedition. The British First Sea Lord, Winston Churchill, noticed that the British army was stuck in the trenches in France and was determined to achieve a military victory by the sheer force of the powerful British Navy. His ambition was to conquer the straits of the Bosphorus and the Dardanelles by the force of his battleships, to seize Constantinople (now Istanbul), the capital of the Turkish empire, and to force the sultan to capitulate – and in the process, to secure a supply route to the Russian ally.

This military adventure resulted in a fiasco. Having suffered heavy losses against floating mines while attempting to cross the straits, the French and British navies had to abandon their attack and leave it to the troops on the ground who were trying to seize the heights bordering the straits. There too, the Franco-British troops met with a bitter failure and suffered heavy losses in the face of Turkish troops who proved to be formidable fighters, defending their national territory. At the end of 1915, the decision was taken in London and Paris to stop the losses and to bring back these troops.

But the French Minister of Foreign Affairs, Aristide Briand, who on October 29, 1915 became the head of the government, had an idea: to use the troops of the Dardanelles to go and help Serbia, which since the beginning of the war was still valiantly resisting the assaults of the Austro-Hungarian Empire. It was with this aim that the French government created the Armée d'Orient (Orient Army), whose command was given to General Sarrail, an officer that Joffre, the head of the French army, had dismissed on the pretext of the excessive losses that he had caused by his military operations in 1914. Sarrail was in fact neither worse nor better than the other generals; but he was a declared radical-socialist (republican center-left party of the French 3rd Republic) sympathizer who had such strong political support that he was a potential candidate to replace Joffre himself... It is therefore not surprising that the government with a radical-socialist majority offered him the position of head of the new Orient Army. It is not surprising either that Joffre had no problem letting him leave for this galley, delighted to get rid of him.

For it was indeed a galley. To rescue Serbia, it was

Below: View of the city of Salonica (Paul Andrillon)

Above: Aerial view of the city of Salonica (Charles Denti)

Below: Surveillance of the Orient Front: a landscape of arid mountains, very different from the trenches of France (Collection Laurent Winer)

Salonica During the First World War

As the rear base of the Orient Army, the city of Salonica (Salonique in French, now called Thessaloniki) is a cosmopolitan city that was for a long time a city of the Ottoman Empire with a strong economic and cultural influence, and that only passed under the sovereignty of the Greek government in June 1913. The Greeks were only a minority here, as according to the 1913 census the city has 157,000 inhabitants, of which 61,000 are Jews, 45,000 are Turks, 40,000 are Greeks, 6,000 are Bulgarians, 3,000 are Aromanians (or Vlachs) and 2,000 are Levantins (descendants of western Europeans from the crusades).

The presence of this important Jewish community, which constitutes nearly 40% of the population, comes from the fact of the expulsion of the Jews from Spain at the end of the 15th century – many of them found refuge in this city of the Ottoman Empire where a Judeo-Spanish dialect is spoken. Salonica is then an economic center of very first order which radiates in all the Balkans. Its railroad, which goes up the Vardar river, is the only true maritime frontage of Serbia. In the 19th century, Moïse Allatini (1809–1882), a Jew from Italy and ancestor of French aircraft constructor Marcel Dassault, developed the city considerably by founding a number of industries, flour mills, food industries, brick factories, and tobacco processing plants. He became the third richest man in the Ottoman Empire and financed a number of charities and opened several schools, which aimed to educate pupils from kindergarten onwards to the evolutions of the modern world and to make them fit for the labor market of a society in the process of industrialization. At the beginning of the 20th century, the population of the city of Salonica was the most educated in the Ottoman Empire, and the French language, widely taught, was commonly spoken among the population to the point that several daily newspapers were published in French.

The city of Salonica had known several centuries of Ottoman sovereignty, until the first Balkan war of October 1912 which saw the Ottoman Empire lose its last European possessions following a combined attack by Greece, Serbia, Montenegro, and Bulgaria. It is Bulgaria which is attributed the sovereignty of the city but this division does not suit the other Balkan states which in June 1913 allied themselves with Turkey and Romania to attack Bulgaria and take back its share of the cake... The city of Salonica then officially passes under Greek sovereignty. The Greek administration was only two years old when the Allied troops arrived at the end of 1915. This event will largely upset the balance of the city of 157,000 inhabitants which sees nearly 300,000 Allied soldiers settle in huge tented camps around the city. These soldiers represented a host of nationalities: the French army, which constituted the largest share, brought all the ethnic groups of its colonial empire. In addition to the poilus from metropolitan France, there were North Africans, soldiers from Black Africa (Senegalese riflemen), and Annamite troops from Indochina. The army of the British Empire was not left out: soldiers from England, Scottish nurses, Hindus... The Serbian army, which was forced to flee its country, was taken in by the French navy and brought back to Salonica to be rearmed in early 1916. There was also a contingent of Italian troops and some Russian troops, and even, in the port, perhaps some sailors from the Japanese fleet which had sent a squadron to the Mediterranean to help escort naval convoys! In June 1917, Greece officially entered the war on the side of the Allies when French troops forced King Constantine I of Greece to abdicate. A new Greek government led by the former Prime Minister Venizelos then engaged the country in the war despite the opposition of many army high-ranking officers who were admirers of Germany.

In 1917, the Allied troops, which had grown to 650,000 men (of which a large third were French), were for the most part on the front, but many of them remained in the city, which served as the main logistical base for the Orient Front. The port of Salonica never stopped receiving ships unloading material, arms, and ammunition as well as reinforcements. They never left empty and brought back many medical evacuees: endemic malaria was rife in the northern regions of Greece and nearly one soldier in two was affected by the disease. The city of Salonica received all the volunteers from all the armies returning from the front; the restaurants were full, as were the theaters and even less the brothels...

One can easily imagine the work of the French gendarmerie in trying to maintain order in this gigantic Tower of Babel. In 1917, two events made things even worse. First of all, the Russian Revolution, which led the few Russian troops to refuse to fight, which in itself was not too serious from a military point of view because their number was quite limited. What is much more serious is that many of them are won over by Bolshevik revolutionary ideas and the staff is very much afraid that they will "contaminate" other soldiers, because we are at a time when mutinies appear in the troops in France following the failed offensive of the Chemin des Dames in April 1917. The Bolshevik agitators were hunted down without mercy and the spreading of their propaganda led immediately to the council of war.

The other dramatic event was a gigantic fire, which started on Saturday, August 18, 1917, in the kitchen of a house in the city center. A spark touched a bale of straw which ignited. The lack of water and the indifference of the neighbors prevented the fire from being quickly extinguished. In addition, a strong wind moved the fire to neighboring houses and then to the entire town. Water was in short supply because most of the water was used by the Allied military camps on the outskirts of the city. Moreover, the city did not have a fire department... The fire ravaged the entire city center and was not extinguished until 32 hours later, after destroying 9,500 buildings and leaving nearly 70,000 people homeless. Among the buildings burned, we can note the post office, the town hall, the gas and water companies, the Ottoman Bank, the National Bank, the deposits of the Bank of Greece, part of the Church of Saint Demetrios, two other Orthodox churches, twelve mosques, the seat of the Chief Rabbi and its archives and 16 of the 33 synagogues. The printing houses of most of the newspapers (Salonica used to have the largest number of newspapers in Greece), most of which cannot be reissued. About 4,096 of the 7,695 stores in the city were also destroyed, leaving 70% of the inhabitants without work. A large part of the population, mostly Jewish, then left to settle in refugee camps that the French and British armies tried to organize around the city. Additional tension: the Greek Prime Minister Venizélos forbids the access to the ruins for the refugees who cannot rebuild their houses, because he prepares a reconstruction according to a modern urbanism plan.

In spite of this chaos, the city remained throughout the war a synonym for rest and entertainment for the airmen of the Orient Front, who, not without displeasure, took turns to fly from their airfields for the mail service and landed their aircraft in the park of the Allatini camp in the south of the city, leaving only after a stopover in the bars, restaurants or other establishments of the port, of which the white tower is the symbol...

necessary to disembark in Greece, a neutral country whose prime minister, Eleftherios Venizelos, was in favour of the Franco-British Entente, but whose King Constantine I, who held the army, was an admirer of Germany and a brother-in-law of the Kaiser. Venizelos gave the French troops permission to land in Salonica, but the King dismissed him from his post. When Sarrail and his men set foot on the quays of Salonica, they were in a neutral country where they were not welcome but where they imposed themselves by force of arms. The British also landed troops there, but they were instructed not to leave the city.

Above: Urban landscape in ruins after the fire of Salonica on August 18th 1917. (Coll Boris Ciglic)

Above: General Maurice Sarrail (1856–1929), commander in chief of the Orient Army. (Rol)

5.2 - The Vardar Expedition (October–December 1915)

The Vardar Expedition, the name of the river that flows into Salonica and has its sources in Serbia, can begin. It was already too late: Serbia had succumbed to the onslaught of German troops who had supported the Austro-Hungarian troops who had been unable to conquer the country since 1914. Bulgaria participated in the curse by attacking simultaneously at its border. The French troops set off as winter began in the Vardar valley, whose course was bordered by a railroad linking Salonica to Serbia, which was then the only means of access to a port controlled by the Entente powers.

With the French soldiers, three squadrons of planes landed in Salonica, initially to reinforce the Serbian aviation: squadron C 89 S (S for Serbia, with 6 Caudron G.3), V 90 S (6 Voisin LAS) and N 91 S (6 Nieuport 10 two-seaters). The head of Orient Aviation, Major Victor Denain, had installed his men on a vast maneuvering field south of Salonica, at Camp Allatini, where the planes were brought up. The first of them, a Caudron G.3, took off on October 26, 1915 in front of a crowd of Greek civilians and soldiers who had come to watch the show.

The Caudron soon followed the progression of the French troops along the Vardar; their northernmost point being the village of Gradsko – the three squadrons then met on an improvised airfield in the valley. But General Sarrail stopped his progression: in the meantime, Serbia collapsed and the remnants of his army fled the country via Albania. Bulgarian troops appeared on the mountains of the northern bank of the Vardar and fired with their artillery on the French troops. Fighting took place and C.89 and V.90 squadrons executed bombing missions on the Bulgarian positions before the order to retreat was given (See See testimony of René Vicaire, page 118). The French troops withdrew in good condition to the city of Salonica, despite the snow and the polar temperatures that swept the region. French airmen encountered German aircraft in the air for the first time and even attempted an interception. Despite the fog and the snowfall, they all managed to return to the city of Salonica on 15 December 1915.

5.3 - Air Battles on the Entrenched Camp of Salonica (January – April 1916)

After the Vardar campaign, the French army withdrew to the city of Salonica, in a curious political situation: the German-Bulgarian army stopped at the Greek border. Between this border and the city of Salonica, Greek troops were stationed, officially neutral, but whose leadership was clearly pro-

Above: This Caudron G.3 of the C 89 S squadron is one of the very first operational aircraft of the Orient Air Force, accompanying the French troops during the Vardar expedition. (Coll. Albin Denis)

Below: Nieuport 10 two-seater (n°377) belonging to the N 91 S squadron which participated in the Vardar expedition (Coll. Paschalis Palavouzis)

Above: The squadrons of the army corps, intended for observation and artillery adjustment, are equipped at the beginning of 1916 with Farman MF 11bis with 130 hp engine. (Coll Boris Ciglic)

Facing Page, Above: First victorious airmen of the Orient Army, sergeants Henri Astor (machine gunner, left, hands behind his back) and Alain Terme (in fur) pose in front of the German aircraft they shot down and captured on February 1, 1916, which is being dragged down a street in Salonica where it will be displayed in front of the White Tower. This two-seater Nieuport 10 crew won a total of three victories in the Orient, with Astor adding a fourth as gunner to another pilot. (Coll Boris Ciglic).eater (n°377) belonging to the N 91 S squadron which participated in the Vardar expedition (Coll. Paschalis Palavouzis)

Facing Page, Below: Terme and Astor's third aerial victory, an Albatros B.II shot down east of Lake Amatovo on April 25, 1916, is also on display near the White Tower of Salonica for the pleasure of the curious. (Coll Boris Ciglic)

German. This did not prevent the British and French troops from taking precautions and establishing a fortified camp around the city. Even though the military situation on the ground was frozen, an aerial war began: on December 30, 1915, bombs were dropped on Salonica by German planes, causing little damage and killing an unfortunate Greek shepherd in a field. This raid was the first one, others were to follow... Five other French squadrons were landed in Salonica during the Vardar expedition: the N 87 (Nieuport 10 and 11 aircraft), the V 83 and V 84 (on Voisin LAS bombers), the MF 83 and MF 86 (on Maurice Farman MF 11 and MF 11bis reconnaissance aircraft), as well as a maritime aviation center equipped with 100 hp FBAs, which kept watch over the bay of Salonica for enemy submarines. This air force of 8 squadrons of ground planes was divided between three improvised airfields around Salonica, at Topcin, Sedes, and Samli – the latter receiving the 3 squadrons of Voisin bombers, the V 83, V 84, and V 90, which were to form the Groupe de Bombardement d'Orient (GBO), directed by Captain de Serre.

The French airmen were the only soldiers of the Entente to actually fight against the enemy in early 1916. The GBO started operation on January 28, 1916, against the bivouacs of German-Bulgarian troops near Lake Doiran. During the night of 31 January to 1 February 1916, bombs fell on the port and military warehouses of Salonica, dropped by

J. ISRAEL & I. ARDITTI
Z I
TANGYES LIMITED BIRMINGHAM

Above: At the beginning of 1916, the French airmen settled on three airfields around Salonica. Here, the one of Sedes, arranged around the building of the agricultural school in the south of the city. (Coll Boris Ciglic)

Facing Page, Above: Aerial view of the Samli airfield located north of Salonica, flown over by a two-seater Nieuport. (René Vicaire)

Facing Page, Below: Meals in the shade for Serbian and French airmen near this Nieuport 10 (n°767) of the N 91 squadron in the spring of 1916. (Coll. Albin Denis)

Zeppelin LZ 85, which returned to its base in Temesvar (Hungary; now Timisoara in Romania) without being disturbed. The emotion was great in the city and Commander Denain had to react: he sent the GBO for a bombing mission against military camps in Bulgaria the next day. While the Voisins returned unharmed, a two-seater Nieuport 10 of the N 91 S squadron, piloted by Sergeant Alain Terme with Sergeant Henri Astor as gunner, scored the first air victory of the Orient Air Force. The two men took off from Topcin to intercept a German two-seater reconnaissance plane flying over Salonica and cut it off on its way back. After a few shots were fired, they cut off its fuel line and forced it to make an emergency landing in the no-man's land between Salonica and the border. Terme and Astor landed next to capture their prey whose crew surrendered without resistance; Greek soldiers rushed to the scene to imprison the four airmen. Astor managed to escape and called for reinforcements: a detachment of French cavalry arrived at full gallop and forced the Greek soldiers back into their barracks... The German plane was dismantled and exposed to the public near the White Tower of Salonica, to demonstrate to the population the power of French aviation.

Other shot down German planes would be in turn exposed on the square of the White Tower the following days. While the GBO repeated its bombing missions on the German-Bulgarian lines, the Germans made several flights to Salonica, which were intercepted by Nieuport fighters. On 17 February 1916, one of them was shot down and captured – it immediately replaced the previous one on the White Tower square. On 23 March, the GBO's Voisin bombers encountered aerial opposition for the first time during their mission in the form of two Pfalz E's from the Hudova-based Armeeflugpark 13 fighter patrol, which did not do much harm, although an escort Nieuport and a Voisin were hit by flak – the crew of the former escaped unharmed, while the latter was killed.

The Germans retaliated against the GBO raids by launching an air raid on Salonica at dawn on 27 March 1916, using all their available aircraft. All the French fighters took off for some tough explanations with their machine guns and 37 mm cannons which equipped some Voisin bombers. A new trophy reached the place of the White Tower, still thanks to the Terme / Astor crew who won their second victory. The Germans stopped their bombing raids on

Above: Nieuport 10 two-seater (n°758) photographed at the Salonica park where it was reassembled on 28 May 1916. Having suffered from its sea voyage from France, it had to be entirely repaired by the local means of the park on June 22, 1916, from where it could be delivered to the N 390 squadron on August 7, 1916. Once again recast in the park, it had a second career at the N 391 where it ended its operational life at the end of 1916. (SHD)

Below: This photo taken at Sedès airfield in early 1916 at N 87 squadron shows the different colors of the squadron's Nieuport 10 two-seaters - combat reports speak of "blue" Nieuports (probably silver canvas), and "yellow" Nieuports (unbleached canvas). (Coll. Boris Ciglic)

Above: Nieuport 10 two-seaters photographed on the Topcin airfield on March 28, 1916. (Coll Boris Ciglic)

Below: A Farman MF 11bis flying over the Salonica park located at Camp Allatini in early 1916. (DR)

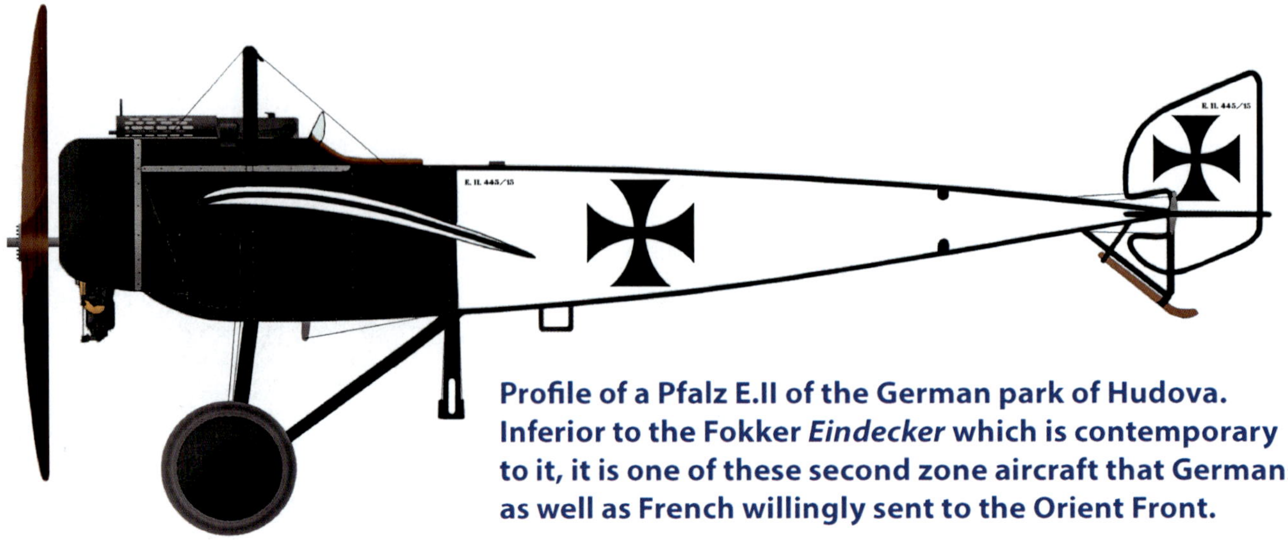

Profile of a Pfalz E.II of the German park of Hudova. Inferior to the Fokker *Eindecker* which is contemporary to it, it is one of these second zone aircraft that Germans as well as French willingly sent to the Orient Front.

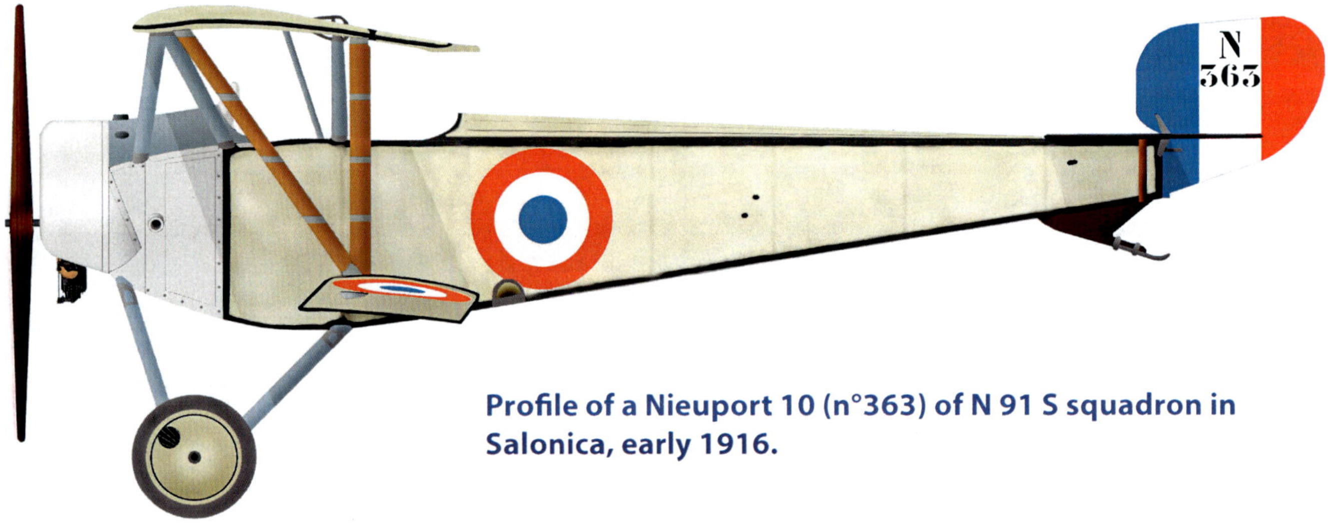

Profile of a Nieuport 10 (n°363) of N 91 S squadron in Salonica, early 1916.

Salonica for a while, but their two-seaters continued to carry out solo reconnaissance missions, such as the one on April 25, when Terme and Astor shot down their third German plane that fell east of Lake Amatovo. The German crew escaped but the plane was captured and (again ...) exposed on the square of the white tower.

It would be replaced on May 5, 1916 by a trophy of choice: the carcass of the Zeppelin LZ 85! Back in Salonica for a bombing mission in the night of the 4th to the 5th, the airship was spotted by the radio listening system of the French and the alert was given in the night. It was attacked by two Voisin-Canon of the V 90 squadron (crew: Sgt Guy Deloupy / Gunner Robert Lelong and S/Lt Albert Mézergues / Gunner Martin). These planes managed to fire respectively 2 and 5 shells of 37 mm respectively before the guns of the navy ships in the port took over. Hit, the Zeppelin fell into the marshes at Vardar's Mouth and its crew was captured. The carcass would be dismantled and transported to Salonica for the great pleasure of photographers.

5.4 - The Fixation of the Orient Front (May – December 1916)

During the month of May, the French and British troops left the entrenched camp of Salonica to make jumps on lines further and further away from the city, approaching the border where the German-Bulgarian troops were. GBO carried out its most daring raid on 24 May 1916 by bombing the town of Xanthi in Bulgarian Thrace, where spies reported the presence of General Mackensen, who was supposed to inspect the troops. It was a 180 km raid for 12 Voisin bombers, escorted by 5 Cannon-Voisin and 4 Nieuport 10, which refueled on a beach close to Orfano on a beach in the Struma Mouth, without suffering any losses. The ultimate mission of the GBO took place on June 6, 1916 against the city of Petric: only 4 bombers out of 12 were able to complete the mission, the planes being at their operational limits. The GBO, last operational unit to use the Voisin LAS in combat, had to be disbanded. The V 83 and V 84 squadrons were reformed on Farman aircraft while the V 90 was re-equipped with Nieuport fighters.

Above: This Nieuport 10 photographed in the park of Salonica carries an original personal decoration. (Coll. Paschalis Palavouzis)

Below: Handling of bombs on the Salonica airfield in 1916. Detail specific to the Orient Front: the colonial helmet worn by the mechanics. (Rol)

Above: General Sarrail himself inspects the frontline by flying as observer in this Farman MF 11bis.

While Prime Minister Aristide Briand succeeded through diplomatic maneuvers in getting Romania to enter the war on the side of the Entente, General Sarrail was ordered to prepare a major offensive on 20 August 1916, seven days before the planned date of Romania's mobilization. He could not do much because Joffre had refused him the military resources he had requested. The German-Bulgarian troops took the initiative of the assault by leaving their positions on August 17th 1916. They advanced to the east of the British troops positioned along the Struma River. To the west, the Bulgarians seized the Greek town of Florina, occupied by a French detachment, and moved into the mountains of Macedonia in the direction of Salonica.

Sarrail ordered a vigorous counter-attack and used the troops of the new Serbian army reformed in Salonica. The aviation of the Orient was also expanded and now included three fighter squadrons on the single-seater Nieuport 21 (N 87, 90, and 91), one bombing/reconnaissance squadron on the Caudron G 4 (C 89), and five observation/artillery adjustment squadrons on the Maurice Farman F.40 (MF 83, MF 84, MF 85, MF 86, and MF 88), reinforced by three others: the MF 98 back from Dardanelles in March 1916, the MF 82 back from Corfu in April, and the new MF 99, which fought in Serbia at the beginning of 1915 and was reconstituted in France in June 1916.

Some of these air units were reserved for the new Serbian army: N 87 (fighter), plus 4 artillery adjustment squadrons: MF 82, MF 84, MF 98, and MF 99. At the request of the Serbian government, Major Roger Vitrat, former head of the MF 99 S in 1915, was appointed head of the new Serbian air force, which became autonomous in relation to Major Denain's Orient Air Force. Several Serbian officers served in these units as observers and some of them as pilots, even if the majority of the pilots were French. Finally, in July 1916, the British landed in Salonica their first aviation unit, the N°17 Squadron, a mixed unit composed of 12 BE2c reconnaissance aircraft and some fighters (2 Bristol Scouts, 3 DH 2) to be established in the British sector on the Struma valley.

In the enemy lines, the aviation of the central powers counted only 6 squadrons: the German Feldfliegerabteilungen 20, 28, 30, 66, and 69, as well as the 1st Bulgarian squadron, all equipped with two-seater reconnaissance planes but reinforced by some single-seater fighters of class E – Pfalz and Fokker Eindeckers. It seems that the Franco-British had an air superiority of 2 against 1.

The Serbian troops would move on the railway from Salonica to Monastir to face the Bulgarian troops. The Serbian air force won four aerial victories, notably on August 17, 1916, when S/Lt Marcel Thirouin, veteran of the Serbian

Above: Fokker D.II of the German Air Force photographed at Drama in 1916. (DR)

Below: Farman F.40 of squadron F.386 with the unit symbol on the front of the aircraft, a blue and red Ying-yang. (Coll Boris Ciglic)

Above: The city of Florina, one of the rear bases of the French Army of the East holding the basin of Monastir.

campaign of 1915, claimed his second personal success, but the victory was not confimed. The Bulgarian advance was halted on 22 August 1916 near Lake Ostrovo; a few days later, on 25 August, Thirouin scored his second official victory in his Nieuport 21 against a German shot down in the mountains near Tresina. The town of Florina was retaken by Franco-Serbian troops on 17 September 1916, who then moved north to try to seize the town of Monastir, where Bulgarian resistance had been reinforced. The troops and squadrons of the Serbian army repositioned themselves around the city, which fell on November 19, 1916, after heavy fighting that cost nearly 32,000 men. The enemy withdrew to the heights above the city, which remained under artillery threat until the armistice.

After the battle of Monastir, the front remained unchanged until September 1918. The French squadrons, renumbered by adding 300 to their number, took their positions which remained practically unchanged until the end of the war: in the Monastir valley and the loop of the Cerna River, the squadrons of the Armée Française d'Orient (AFO) were stationed, composed of the N 390 (Nieuport fighter), C 389, MF 383, MF 383bis (or heavy artillery section no. 1), MF 386, and MF 388, all of which were observation and artillery reconnaissance aircraft. In the center of the front, facing the mountains of Macedonia, the Serbian aviation (N 387, MF 382, MF 384, MF 398, and MF 399) is distributed on 3 airfields of Vrbeni, Ostrovo, and Vertekop. Remaining in Salonica and on the airfield of Gorgop, along the Vardar, the reserve aviation of the Allied Army Command (CAA) composed of the squadron N 391 (fighters) and the MF 385 (observation). The British army now had two Squadrons, N°17 and 47, divided into several patrols on the airfields of Janes, Kukus, and Lahana.

The German-Bulgarian aviation, always outnumbered, would, however, end the year 1916 with a qualitative superiority. On November 28, 1916, Jasta 25 was created at Prilep airfield, with Fokker D II and Halberstadt D fighters, which were rather mediocre according to the criteria of the Western Front, which was still a priority for the German high command for deliveries of the best equipment. But this was also the case for the French: at the end of 1916, the Orient Front fighter units only had Nieuport 21s, while the SPAD VII and Nieuport 17 were flying in France... Equipped with an 90 hp engine and a non-synchronized machine gun on the upper wing, the Nieuport 21 was clearly outdated and unable to catch the German fighters that came and went as they pleased over the Allied lines and inflicted increasingly important losses on the Farman observation planes. The Nieuport 21 could hardly intercept the German reconnaissance planes because their speed differential was

Above: Another view of Florina.

Below: The Nieuport 21 was the standard fighter of the French Orient Army during the year 1916. Very maneuverable, it was underpowered with its 90 hp engine and proved to be very insufficient to intercept German aircraft. (Basile Sauné)

Above: French and Serbian pilots of the N 387 squadron photographed at Vrbeni in November 1916. In the center, under the propeller, second lieutenant Marcel Thirouin, one of the best pilots of the orient aviation, holder of 3 confirmed aerial victories, including one won in Serbia in 1915. (Coll Boris Ciglic).

Below: The repair of an aircraft landed in the wilderness takes on the aspect of an expedition under the Orient front, like this Nieuport 21 brought back by Serbian mechanics. (Coll Boris Ciglic)

Above: Serbian pilot Milorad Jefremovic (right) and his mechanic next to his Nieuport 11 n°1183 at Vrbeni airfield in 1916. (Collection Boris Ciglic)

low. And when it managed to catch up with one of them and get into firing position, it still had to shoot it down, which was quite difficult with only 47 cartridges feeding its Lewis gun... when it did not jam! After these 47 rounds, the magazine must be changed, which was impossible to do in combat. Most of the time, the enemy plane does not have the courtesy to wait. Captain Hamoir, commander of the AFO, kept statistics: from 16 to 31 December 1916, his Nieuport 21s made 22 interception attempts and were lost 14 times to their prey. Eight interceptions were successful: one was aborted due to a machine gun jam, six resulted in a fruitless exchange of fire, and only one saw the enemy hit! The French pilot lost sight of his prey while reloading his weapon, the victory was not confirmed to him... Not surprisingly, no aces emerged from Orient Aviation, which had only 15 confirmed victories in the entire year of 1916.

But for the pilots of Orient Aviation, these material problems were the least of their worries. Their own state of health was more worrying. Like all the soldiers of the Army of the East, they were victims of malaria, which was endemic in Greece at the time, and which caused many more victims than the German air force and forced many pilots to fly in a state of great weakness. As shown in an official report written by Commander Vitrat on October 31, 1916, the five squadrons of the Serbian Air Force had more than 300 sick people in their ranks – pilots, observers, mechanics, and other ground personnel – representing between 30 and 50% of these formations left in a state of chronic understaffing. Another concern for the pilots was their living conditions, which were much more precarious than on the French front. Some airmen assigned to the Army of the East were sent there on orders. Others were volunteers, attracted by adventure, by supposedly better weather conditions, or by a front where enemy encounters were much less frequent. But all of them were disappointed to discover a very poor country where everything is lacking and where the temperatures are extreme: very harsh winters and very hot summers. In this context, one can imagine that in each squadron, there are many volunteers to fly the mail to Salonica, the back city synonymous with pleasures where the military administration already regulates the brothels for the soldiers. As in Paris, the restaurants and cabarets were full of soldiers of many nationalities, and the airmen distinguished themselves by their extravagance. But for the frontline pilots, daily life was much more difficult than in France, living in the packing crates of their planes arranged in barracks, which were

Above: Map of the Orient Front, unchanged from the end of 1916 to September 1918.

quickly invaded by fleas and other insects of the local fauna. Only alcohol and the Scottish nurses of the British military mission could entertain the aviators of the Oriental Army who, until the end of the war, would remain the forgotten pilots of the French air force.

5.5 - The German Air Offensive

The arrival of the winter of 1916–1917 froze flights on both sides. The head of the French air force, Major Denain, took advantage of this to request reinforcements and obtain a few Nieuport 23 fighters equipped with a 110 hp engine. They were not enough to face the German air force, which launched an unprecedented air offensive on 15 February 1917, as soon as Kaghol 1 arrived on the Macedonian front.

The Kampfgeschwader der Obersten Heeresleitung 1, that is to say the Combat Group 1 of the Supreme Army Command, was a German bombing group composed of 3 squadrons each equipped with 6 powerful twin-engine AEG, Friedrischafen, and Gotha bombers. Arriving from Romania by train, the German formation, which was also equipped with its own escort of Halberstadt D II fighters, settled at the Hudova airfield in early 1917. Despite the cold and bad weather, it carried out its first raid on 15 February 1917 on the British airfield of Janes, which received a deluge of bombs without understanding what was happening to it. The second action of the Kaghol 1, two raids carried out on 26 February because of the weather, made the Allies understand the extent of the problem. In the morning, the personnel of the MF 385 squadron at Gorgop heard the hum of the engines of 8 bombers which dropped their cargo of bombs on the airfield. Balance sheet: 3 hangars destroyed, all the aircraft are hit, and 2 airmen are wounded. Still at dawn, the city of Salonica was targeted. The Nieuports of the N 391 squadron took off and confronted the bombers, while the anti-aircraft of the port fired a heavy fire, damaging some aircraft, which withdrew having made some 330 victims killed or wounded in camps and hospitals in the west of Salonica. Harassed on the way back by the British fighters (including Captain Murliss-Green, who courageously went to the attack alone on his BE 12), the Germans lost one of their Halberstadt D II escort fighters hit by the flak, whose pilot was captured and interrogated by the French Intelligence officers; this allowed Commander Denain to realize the extent of the threat. The Farmans of MF 385 being all destroyed or damaged, he could only organize a symbolic raid against Hudova by night on a single Farman taken out of storage and piloted by a hothead, a certain Dieudonné Costes.

On March 1st Kaghol 1 returned to Salonica and found only two French Nieuports of the MF 385 squadron, as well as two British two-seaters and one single-seater of the N°47 Squadron. Denain, who had only 24 French Nieuport fighters at his disposal out of the entire Orient Army (two thirds of which were Nieuport 21s) and 9 British fighters, decided to order N 387 to join forces with N 390 to block an incursion into the Monastir valley. The sector of the mountains held by the Serbian army was left to the protection of the flak alone, while the N 391 squadron of

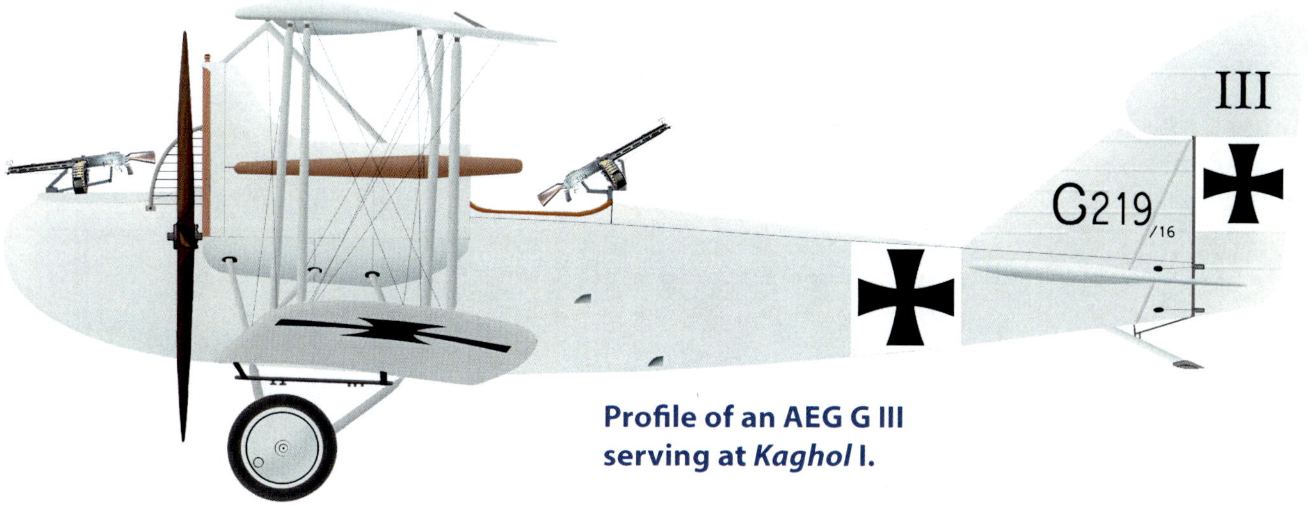

Profile of an AEG G III serving at *Kaghol* I.

Topcin and the British air force of Janes were supposed to group their efforts to block the incursions on Salonica.

The bad weather brought a short respite until March 12, 1917, when Kaghol 1 went on the attack. All the air defense of Salonica is active but unfortunately the German bombers obliqued to Vertekop, in the Serbian sector of the mountains, precisely abandoned to its only anti-aircraft artillery, and to a small patrol of two Nieuport 21, piloted by Captain Vukosavljevic and Second Lieutenant Miletic which could not make miracles against the armada of 18 bombers and 6 escort fighters. The bombs fall and volatilize the ammunition depot of the 2nd Serbian army in a tremendous explosion. Several bombs fall on the nearby British field hospitals and kill some patients in their beds as well as medical staff – the Serbs will accuse the Germans of having deliberately violated international conventions by having attacked a target very clearly marked with the red cross, while the Germans will defend the proximity of the hospital to a military target. The British organized a large-scale response by mounting a raid the same day with all their aircraft (13 bombers escorted by 7 fighters) on the airfield of Hudova to try to destroy Kaghol 1 on the ground. But the Germans, warned by their alert network, took off their fighters and bombers before the arrival of the British. A melee ensued during which a British

Above: Map of the Kaghol 1 attacks.

Above & Below: For the allied troops, the awakening is brutal. These two pictures show the result of the attack of *Kaghol* 1 around Vertekop on March 12th 1917.

Above: BE 12 of N°17 Squadron of the British Army. Heavy and unmaneuverable, this failed fighter that the British relegated to the Orient Front was equipped with an RAF 4a 140 hp engine, and was faster than the 90 hp Nieuport 21s of the French squadrons. (DR)

aircraft was shot down by a Halberstadt escort.

In the following days, the German bombers varied their objectives and struck the Salonica region as well as the rear of the Serbian sector with impunity – the victims on the ground of each raid were counted in tens. On March 18, 1917, three bombers were claimed by the British, but no tangible losses were recorded on the personnel of the Kaghol 1...

On April 3, 1917, Kaghol 1 returned to Vertekop for the second time. But this time, the Serbian fighters of the local detachment of the N 387 hold their revenge since the Lt Miodrag Miletic, very in form since it downed the day before a German plane of recognition, launches itself in the swarm of German twin-engines and frames one of them which it does not release any more, leaving it disabled at low altitude north of the Dobropolje massif where it may have been shot down – the German archives do not allow to confirm the victory which is nevertheless duly certified to the Serbian pilot, who will end the war at the head of the best fighters of his country with 3 successes. It is on the sector of Doiran that on April 8, 1917 the Allies gain their first indisputable success against the German twin-engine bombers by shooting down a Friedrischafen G III which is pulled by four British two-seaters and falls to Paturos near Snevce – the plane is captured damaged and is confirmed to the Sopwith 1½ Strutter of the Flight Lieutnant E.T. Bradley (pilot) and 2nd Lt. C.T. Repton (gunner), of the Royal Navy Air Service.

On April 15, 1917, noting that the target of the German bombers was now in the Monastir valley, Major Denain ordered 391 Squadron to reinforce 390 Squadron in its sector to the west of the front, leaving a small detachment under Lt Noël to defend Salonica. The French fighters thus reinforced, which received their first SPAD VII, were able to attack the bombers whose route was previously "cleaned" by the Halberstadt D II fighters. On April 24, the 390–391 group pursued the bombers in the direction of Prilep but could not join them. It was only on 29 April that the French fighters had their first success, during a raid on the village of Skocivir (south bank of the Cerna loop), which was hit with 2300 kg of bombs. The Nieuports were this time in a good position and intercepted the bombers over the mountains, but had to face first the black cross fighters. MdL Christmas, of N 390, managed to slip his red-nosed Nieuport 21 between the fighters and reach a bomber on which he emptied – without result – the 100-round magazine of his Lewis gun. Adjudant Dieudonné Costes, flying his Nieuport 23, was luckier (and also better armed) and attacked a bomber which fired a distress flare to request help and continued on its way while losing altitude. Costes was then taken on by two fighters and had to

Above: Nieuport fighter on patrol over the mountainous landscape of the Orient Front (SHD).

Below: Friedrischafen G.III bomber shot down by the British Air Force on April 8, 1917. (Coll Benoit Henriet)

Above: Embarkation on the port of Marseille of aircraft in crates for the aviation of the East. The bad conditions of transport damaged the aircraft, giving an important extra work to the park of Salonique to put them in order.

drop his prey, although he was awarded the victory (his first). The success of the fighters was, however, relative, as on the ground 12 men and 50 horses were killed or wounded... The next day, 30 April, the village of Vodena, in the Serbian sector of the N 387, was hit and there were still 41 victims.

May 1st 1917 saw the last battle of the twin-engine bombers of Kaghol 1, which bombed the village of Batch where a French squadron was stationed, losing 2 Farman F.40 on the ground and its fuel depot. The French fighters succeeded in intercepting the aircraft and engaged in combat, which caused no losses on either side. Kaghol 1, whose aircraft were vulnerable on the ground in Hudova, increasingly harassed by the Franco-British air force, was then transferred by the German staff to Flanders to carry out raids against England, after having led nearly two and a half months of raids that had shaken all the rear of the Macedonian front.

It was precisely at the critical moment of the air war against Kaghol 1, that General Sarrail, obviously quite ignorant of things in the air, launched a ground offensive in the Monastir valley. He was pressured into this by Paris, which refused him the heavy artillery reinforcements he demanded. He attacked in spite of everything on 22 April and 9 May, but neglected an essential fact: in this artillery war, control of the sky was essential to allow his air force to adjust its own fire and prevent the enemy air force from doing the same. However, all the French fighters in the sector, which were few in number, were used to hunt twin-engine bombers... The German fighters of Jasta 25 of Prilep had a field day. Thus, the MF 386 squadron, which until then had only known losses due to accidents, experienced a black series: from mid-February 1917 to the end of March, the interceptions of the German fighters made it lose 4 pilots out of 8 (3 killed, 1 wounded) and 3 observers out of 6 (all killed). In addition to the bad weather, which did not help, all observation and artillery adjustment flights were almost systematically attacked by fighter patrols and had to interrupt their mission.

The failure of Sarrail's offensive increased his disgrace in Paris and led to his replacement at the end of 1917 by General Guillaumat. The Orient Front was to experience mutinies among the infantry units. However, in the air, Commander Denain took advantage of the lost battle to patiently rebuild his air force, to make it accepted by the high-ranking officers and turn it into a decisive tool for a victorious offensive.

Above: Following the German offensive, Denain obtained modern equipment from France, including Nieuport 23 and 24 fighters with 110 hp engines, representing an improvement over the small Nieuport 21. Here, a Nieuport 24 of the 504 squadron, named "Routchichoune" by its pilot. (SHD).

Facing Page, Above: Farman F.40 damaged. This photograph symbolizes all the problem of the aviation of the East, which systematically receives outdated apparatuses compared to those put in service on the French front.

Facing Page, Below: After having launched an appeal in France in February 1917, Denain obtained only one experienced fighter pilot to volunteer for the Orient Front: Maréchal des logis Gabriel Hébert, who is seen in white uniform (7th from the left) and smiling with his decorations among the pilots of the N 390 squadron who were waiting for him "like the messiah". If Denain was satisfied to have an experienced manager to instruct his fighter pilots, he was to be somewhat disappointed: Hébert was to indulge in pimping in Salonica, which led to his disciplinary dismissal in the infantry. This punishment was soon lifted by Denain, who took revenge by refusing to certify him one of the three victories he claimed in the Orient front, in addition to the two he had won in France. (Coll Cross and Cockade)

5.6 - Denain Reforms the French Air Force

Denain learnt from this failure and we find in the archives a document dated 9 May 1917 which summarizes the measures he undertook.

First of all, the Serbian air force was in the hot seat: it was criticized for its autonomous air force, which would benefit from being merged with that of the French air force in the East into one large air force. Moreover, its chief, Major Vitrat, protested a little too loudly against the reinforcements that were not arriving, both in terms of aircraft and pilots. He accused the French units of being the first to use the reinforcements coming from France and of leaving only the crumbs to the Serbian air force... True or false, this assertion discredited him and Sarrail dismissed him without mercy on August 28, 1917, writing to the general headquarters that he was "an officer with no personal drive, who flies little or nothing, and who has committed errors of command." The statement is unfair, but Vitrat paid for being too close to the Serbs with whom he had shared the suffering in the Serbian campaign at the head of the MF 99 S. His deputies shared his disgrace (including Capt. Tulasne, also a veteran of the MF 99 S) and were all replaced at the head of the Serbian air force by officers who were better disposed towards Sarrail and, above all, were now officially placed under the orders of Major Denain, as were the two squadrons of the Royal Flying Corps in Salonica and the Italian Farman squadron (47th Squadriglia), reinforced by a fighter patrol which reached the Salonica front. As for the Serbian fleet, it was purely and simply dissolved despite the protests of the head of the Serbian army, General Petar Bojović.

Then came the question of equipment. The obsolescence of the Farman F.40 is clearly demonstrated by events and its

Above: One of the very first SPAD VII fighters delivered to the Orient Air force, at the beginning of 1917.

replacement in all the Orient squadrons by the Dorand AR 1 is finally accepted, the first ones winning the Eastern during the summer of 1917. Concerning the fighters, the Nieuport 21 (90 hp) were already replaced by the Nieuport 23 (110 hp), then 24 and 24bis. But Sarrail expressly requested – and obtained – SPAD VIIs, albeit in small quantities, but which were entrusted to the best fighter pilots.

The question of fighter pilots was precisely the recurring concern that Major Denain had to face, and of which he had personally alerted Paris as early as April 15, 1917, requesting that experienced pilots from the French front who had shot down enemy aircraft be seconded to him, if only for a month, so that they could pass on their experience. Volunteers were requested from all the squadrons on the front... who were not in a hurry to go to Greece. They thought of assigning the ace Jean Navarre to Salonica, mainly to get rid of him. This project was short-lived. Throughout the war, pilots were transferred to the Orient directly from the flying schools, and although some of them left because of a taste for adventure, a good number of them volunteered to fly in a sector where German aircraft were scarcer. Denain received only one pilot who had already won victories in France, Maréchal des Logis René Hébert, who had already won four victories, two of which were homologated at SPA 62, and who would win as many in the Orient. But Hébert took wide liberties with the discipline and met a French prostitute in Florina, whom he made his mistress and whose pimp he became...! Denain had him transferred to the Salonica hunt, but Hébert stole a squadron car to join his mistress, whom he put back on the street in Salonica itself! Denain expelled him from the air force, but because of the lack of experienced pilots, he had no other choice than to reinstate him...

The last point on which Denain was working was the improvement of the logistics. The new equipment obtained raised the question of delivery methods to Salonica. Until now, all the planes were delivered in crates by the French Navy to the port of Salonica – many of them being damaged during the journey. Denain then asked for and obtained the creation of an assembly factory in Salonica itself, to assemble aircraft from elements machined in France. The gain in space in the transport ships was considerable compared to sending dismantled aircraft in crates. The factory, which took a long time to set up, was not fully operational until 1918, and by the end of the war it was producing aircraft stamped "US" for Usine de Salonique (Salonica factory).

More immediately, Denain obtained the creation of an assembly center in Otranto, the heel of the Italian boot, where the dismantled planes were delivered by train from France, reassembled on site and delivered directly in flight to the Salonica front. This route was not an easy one at the time, as it involved crossing the mountains of Albania and Macedonia, in an area quite close to the front. Grouped convoy flights were organized in groups of seven aircraft, led by a team of specialized convoy pilots. The first flights started on August 25, 1917 and the archive documents allow us to find proof of 72 aircraft (Dorand AR 1 for the most part) delivered to Salonica in nine trips, during which five aircraft were lost (7% of the total). It should be noted that the Otranto center, whose importance the Austro-Hungarian command understood (even if it took it for a British center), was attacked by an air raid during the night of July 24–25, 1918.

Above: In 1917, the Dorand AR becomes the standard aircraft of the army aviation for the Orient Army. Although it represented a definite improvement over the old Farman F.40, it was far from satisfactory. It is an aircraft that the military aeronautics of the French front is getting rid of... Here, an aircraft of the heavy artillery section n°1, which becomes the 508 squadron in June 1917. (Coll Albin Denis)

Below: Alignment of Dorand AR of the Orient Army in 1917. Many of these aircraft were delivered directly by air from the assembly center in Otranto, in the heel of the Italian boot, and flew across the Adriatic and the mountains of Greece. A rather perilous journey for the time. (SHD)

Above: In profile, Major Victor Denain inspects with the regent of Serbia the assembly plant of Salonica which was only operational at the very end of the war and assembled aircraft from spare parts delivered from France. The registration of the Breguet 14 can be seen on the tail, bearing the number US 1 (Salonica factory n°1). (DR)

5.7 - Securing the Naval Links

If many planes were delivered in flight from Otranto, many more remained in boxes in this port and were sent to Greece by sea. This is where the French maritime aviation comes in.

Since the beginning of 1916, the French Navy had had a Maritime Aviation Center (CAM) in Corfu, where its seaplanes provided part of the coverage of the Otranto dam, a gigantic floating net that was supposed to block the exit of the Adriatic Sea to German and Austrian submarines. This CAM, which at the end of 1916 consisted of about thirty aircraft of various types, was to be considerably strengthened to reach about fifty aircraft on the line at the end of the war, including in particular Sopwith Baby fighter seaplanes. In parallel to the seaplanes, a fighter squadron, the N 562, was created in Govino to ensure the coverage of the island.

Two other naval stations (CAM) existed at the end of 1916 in Greece: one in Salonica (12 aircraft, eight of which were armed), covering the approaches to the port, and another in Argostoli to cover the entrance to the Corinth Canal. During the war, the CAM Salonica increased its number of aircraft to 24, of which 12 were armed. The CAM Argostoli would benefit from a similar increase in power but malaria would seriously limit its effectiveness: the airfield will have to move to Mytika on the mainland, then, once again chased by mosquitoes, will finally settle in Platali in 1918.

Another CAM was created in 1918 on the island of Milo with 16 seaplanes, 12 of which were armed. Its patrol area was comprised between Crete and the Peloponnese. In parallel to the action of the maritime aviation, military aeronautics would create two naval patrol squadrons on land aircraft in 1918. The first one was the G 488 squadron at Kourtesis, in the north of the Peloponnese, to monitor the Corinth Canal. The second one was an unofficial squadron (which did not receive a number) created on the initiative of Major Denain, with Farman aircraft decommissioned from the Oriental Air Force, which he used to patrol the approaches to the port of Salonica from the village of Aponomi.

The activity of these CAM and coastal squadrons was not limited to uneventful maritime patrols. The CAM Corfu was in direct contact with the Austrian air force and took part in attacks on the ports of Durazzo and Kumbor – and participated twice in interceptions of Austrian seaplanes over

Above: Nieuport seaplane launched at the Naval Aviation Center (CAM) of Argostoli, created in May 1916. (Coll. ARDHAN)

Profile of a Nieuport seaplane of the Argostoli Naval Aviation Center coded N 19. Very resistant, these aircraft produced in 1914 were only withdrawn from service at the beginning of 1917.

Corfu on February 11, 1917, May 26, 1917, and June 6, 1918, but without any success. It was mainly in naval attacks that the main part of its activity was summarized: if in 1916 it only participated in one U-Boat attack, it carried out 13 in 1917 and two in 1918, and also took part in attacks against Austrian surface ships during Admiral Horthy's raid against the Otranto dam on May 15th 1917.

The CAM Salonica ranks second in terms of activity, since it carried out three U-boat attacks in 1917 and five in 1918. The CAM Argostoli/Mytika/Plateali carried out one attack in 1917 and another one in 1918. But most importantly, it actually managed to sink a U-Boat on 14 December 1917. The UC 38 of Oblt z.S. Hans Hermann Wendlandt torpedoed and sank the French cruiser *Chateaurenault*, which was on a troop transport mission between Taranto and the Greek port of Itea. Hit by 2 torpedoes between the peninsula of Leucate and the island of Argostoli, it sank while most of its passengers were rescued. But the submarine was spotted by the destroyers of the escort, the *Mameluk* and the *Lansquenet*, which shot it and damaged it a first time. The submarine was then finished off by the bombs launched by two FBA 100 hp seaplanes (type C) and more particularly the "S 11" of the Chief Petty Officer Jeanblanc and the sailor Trouillet. Before sinking, the UC 38 was able to surface and was evacuated by its crew, which was captured – it was one of the only two U-boats put out of action by the French naval aviation during the First World War.

As for the CAM Milo and the coastal squadrons of

Above: In The Argostoli Airfield also received in 1916 FBA seaplanes with a 100 hp Clerget engine, called "type C" by their manufacturer. (Coll ARDHAN)

Below: FBA 100 hp Clerget n°219 of the Argostoli CAM at the end of 1916. The aircraft has an original decoration. (Coll. ARDHAN)

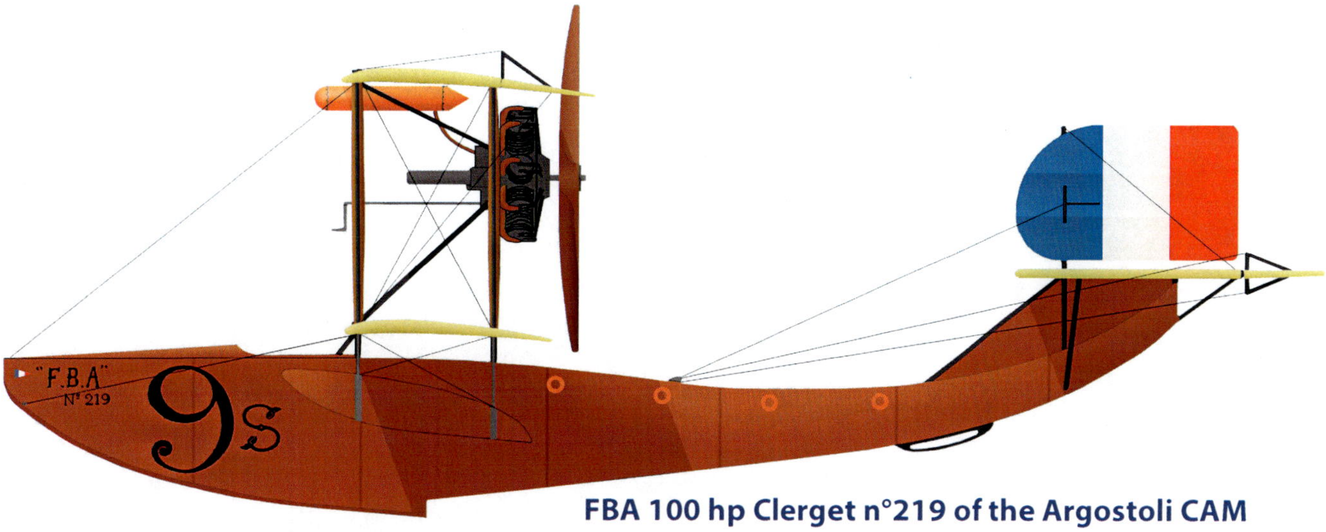

FBA 100 hp Clerget n°219 of the Argostoli CAM

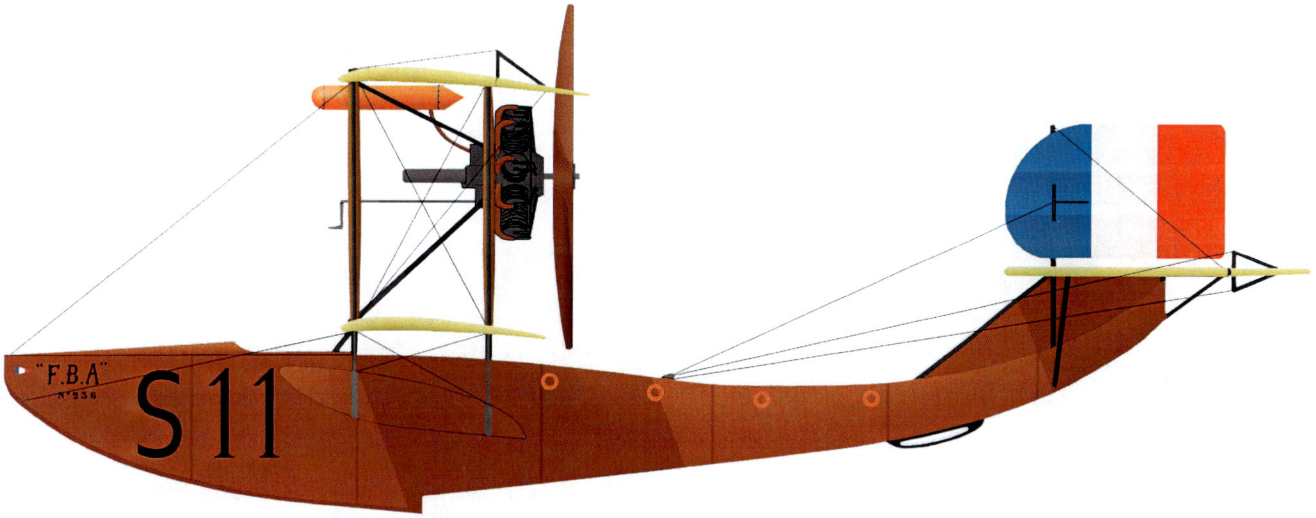

Theoretical profile of the FBA 100 ch Clerget of the second master Jeanblanc and the sailor Trouillet, victorious crew having contributed to sink the submarine UC 38 at the north of the island of Argostoli on December 14, 1917

Panomi and Kourtesis, these units did not encounter any trace of enemy submarines.

5.8 - Fifteen Months of Air Warfare on a Static Front from May 1917 to August 1918

After the aborted offensive of May 1917, no further large-scale operations were undertaken on the Orient Front. The German air force was now clearly reduced to the defensive: it no longer had any bombing capacity apart from symbolic attacks with a few bombs dropped at high altitude by its isolated reconnaissance aircraft, which rarely ventured onto the Allied lines. The fights were only a series of skirmishes against reconnaissance planes and some clashes between fighters, from which some aces in both camps emerged.

Nevertheless, a few events should be noted. As early as the summer of 1917, while the replacement Dorand AR 1s were not yet delivered, Major Denain wanted to instill a certain offensive spirit in his men and employed Farman F.40s in bombing raids in the Monastir sector. Each AFO squadron had to send several Farmans as well as its Nieuports to join the air armada that was going to bomb a specific target – Denain did not hesitate to call in a detachment of British air force escorted by a detachment of the 507 squadron. Thus, on 7 August 1917, 16 Farmans belonging to eight squadrons left to attack the headquarters of the 47th Austrian Infantry Division and military camps, escorted by 13 Nieuports for close protection, plus four Nieuports and SPADs from 506 for group protection. On August 16, 11 Farmans escorted by 12 Nieuports attacked military camps and boats on the Presba lake, while five fighters of the 506 made a blockade further north. At the same time, in order to saturate the enemy defenses, the British detachment escorted by fighters of the 507 attacked the Kanatlarci airfield – a similar raid was organized on 18 August and more generally all the AFO air force got into the habit of pooling its resources.

However, it would be wrong to believe that the German fighters were standing on the ground. In June 1917, receiving

Above: Created in May 1916 at the same time as the CAM Argostoli, the CAM Corfu became one of the most important French naval air bases of the Great War, whose seaplanes patrolled the southern Adriatic. Here, a Donnet-Dennaut 160 hp Canton-Unné on a stopover at the Italian base of Otranto. (Coll Paolo Varriale)
Below: The CAM Corfu even had a squadron of fighter seaplanes. Sopwith "Baby" with a 130 hp Clerget engine. However, they hardly succeeded in intercepting any enemy aircraft, due to their poor performance and the low number of incursions of the Austro-Hungarian naval aviation on the island of Corfu. (Coll ARDHAN)

Sopwith Baby with 130 hp Clerget engine of the CAM Corfu, in 1917.

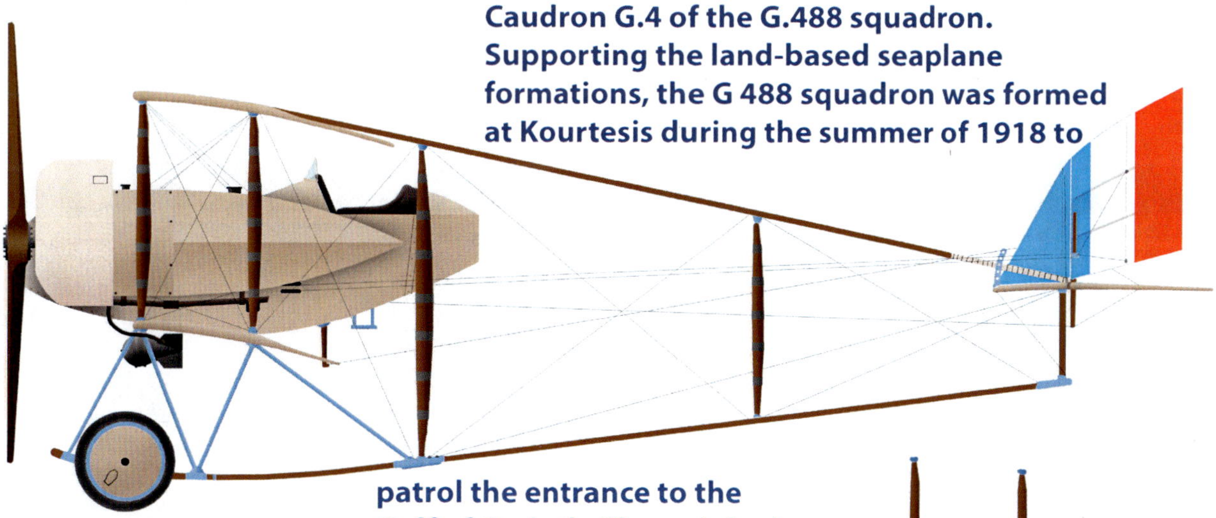

Caudron G.4 of the G.488 squadron. Supporting the land-based seaplane formations, the G 488 squadron was formed at Kourtesis during the summer of 1918 to patrol the entrance to the Gulf of Corinth. The unit had very low performance due to a relentless aerial opponent: the mosquito, which transmitted malaria to many of the pilots and mechanics.

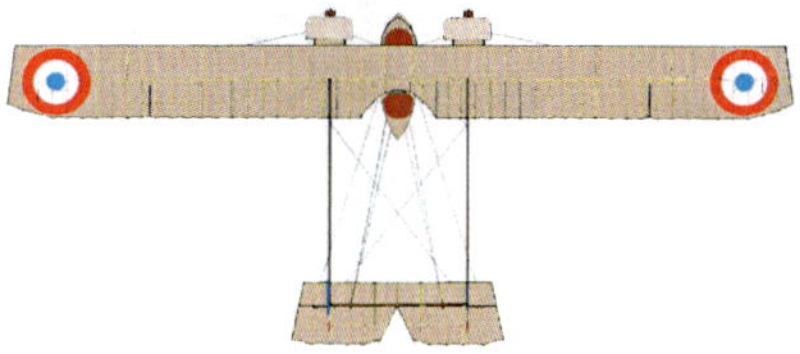

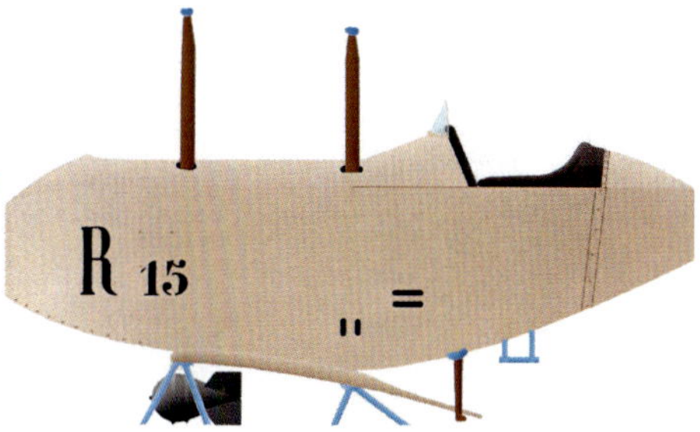

Roland D IIs, the Germans created a second fighter squadron in Hudova, Jasta 38, which covered the Vardar front. Re-equipped and reinforced, the German fighters were now going to take their due with regularity on the Allied aircraft flying over their lines, even if most of the results were obtained by two exceptional pilots, Vicefeldwebel Gerhard Fieseler, who won his first confirmed victory in June 1917, and Leutnant Rudolf Von Eschwege, who was practically the only fighter pilot in his sector, the Struma front, facing the British airmen. Based in Drama and in charge of escorting the observation planes of the sector, Von Eschwege shot down day after day the British planes passing within his range, acquiring a reputation in Salonica comparable to that of the Red Baron in France. The British finally shot him down by means of a war ruse on 21 November 1917, stuffing the gondola of an observation balloon with explosives and blowing it up when the German ace dived to attack him.

When the winter of 1917–1918 came, the air deliveries from Otranto stopped and the planes in crates piled up on the docks. It was at this time that the Greek air force was created and developed, as it had officially entered the war on the side of the Entente. A corps squadron, the 532, was created in December 1917. The Greek Prime Minister, a skillful negotiator, offered the assistance of his merchant navy to fetch the aircraft from Otranto, in exchange for the creation of new squadrons. Two new corps squadrons were thus created by the end of the war. But above all, a fighter squadron, the 531, led by S/Lt Alexandros Zannas, the only

Above: FBA 200 hp Hispano seaplane (called type S) at the CAM Salonica in 1918.

Below: A Dorand AR of the 524 squadron (Serbian army) flies over the snow-covered mountains of the front, during the winter of 1917–1918 (Coll Boris Ciglic).

Above: Assigned to the n°501 squadron at the western end of the front, in the Albanian mountains, the pilot, Georges Baudeuf, does not seem to be satisfied with his assignment and expresses this on his Nieuport 23. (René Germain, coll. Albin Denis)

Greek fighter pilot, was set up and gathered the best French fighter pilots. The unit had some success in the spring of 1918 when it was engaged on the Vardar, participating in particular in the coverage of the Greek troops' assault on the Serka Di Legen mountain at the end of May 1918. Two of its pilots, Dieudonné Costes and Basile Sauné, became aces the following month, winning their 5th confirmed victory together. They were then detached to the Monastir sector to boost the local fighter squadron. But a tragedy occurred on June 20, 1918: At the time of a patrol flight gathering Costes, Sauné, and another pilot named Cochet-Forthuny, Costes' plane did not start. Sauné nevertheless took off and started the patrol with two aircraft. He and his partner surprised a patrol of three German fighters and attacked them from behind. But one of the Germans was none other than Gerhardt Fieseler who did not let himself be surprised. He engaged in a furious aerial combat, which he described in his memoirs, and finally shot down Sauné's SPAD at ground level, while Cochet Forthuny, with his engine in smoke, managed to reach the French lines in extremis. Despite the evidence, the officers of the group did not attribute the victory to Fieseler but to one of them, Ltn Rose.

In the east of the front, in the Struma sector held by the British, the latter reinforced their air force by creating the N°150 Squadron specialized in hunting, leaving the N°17 and 47 Squadrons to the reconnaissance missions. The British fighters, equipped with Camels, SE5a, and Bristol Monoplanes, showed a certain dynamism and claimed (if not really shot down...) many victories, the best of them being Canadian Captain Gerald Gordon Bell who won 13 victories in the Orient front on the SE 5a, to which three others were added in France.

Let's also mention the arrival of an additional squadron for the French air force, the 510, which landed in Salonica in January 1918 equipped with Breguet 14s, and which was to play a decisive role.

5.9 - The Victory Offensive (September 1918)

In August 1918, with the continuous arrival of reinforcements, the number of squadrons on the line continued to grow for the Entente forces.

The Aviation Française d'Orient (AFO), compared to the

A mon ami et camarade
Viallet -
Affectueux souvenir de notre
séjour d'Orient -
Salonique le 20 Septembre 1917

This Page & Facing Page: An acrobat at his leisure, Maréchal des Logis Georges Viallet, of squadron n°506 (formerly 390), was photographed in his Nieuport 23 by the Farman he was escorting, during the late summer of 1917. This pilot won a confirmed aerial victory on February 6, 1917 in the Orient Army... (Coll. Jean Vaissade)

ТУТРАКА

Facing Page, Above: German two-seater Albatros C.III flying over the Orient Front. (Coll. Boris Ciglic)

Facing Page, Below: Crew of an Albatros C.III of one of the two Bulgarian squadrons present on the Orient Front, which nothing, except the personal markings, makes it possible to differentiate from the aircraft of the four German reconnaissance squadrons present on this front. (DR)

end of 1916, saw the transformation of two heavy artillery sections into full-fledged squadrons. Its six corps squadrons all had a fighter section (Nieuport 24 and 24 bis) to escort their Dorand AR 1 two-seaters, which were beginning to be replaced by Breguet 14s. The AFO's fighter squadron, 506, was entirely equipped with SPAD VIIs, in addition to a section of Breguet 14s for major reconnaissance missions.

The Allied Air Command (CAA) reserved aviation had a corps squadron, a fighter squadron defending Salonica, and the 510 squadron with its Breguet 14 bombers.

The Serbian air force, with four corps squadrons and one fighter squadron, saw an influx of Serbian pilots and observers from French schools, which allowed the transformation of two corps squadrons into 100% Serbian squadrons - one squadron, 522, being dissolved to gather the Serbian personnel.

The Greek air force had one fighter squadron and four corps squadrons. The British had one fighter and two corps squadrons, and the Italians had one fighter and one corps squadron.

In total, at the end of the war, there were 25 squadrons, including six fighter squadrons, 18 corps squadrons, and one bomber squadron... As for the German air force, a report from French Intelligence dated 25 June 1918 lists two fighter squadrons (Jastas 25 and 38), six observation squadrons (Fliegerabteilung 20, 30, 34, 246, and two Bulgarian squadrons) plus a maritime squadron, i.e. around 85 aircraft representing a force three times smaller than that of the Entente.

The summer of 1918 saw intense activity above and behind the lines. General Guillaumat was recalled to France following the German offensives that threatened Paris; he was replaced on June 6 by General Franchet d'Esperey. The latter continued and amplified his preparations for the offensive by carrying out infrastructure work in the mountains to allow the deployment of artillery and prepared the organization of a cavalry division to exploit the future breakthrough. A deceptive calm reigned in the trenches but the war continued in the air. From its base at Samli near Salonica, the 510 Squadron renewed the exploits of the GBO two years earlier by launching more than three tons of bombs at each sortie on the enemy's rear, unable to intercept the Breguet 14s which flew fast and climbed up to 6,000 m, out of range of the fighters; neither did it attempt to launch retaliatory raids because it no longer had any bombers. Raids on enemy communication nodes with up to 10 aircraft multiplied and the only loss in combat was due to French flak. At the beginning of September 1918, the 510 intensified its efforts and, on the 2nd of the month, attacked the enemy airfield of Hudova, claiming a victory against a German fighter shot down by the crossfire of the Breguet gunners. On the ground, the troops were ready for the attack planned by Franchet d'Esperey, who intended to break through where the enemy least expected it, in the Serbian sector, lined with mountains that were a priori impregnable, where the Bulgarian troops were entrenched. In the Serbian lines, artillery pieces were hoisted on human or animal backs on the top of the reliefs on paths dug in the rock for many months. In the second line, the cavalry was ready to attack.

Above: Leutnant Rudolf von Eschwege, ace of the aces of the Orient Front in 1917, who would total 19 certified victories. (DR)

At dawn on September 14, 1918, the infantrymen could see many planes belonging to all the Entente's air

Above: British fighter BE 12 descended on the Struma front and surrounded by Bulgarian soldiers. The British air force was technically inferior to the German air force fighters. (DR)

Above: View of a German squadron on the Orient Front, towards a DFW C.V in the foreground, a two-seater adapted to high altitude reconnaissance. In the background, a Halberstadt D.V fighter, an aircraft on which von Eschwege won most of his 19 victories. (DR)

Above & Above Right: Reconstruction of the dummy used to trap von Eschwege. (Coll Cross and Cockade)

Below: All that remains of von Eschwege's Albatros D.III is a pile of wood and linen, blown up by the explosion of the balloon bomb. (Coll Cross and Cockade)

German and Bulgarian airmen in front of a captured British Armstrong-Whitworth FK3. (DR)

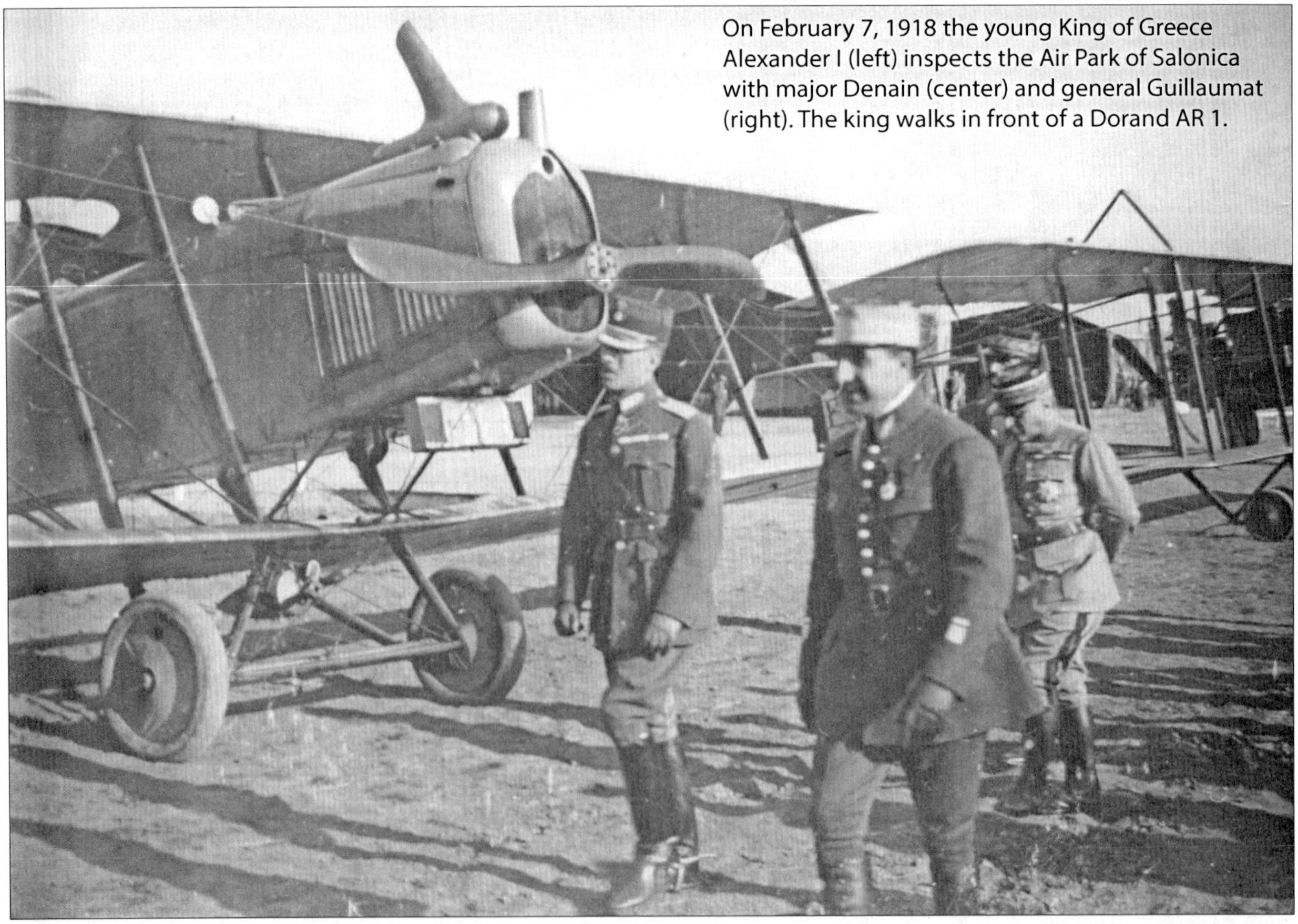

On February 7, 1918 the young King of Greece Alexander I (left) inspects the Air Park of Salonica with major Denain (center) and general Guillaumat (right). The king walks in front of a Dorand AR 1.

Above: A Dorand AR flying over the front in 1918. (Coll Boris Ciglic)

forces humming above their heads as they set off for the preparatory bomb and machine-gun attacks on the enemy's rear. D-Day was September 15, when, after a brief and violent artillery preparation, the Serbian troops, with the help of two French divisions, captured the Dobropolje mountain (1850 m) with great difficulty – the French cavalrymen then rode on the heels of the routed Bulgarians. In the air, French Orient aviation as well as Serbian aviation supported with all their potential, reinforced by the squadrons 505, 507, 510, and a detachment of the 531 in which participated the now Second Lieutenant Dieudonné Costes whose SPAD was based in Vertekop. 78 French pilots and observers, 20 Serbians, and four Russians carried out not less than 90 sorties during the day – the rhythm of the sorties was the same during the following days. Threatened in their rear, the Bulgarian troops and the few German detachments that accompanied them had to retreat but came up against the natural barrier of the Cerna that they could only cross on rare bridges. The bridge at Rasimbej was an obligatory passage to the rear and became an important objective for the French air force, which lost three Nieuports in its vicinity – one by a fighter, one by flak, and one missing.

The German fighter squadron fought its last battle in the East and Gerhard Fieseler won the last five of his 19 certified victories – the last one on September 20 at the expense of a Breguet 14 of the 510, the only one that this squadron would lose by fighters. On September 21, with the evacuation of the Kanatlarci and Prilep airfields threatened by Entente troops, the German air force disappeared from circulation and the following days were limited to attacks by routed Bulgarian troops. On September 28, 1918, the Franco-Greek 531 squadron landed on the airfield of Hudova abandoned by the enemy and, the next day, a French reconnaissance plane discovered that the French cavalry was occupying the town of Uskub. The Bulgarian army being cut in two, the Tsar of Bulgaria understood that he lost the game and signed the armistice on September 30, putting an end to hostilities on the Orient Front. In September 1918, the French, Serbian, and Greek air forces lost no less than 10 aircraft due to air combat, seven due to flak and three due to accidents – a level never achieved during the entire war, even during the most dramatic months of confrontations against Kaghol 1. But it is indicative of the level of air activity and the importance of ground attacks, which made a definite contribution to this great victory, later cited as the best example of the breaking of a front and its exploitation in the military training manuals of the German army...

Above: In front of his Albatros D.III, Gerhard Fieseler of Jasta 25, top ace of German aviation in the Orient front, winner of 19 victories mostly confirmed by French archives... A one-man airforce.
Below: In 1918, the Italian Air Force sent two squadrons to the Orient Front, the 73a (fighter) and 111a (reconnaissance) Squadriglia. Here, Nieuport 27s of the 73a Squadriglia. (Coll. Paolo Varriale)

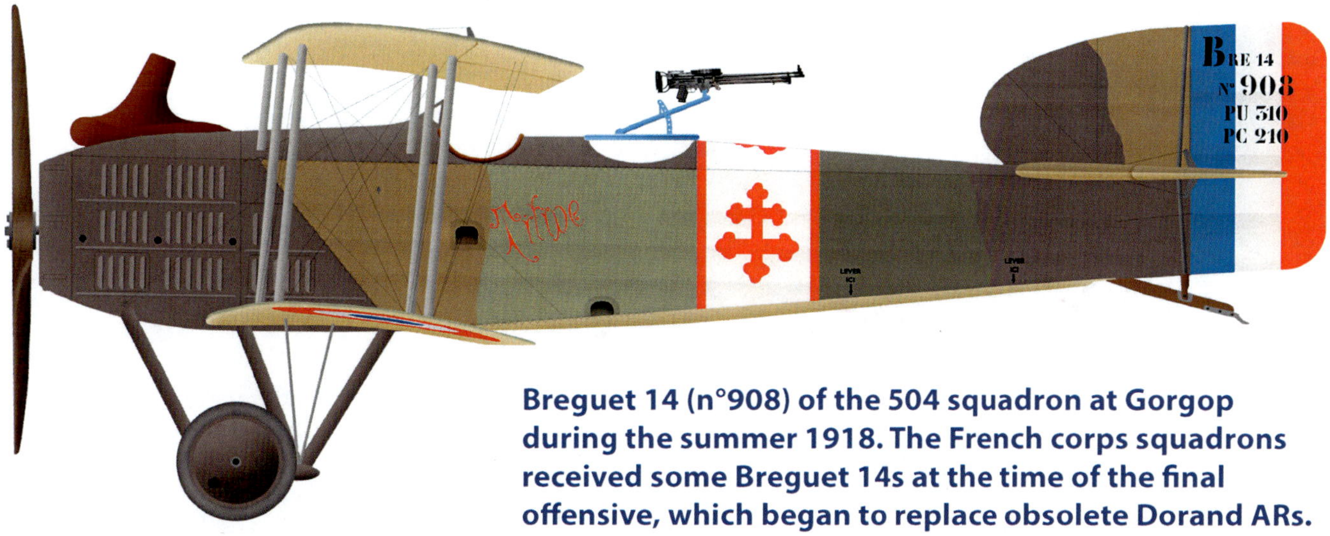

Breguet 14 (n°908) of the 504 squadron at Gorgop during the summer 1918. The French corps squadrons received some Breguet 14s at the time of the final offensive, which began to replace obsolete Dorand ARs.

Above & Facing Page, Above: Albatros D.III captured on the Orient Front by Serbian troops after the final offensive. (Coll. Boris Ciglic)

Chapter 6: Against the Bolsheviks in 1919

Above: This group of Cossacks is none other than the Breguet 14 pilots of escadrille 510, photographed in Bucharest on February 17, 1919, shortly before their engagement on the Dniester front. The pilot seated in the middle of first rank is probably the squadron leader, Lt Raoul Bergeaux. (Coll Albin Denis)

6.1 - The Upheavals of the Great War

Georges Clémenceau came to power in France on November 16, 1917, when he was appointed Prime Minister (in this time, the French President only had a symbolic role). He had the heavy responsibility of leading a country exhausted by the war and whose political representatives were advocating a compromise peace with the Central Empires. The inflexible politician, nicknamed "Le Tigre" (The Tiger) since his time in government during the years 1906–1909, acquired the new nickname of "Father of Victory" by imposing the continuation of the struggle on a nation won by doubt, until the armistice of November 11, 1918.

Although this date marked the end of the First World War, only the soldiers in the trenches from Flanders to the Swiss border were able to return home quickly. For all those in the Orient Army which defeated the German-Bulgarian troops in Greece and now occupied a vast portion of the Balkans from Yugoslavia to Romania, they would have to wait, because a new enemy was emerging from the plains of Russia: Bolshevism. The Russian government led by Lenin was currently engaged in a civil war with the White armies that were rising in every corner of the former empire of the Tsars.

Georges Clémenceau had strong ideas about the new occupants of the Kremlin, whose exit from the war in March 1918 with the signing of the Treaty of Brest-Litovsk, which allowed Germany to redeploy a large number of troops to the west and narrowly miss breaking through the French front in the spring of 1918, was not forgivable. He considered the Bolshevik leaders as "*a gang of bandits*", and wrote about them: "*Russia represents the biggest problem we have to face. The Bear is not a threat at the moment, but he is big and strong, and may cause us trouble in the years to come (...) The fixed idea of the Russians is that communism will flood the world under the leadership of the Kremlin. Therefore, in the years to come (...) I think that the Soviet government, having solidified its positions at home, will have no limits. It will trample the weak nations. It will plunder them; it will sign solemn agreements if they serve its purposes – and it will deliberately violate them.*"

As soon as the victory against Germany was achieved, even before the armistice was signed, the order was sent to the Armée d'Orient to engage in the fight against the

Above: A Breguet 14 B2 given to Romania in 1919. Barely noticeable is the serial number "US", coming directly from the Salonica Factory.

Bolsheviks. With 18 squadrons, the aviation of the Armée d'Orient was an important force that accompanied the French armies advancing in the Balkans, which were evacuated by the German, Austro-Hungarian, and Bulgarian troops. A new "front line" was taking shape around the Danube, in Romania against Bolshevik Russia, but also in Hungary where a communist government had taken power.

But in the chaos of the wreckage of the Great War, there will be no new battle for the French air force. For lack of opponents, as well as the will to fight.

6.2 - Skirmishes on the Dniester (April–July 1919)

The intervention decided by Clémenceau was carried out in December 1918, with the help of Greek troops whose government hoped for French support for the annexation of Turkish territories. In a great improvisation, the French navy crossed the Bosphorus strait and landed troops in Odessa, Sevastopol, Kherson, Kerch, and Mariupol. The French Prime Minister largely underestimated the demoralization of the French troops, in a hurry to return home and with little motivation to leave for a new war. Logistics were also lacking, as the merchant navy was unable to supply these troops, who had to face a very cold winter, and who found real chaos on the spot. The German troops that were supposed to occupy the country and maintain order had vanished; refugees were cluttering up the cities, and the white authorities that were supposed to control them had no authority, when they were not hostile to those who came to help them.

When the Bolshevik troops approached in March 1919, the French garrisons were very weak and poorly supplied. Several units refused to fire on the Bolsheviks. Clémenceau had no choice but to decide to evacuate all the Ukrainian cities (the first one, Kherson, on March 9) and to withdraw them south of the Dniestr river, in Bessarabia (now Moldavia) annexed by the allied Romania and defended by its troops. Odessa was evacuated on April 6, 1919 and the port of Sevastopol, where French ships were mutinous, was evacuated on April 29.

It was in this context that the French air force intervened. At the end of 1918, the aviation of Armée d'Orient had melted away with the demobilizations and five squadrons had to be disbanded.

Escadrille 501: Dissolution
Escadrille 502: Dissolution
Escadrille 503: Dissolution
Escadrille 504: Romania
Escadrille 505: Occupation of Constantinople
Escadrille 506: Dissolution
Escadrille 507: Romania (Galatz)
Escadrille 508: Serbia (Novi Sad)
Escadrille 509: Occupation of Constantinople
Escadrille 510: Romania
Escadrilles 521, 523, 524, 525: Serbian army. 521 has only Serbian pilots, 525 is dissolved.
Escadrilles 531, 532, 533, 534: Greek army. 531 is dissolved.

With the four Franco-Greek squadrons which pass under the control of the Hellenic government, and three to the Serbians, it leaves 6 squadrons at the disposal of the

Above: German DFW C.V captured by the Red Army and used by the 24th RAO, pictured near Kiev in February 1919. The aircraft is decorated with red roundels - not stars - hastily repainted on the German markings. The Red Army Air Force was able to field an assortment of outdated and patched-up aircraft, using any means available due to the lack of an aviation industry. (DR)

army of the East. Two (505 and 509) are part of the troops of occupation in Constantinople, one (508) stays in Serbia. Three others will be sent to support the action of General Berthelot's troops in Ukraine: 504 (6 Breguet 14 + 3 SPAD), 507 (12 SPAD), and 510 (12 Breguet 14). The Greek government brought its support with the 534 squadron (6 Breguet + 3 SPAD) which joined the expedition.

The lack of logistic means delayed the departure of these squadrons from Salonica to Odessa. It was only on March 12, 1919, that a ship left Salonica with the equipment of the aviation park intended to settle in Odessa. After a stopover in Constantinople on the 15th, the French equipment arrived in Odessa on March 17, where the landing took place in the following three days. The 504 squadron, incomplete, was already there, as well as a Polish squadron trained by the French, plus three squadrons of White Army aircraft and a seaplane squadron. A first SPAD XIII was reassembled on April 1st 1919 and entrusted to the Greek 534 squadron.

There was no other one: the Bolsheviks advanced and the order to evacuate the city fell on April 3. The French mechanics had to pack up the equipment just unpacked. They improvised a convoy to the station with recovered German trucks. But the pro-Bolshevik railway workers were on strike and they had to go to the port. The air force evacuated by boats which deposited their cargo in Constanza (Romania) and Constantinople. On April 14, the head of the Park received the order to go to Galatz (currently Galați, in Romania, on the Danube) where he disembarked with his equipment on April 17, supplemented by a new shipment of technical equipment from Salonica allowing the squadrons to operate effectively. These settled in two localities near the Dniester: Arcis for the 504 (currently Artsyz, Ukraine), and Kishinev (currently Chişinău, capital of Moldova) for the 510, while the 534 remained in Galatz. The 507 fighter squadron, countably present, will in fact never be reformed. A Polish squadron in the process of being formed was also dependent on the air force, which was to receive several shipments of new Breguet 14 B2s from France in boxes by rail. The question of equipment supply was therefore resolved for the Danube Army Aviation, whose command was assured from Budapest by Major de Serre.

The only question that worried the staff of the Danube army was the real strength of the Bolshevik army, to know if it planned to cross the Dniester. An indication of the state of the enemy aviation came to them on April 28, 1919 with the capture of a red LVG two-seater that landed in Zalău, Transylvania, where its crew got lost. Having taken off from Proskurov (now Khmelnyskyï, Ukraine) with the aim of reaching Budapest to transmit documents to the Hungarian communist government, the crew got lost in the clouds and was forced to land in Transylvania where they were captured. Questioned separately by the French military, the pilot Rudolf Pier, a former pilot in Kerensky's army, and the observer Bernard Gretz, a journalist, had no difficulty in revealing all they knew about the Red Army, of which they were theoretically a part. The observer even seemed relieved by his fate and indicates that "*the Bolshevik party includes 5% of convinced people, 50% of people attracted by need and 45% by interest. The air force is not frankly Bolshevik and is the object of a certain suspicion. To avoid desertions with the aircraft the principle of collective responsibility was instituted in the squadrons, for a pilot who deserts all the pilots of the squadron are shot.*"

The two men belonged to a special unit of four two-seater LVGs with 230 hp engines and three pilots, set up to provide air links with communist Hungary. On their field of

Above: Profile of the SPAD VII n°11 247 manufactured by the company De Marcay, Lt Julien Monier, squadron 504, Galatz (Romania) July 1919. This aircraft, from a batch produced in October 1918, probably never received the markings of the 504 squadron, namely a red Lorraine cross. It probably performed the last mission of the Danube Air Force on July 14, 1919, performing a reconnaissance over the Dniester River.

Proskurov is stationed the 21st squadron of the Red Army, which seems to have only a relative military value in view of its composition: an LVG (old and in poor condition), a French SPAD captured in Odessa, a very old Caudron, three Nieuport 120 HP (2 mounted), and 1 Farman XXX. There are only four bad pilots in the unit who hardly fly at all, because the gasoline is almost completely lacking. The arrival of the Bolshevik regime plunged the aviation industry into chaos; no new aircraft were produced and the Red Air Force could only rely on the stocks of the former Tsarist army, itself equipped with outdated equipment: during the winter of 1918–1919, the Bolsheviks managed to get their hands on 502 aircraft and 1320 engines in depots in Moscow, Smolensk, Rybinsk, and Yaroslavl, as well as 250 French and English aircraft in the ports of Murmansk and Arkhangelsk. Most of them were Nieuport fighters and Farman or Voisin two-seaters – antiques in 1919. Not to mention the fact that there were hardly any pilots left from the former Tsarist army to fly them: almost all the pilot officers, representing 2/3 of the 871 military pilots listed by the Bolshevik government on December 14, 1917, had defected to the White armies.

There would be no air support for the Red Army, which did not have the strength to launch a war against a regular army such as the Romanian army, supported by French and Greek troops. The few attempts to cross the Dniester River were not renewed and the annexation of Bessarabia was de facto recognized by Lenin's government in spite of the ultimatums he sent to the Romanian government.

The French airmen would quickly realize this. On May 20, 1919, the 510 squadron, flying over the northern part of the Dniester river, noticed the presence of an LVG and a Nieuport 80 hp near the Tiraspol cemetery, one of them even being seen in flight. General Franchet d'Esperey, head of the Armée d'Orient and then Armée du Danube, had already become angry on 16 May about the lack of combat ability of his airmen and had ordered that all enemy aircraft be shot down. Major de Serre followed up on the request of the supreme commander with a note to escadrille 510 indicating that a Breguet 14 B2 was to be assembled at the Park with the English bomb launchers. It was to be entrusted to Lt Maurice Lashermes, the best fighter pilot at the disposal of the French air force in Romania, holder of three certified victories and four probable victories in the Armée d'Orient, where he had been the wingman of the ace of the aces of Salonica, Dieudonné Costes. Lashermes received the order to destroy this red plane, and, as if to motivate him, the commander of Serre specified that he would have the right to a leave to return to France once this mission was accomplished.

He went to work, but on board his SPAD VII fighter. On May 29, 1919, when the Bolsheviks had been repulsed two days earlier while trying to cross over to the neighboring town of Bender, he carried out a reconnaissance of Tiraspol and discovered a battery there. He returned with three bullets in his aircraft from individual weapons fire. It was more or less the same thing as the harvest of Lt Bergeaux, leader of escadrille 510, who on June 2 made a reconnaissance and found a null activity on the whole front. He flew over Tiraspol, and fired about thirty bullets at workers at the northern exit of Slobozia (downstream from Tiraspol). Finally, on June 12, Maurice Lashermes achieved his goal by strafing the Bolshevik plane, which was found between the railway station and the cemetery of Tiraspol. According to the intelligence bulletin of the 30th Infantry Division, "*The*

Above: Aviatik-Berg D.I (92.126) of the Hungarian Communist Air Force.

Bolshevik plane was completely disabled. There were about forty bullets in the engine. The engine was sent to Odessa." Lashermes was rather enthusiastic about his victory and wondered in a postcard addressed to his relatives if this victory will be certified. It seems not, but what is certain is that he has earned his ticket back to France, which he has not seen since June 1917 when he was sent to the Armée d'Orient.

As for the Bolshevik army, the subsequent reconnaissance carried out upstream and downstream of the Dniestr showed that it did not seem willing to cross it. On June 16, 1919, a Breguet 14 flown by S/Lt Arnault with Captain de Moustier as observer carried out a long reconnaissance from the mouth of the river at Ovidopol, flew over the cities of Odessa and Mayaky, then flew upstream to Tiraspol and Kishinev. The two men saw only defensive positions and an airfield tent in Odessa. Only pontoons at Mayaky suggest a river crossing, but the place does not seem to be the best place to conduct an offensive because the area is easily defensible on the south bank.

One may wonder whether the French army would have been able to sustain a new war if the Red Army had been able to attack. At the time of the Black Sea mutinies, when French troops refused to fight the Bolsheviks in the ports of Ukraine, acts of insubordination also spread to the air force, which had never been subject to mutinies during the First World War. On May 12, 1919, Lieutenant Economacos, the leader of the Greek squadron 534, complained about the behavior of a French pilot, S/Lt Jacquart, who takes his service car for personal use by treating him with the greatest contempt. The officer was gently reprimanded by his superiors who lectured him on his lack of tact. The soldiers Core and Muller, who worked at the airfield, were not as lenient: on 17 May, furious at not having had their wine ration, they went on strike and incited their comrades to insurrection. They were immediately put in prison... Generally speaking, the atmosphere was no longer disciplined in the air force, as a staff officer who was on an inspection tour at the end of June 1919 discovered the 510 at Kitchinev "*in the greatest disorder, with neglected equipment and officers living in the city, where they only thought of having fun.*" General Franchet d'Esperey, angry, immediately ordered that Lieutenant Bergeaux be relieved of his command of the 510 squadron.

Bergeaux's dismissal at the head of the 510 squadron turned out to be of little importance for the unit, as it occurred at the time of the withdrawal of the French units, which were relieved by Romanian units. At the end of June, the 510 squadron was withdrawn to the Galatz park, where the chief pilot of the park, Lt Cochet-Fortuny, was killed on the 25th – the only pilot killed in the French air force in Romania. On July 2, the order was given to withdraw to Sofia in Bulgaria. The 504 squadron reached Galatz on July 7, from where on July 14 it left on what seems to be the last combat mission of the Danube Air Force: Lt Julien Monier, on SPAD VII n°11 247, carried out a reconnaissance over the mouth of the Dniester river and concluded it with a laconic "RAS", apart from a few individual gunshots which greeted his passage. On August 1, 1919, the Galatz Park

Above: UFAG CI (161.169) of the Hungarian Communist Air Force.

delivered all its Breguet 14s in stock to the Romanian government and the following month evacuated the country for Constantinople, where its logbook ended on October 11, 1919, putting an end to the aviation of the Danube Army, whose ground troops also left Romania at that time.

6.3 - Facing the Hungarian Communists (March–August 1919)

While the French troops of the Danube army were facing the Russian Bolsheviks along the Dniester, other French troops were facing another communist government in Hungary. This country, one of the two components of the Austro-Hungarian Empire of the Habsburgs, was part of the defeated camp, as the countries of the Entente cruelly reminded it by setting its new borders. On March 20, 1919, Lieutenant-Colonel Vix, the French military attaché in Budapest, handed over a note to the Hungarian republican government setting the new borders. The country was greatly reduced to the benefit of the new Yugoslavia (Banat region), the new Czechoslovakia (through Slovakia, part of which was populated mainly by Hungarians) and especially Romania, which gained Transylvania, which included a large Hungarian minority.

Hungarian government preferred to resign rather than recognize this dictate. A new government, headed by the communist Béla Kun, was formed immediately and rejected the Vix note. This gesture initially won him a certain amount of popular support, which he quickly lost in the face of the abuses committed by the "Red Guards" and the forced collectivization of the country. But this initial position allowed him to win the support of many officers and soldiers of the former Austro-Hungarian army who served in the new Red Army, more out of nationalism than adherence to communist ideas. The military threat of this army in formation was not to be neglected, because the stocks of weapons awere abundant and could lead to a real military power if volunteers ready to use them were organized.

From the aeronautical point of view, the Hungarian Communist government was paradoxically in a much better position than the Russian Communists. The Austro-Hungarian air force fought until 1918, benefited from German technical assistance, and had a certain industrial base to produce new aircraft. As the French intelligence service realized, the Hungarian Red Air Force, led by Lt-Col Petroczi, had about fifty aircraft on line, composed of two-seater Hansa-Brandenburg C.I and UFAG C.I, as well as Aviatik-Berg D.I and Fokker D.VII fighters, divided between eight squadrons of four to six aircraft, including two fighter squadrons stationed in Budapest. A squadron of seaplanes was also stationed in the capital, where there was also a flying school.

The French troops were not far away, occupying the Banat, a region with a very heterogeneous population including a strong Hungarian minority. Their northernmost position was the Hungarian city of Szeged. Led by General de Lobit, the "Hungarian Army" was mainly composed of colonial troops, whose numbers were constantly decreasing due to demobilizations. From an aviation point of view, an air park was set up in Novi-Sad (formerly Neusatz), bringing together the Serbian squadrons and the only French squadron, the 508, which on March 14, 1919 received the order to reach the city of Szeged. It was operational on the 23rd and immediately carried out reconnaissance on Hungarian territory held by the Communists. Its leader, Lt Picot, had only three Breguet 14s in flying condition at his disposal. On March 25, 1919, he flew over the city of Budapest in one of them, with Lt Charmette as observer. What they saw was

Above: SPAD VII and Nieuport 27 fighters of the French-Serbian 523 squadron at Novi-Sad in 1919.

confirmed by the other planes of the squadron: there were no movements towards the French troops in the south of the country, who, with the exception of a few exchanges of fire between outposts, did not have to face the Communist troops.

Bela Kun's government faced a much greater threat: the Romanian government was prepared to take possession by force of the territories allocated to it by the peace treaties and went on the attack at the end of April 1919. The Hungarian troops, disorganized and fighting on a territory mainly populated by Romanians, were defeated and retreated. At the beginning of May, Bela Kun asked for a cessation of hostilities by declaring that he recognized without reserve the territorial claims of the Romanians. The French troops observed this confrontation as spectators, the planes of the 508 squadron, Breguet 14s and even an old Dorand AR, regularly flying over the lines. They were reinforced on 11 May 1919 by the Serbian 523 figher squadron arriving from Novi Sad. It received the unequivocal order to shoot down Hungarian planes that sometimes ventured to Szeged to launch leaflets inviting the population to rise up against the French occupier.

But if the aircraft of the Hungarian army did not have to fight against aircraft bearing the red star, other French aircraft had to do so at the other end of Hungary. Having concluded a cease-fire on the Romanian front, Bela Kun opened another one against the young Czechoslovakia, which he attacked on May 20, 1919 in Slovakia and Ruthenia. The first of these regions was mainly populated by Hungarians in its southern part; the second would make it possible to establish a territorial link with Bolshevik Russia. The study of the air operations in Slovakia and in particular of the French squadron BR 590 which was sent there is beyond the scope of this book which deals with the aviators of the Army of the East. Let us also mention the arrival of French airmen in Poland to train the local aviation and to fight the Bolsheviks.

The conflict with communist Hungary ended with a cease-fire extracted from Bela Kun on June 24, 1919, which the French government threatened with a joint attack by Romanian, French, and Serbian troops in the south and east of the country. The Hungarian communist government failed in its aim to open a corridor to Bolshevik Russia for external aid. Its days were numbered: the terror instituted by the Red Guards in the interior of the country encouraged defections in its army and increased rallies to the counter-revolutionary government in Szeged. Two Red planes voluntarily landed in Szeged, their crews showing the state of decay of the Hungarian forces. In Szeged itself, the only loss of French aviation in Hungary occurred when a Breguet 14 accidentally crashed, killing its crew composed of Lt Robert Gaborit and S/Lt Pierre Paul Marguerite.

The Hungarian regime disintegrated after 133 days of

existence when Romanian troops attacked again in August 1919, crossing the Tisza River to seize the city of Budapest, against the advice of the Entente powers and the French government in particular. Nevertheless, with the end of the communist threat, the French army in Hungary no longer had any reason to exist and its units were quickly disbanded to make way for military missions with much smaller numbers.

Thus ends the action of the French air force in the Front d'Orient, after five years of presence. The action of French airmen in Eastern Europe during 1919 played a significant role in the establishment of the new borders resulting from the Treaty of Versailles and allowed for the provision of substantial military support to Romania and the new nations of Yugoslavia, Czechoslovakia, and Poland. The French military and political influence was to be preponderant in these countries, which were to form the "small entente" against Communist Russia and acquire French aircraft during the 1920s. But as hegemonic as France was after the First World War, its industrial decline was to undermine all its political and military influence, which fell in the face of the rise of Nazi Germany in the 1930s.

Right: Funeral of Lt Cochet-Forthuny on June 27, 1919 in Galati (Romania). Cochet-Forthuny was the only French pilot killed in Armée du Danube.

Appendix 1: Testimonies

The Vardar Expedition and the Groupe de Bombardement d'Orient, by Corporal René Vicaire

Trained as a mechanic, René Vicaire (1891–1985) did his national service when the war began and was assigned to the DO 22 squadron when war was declared, where he flew as a machine gunner for Jules Vedrine on his Blériot "*La Vache*". In 1915 he was promoted to corporal and became a student pilot, obtaining his military licence n°1383 in August. He was then appointed to the V 90 S squadron, which was to rescue Serbia, and embarked for the Orient. Promoted to sergeant in 1916, he was transferred to the Salonica air park from where he was repatriated for health reasons (malaria) in the fall of 1917, ending the war in a technical position in France with the rank of adjudant. He recalls in a testimony preserved by the Service Historique de la Défense his arrival conditions in Orient, his participation in the Vardar expedition as well as in the bombardments of the Groupe de Bombardement d'Orient (GBO).

Above: René Vicaire (1891–1985) (DR)

On the Way to Salonica (October 1915)

Assigned to the V 90 squadron, I left for Lyon-Bron for a meeting. The pilots were Morénas, Carier, Fabre, Breton, Deloupy, and Captain de Thézy. Then we went to Lyon-St-Priest where we were given injections against cholera and other oriental diseases. The first night spent in St-Priest was very agitated because some of us had a tormented mind and we had a real shooting. We never knew why and the next day we could only see the damage: not a single window had been broken. A few days later we were all in Marseille, waiting for the departure. The sea was beautiful but our funds were low, especially after visits to Bassot and Pascal, two famous restaurant owners in Marseille whose prices were not in line with our modest means.

We stayed in a big hotel on the Cannebière and were well treated, but everything has an end. On October 21, 1915, we embarked on the *Gaule*, an ugly old ship whose launch date was so far away that it was forgotten. I will try to describe life on board.

The first few days we enjoyed a flat calm and an oily sea. We are escorted by two destroyers: the *Massue* and the *Arc*. Another boat followed us, the *Mont Cervin*. Others are still part of the convoy and all this small flotilla is heading towards Salonica.

However, after two days at sea, the weather became bad and we started to dance terribly on our old cargo ship: the boat was buckling, with each dive the machine was getting out of control and the noise, always the same, became unbearable. We would hear it until the end of the trip. We are almost all sick. Sergeant Catin has been tied up on the bridge, he refuses to eat and drink, and will remain so until we arrive in Salonica. The mess hall is infectious with its smell and its kerosene lamp that friend Fabre throws in front of us from time to time to show us that the boat is running! This joke forces us to go out to feed the fish, as the sailors say, who, sitting on the deck, with a bucket of wine between their legs, drink their ration and, of course, ours. We have been away for a few days now and we are now in sight of Cape Matapan; it is, it seems, the most dangerous place of the trip. It was there, in fact, that the German submarines were refueling. One evening, the battle was on! Our boat

Above: Greek civilians and soldiers observe a Voisin LAS of the V 90 S squadron mounted at camp Allatini, the maneuvering field located south of the city of Salonica, which will house the entire Oriental Aviation Park. (DR)

Below: The Voisin LAS of the V 90 S squadron stationed in the snow on the field of coast 262, south of Krivolak, during the Vardar expedition in November 1915. (Coll. Albin Denis)

Above: A Voisin LAS with the red markings of the V 90 S squadron, photographed during a mission to bomb enemy camps in early 1916. (DR)

Voisin LAS n°1013 of Corporal René Vicaire, V 90 S squadron, winter 1915–1916.

turned quickly, the destroyers fell in circles around us, a few gunshots and the calm returned. We continued our route.

Finally, after eleven days of crossing, we are in sight of Salonica. We disembarked on 31 October 1915. During the transfer of the equipment we visited the city; we were not too badly received. Then we arrived at the Zetenlik camp: assembly of hangars, planes, and some tests over the city and the harbor. Among the contingent was a friend of Denain, who later became an observer, a prisoner of the Bulgarians and ended the war in Sofia as a tramway driver. He was very rich and a bit of a joker. One evening he invited Morenas and me to dinner in Salonica. He did things well, but demanded that the waiter who served us have a tureen by his side. It's unusual, but here's the explanation: there are three of us at

Above: Map showing the bombings made by GBO.
Right: View of a Voisin-Canon, an aircraft of which a few examples were delivered to the Oriental Air Force and which escorted the Voisin LAS aircraft of the Orient Bombardment Group. (Coll Albin Denis)

the table, and yet he orders four portions at each service. When everyone has been served, he pours the fourth portion into the tureen; the drink follows the same path until the end of the meal. He then calls the waiter, stands up and caps him with the tureen. He then asks for the bill for the dinner and for the suit. The victim did not flinch and we leave the restaurant convinced that our host is a little disturbed.

We spend a few more days in this city and then we take our equipment by rail, except for two or three planes that will join us by air. Fabre and Morenas were on the 1013, the plane that would be assigned to me a few days later. The train was a real wheelbarrow; the track was in very poor condition and in the middle of the first night a car came halfway off the rails. We were all busy putting it back in place. Next to me, the observer Captain Robert ordered me to go under the car; as a pilot, I refused. He then took his revolver to force me to obey him. Major Cheutin, who witnessed the scene, made him holster his weapon and gave him a severe warning; needless to say, our pilot-observer collaboration did not last long!

We arrive at Krivolak: unloading of the equipment. To stretch our legs a little, we leave on foot towards the field, a rough hill in the middle of the rocks. Just halfway up, we

received a few shells, most of which fortunately did not explode; we used the rocks as shelter.

Among the planes that came by air, some did artillery adjustments, in particular on the Kara-Hadgali battery; among them were Morénas and Lt Richepin. That same day, November 23, 1915, I took over the 1013 and went to Négotin. I spent one of the worst moments of my life as a pilot.

Above: Gardening on the airfield (Rol)
Left: Restaurant in Salonica in 1916. Greek restaurant owners know how to adapt to their customers... (DR)

The air camp was set up on a plateau between rocks, ravines, and mountains. The observation flights started, but we were very much hampered by the fog and some pilots joined Guevgueli. I stay here for the last reconnaissance flights, without supplies, but with my shotgun which will be used to supply us. The snow is falling in large flakes. We have to make it disappear from the wings of the plane and shake the tent. Then the temperature dropped: there were 60 cm of snow and it was minus 26 degrees. I sleep with my wetsuit and we make a little fire with gasoline inside our frail house. The next day, with two mechanics, I go to the refueling station in the direction of a small village that we can see very far away, at the end of a snow carpet. How far it is to get there and even further, it seems, to get back! We bring back a small bottle of wine which will solidify on the way. The inhabitants had deposited some piles of manure in the fields. Their tops were above the snow and hundreds of birds came to look for their food. Our need to eat did not allow us to be tender and I shot at these poor larks: 87 fell in five shots! They came from everywhere and my two men, while running

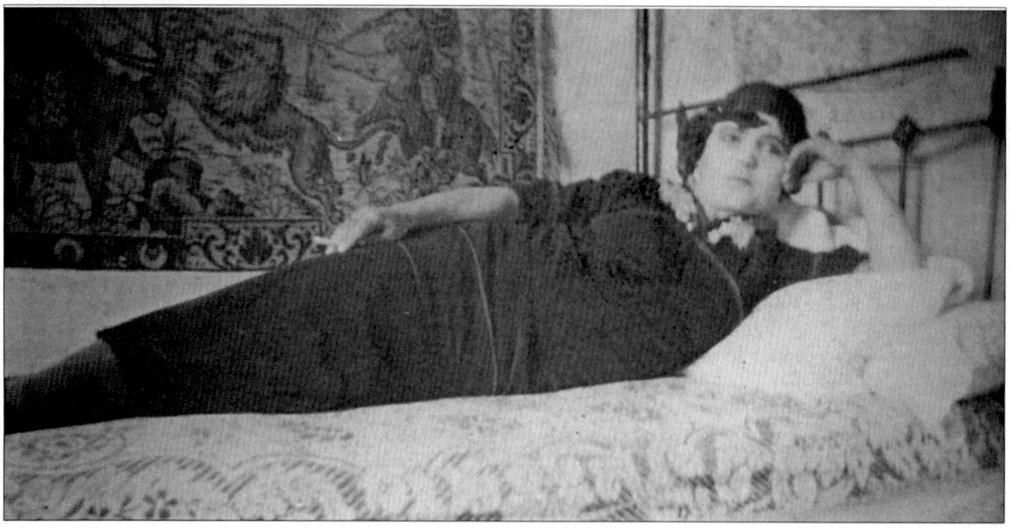

Left: Lili, an Algerian prostitute from Salonica, of whom René Vicaire was a regular client...(René Vicaire)

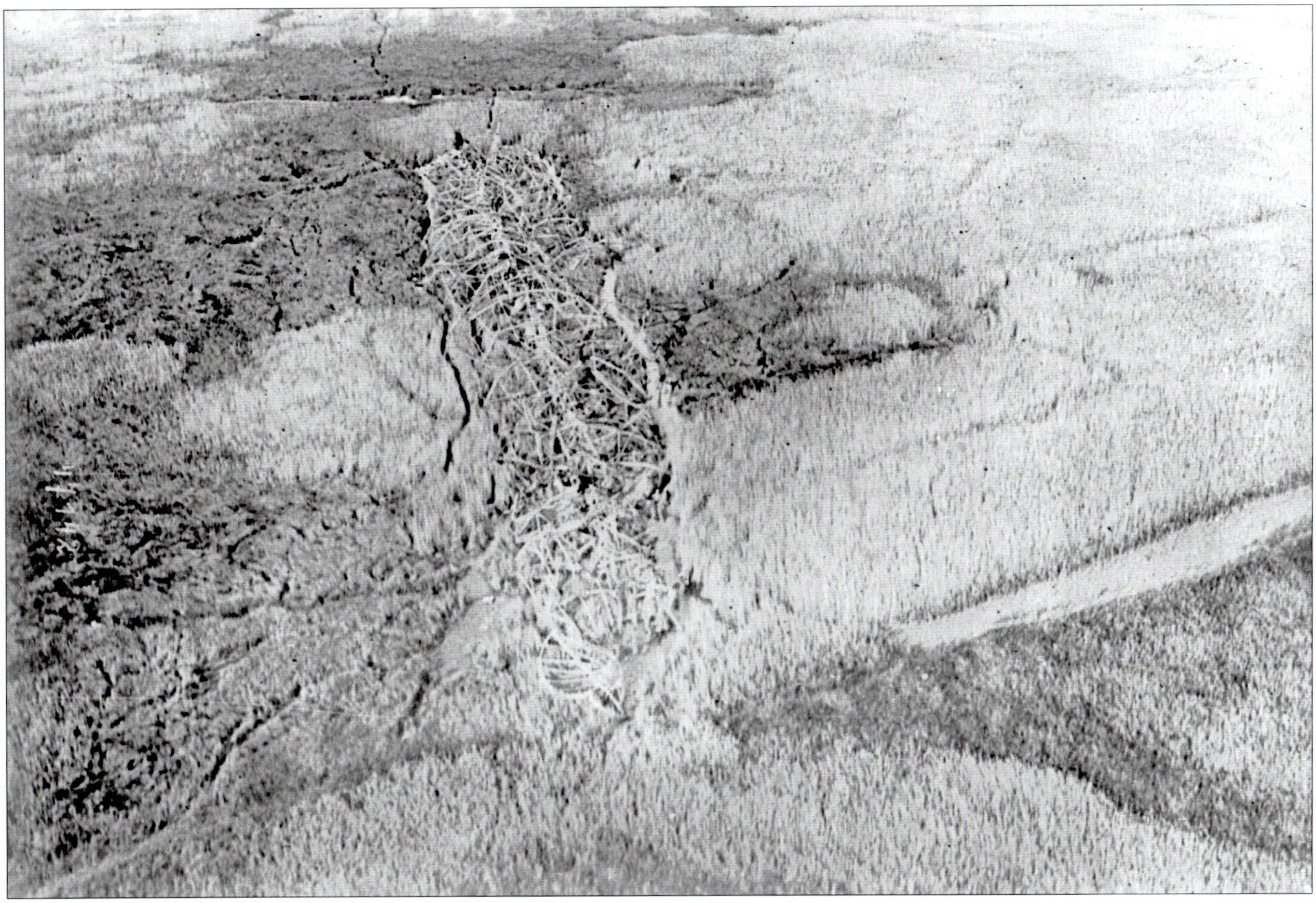

Above: Wreckage of Zeppelin LZ 85 fallen in the Vardar swamps, May 5th 1916. (Rol)

to pick them up, disappeared in huge ditches full of snow. We returned from this expedition in fairly good condition, happy to have something to eat.

Socks and Mechanics

On December 2, 1915, the beginning of the retreat was announced and soon the first elements of the infantry appeared. We had to leave but the terrain was very short, the Canton-Unné engine was frozen, and there was still a lot of snow. So, the radiator was dismantled, repaired, and soldered with a lamp that agreed to work for a few moments and then stopped and we started again. Finally, after a lot of efforts, the radiator was put back in place, but, in spite of my recommendations, my mechanic Lefranc, after having melted the snow to have water, introduces it in the radiator without turning the engine: it solidifies immediately! The only solution: there is nothing left to do but to start again ! After many more problems, the radiator is ready again but this time with the engine running and ready to leave. At the moment of departure I refuse to take my mechanic on board. He starts to cry. But as I am not sure of the take-off, I prefer to be alone to take the risk and, on the small plateau which is used to us as runway, after a new test of the engine at the fixed point, I push the throttle to the bottom and stick the stick to the belly; here I am left in the snow. I lift the plane, but at this moment the engine slows down and I arrive at the end of the runway with, in front of me, a huge hole in which I have to dive. This hole is the accident and probably the death; in this moment and in the time of a flash, many things passed in my head: my parents, my friends, my youth, and I see my end in this immense hole; it comes towards me at a terrible speed, everything is going to collapse and I am going to finish stupidly in this cursed country. But, automatically, by arriving close to the ground, I straighten my plane, which, obediently, lands in an enormous snow pack, which brakes it rather brutally, avoiding me the meeting with the rocks ! I had put my engine at idle during the descent, which I should not have done because it is always necessary, in case of a fall, to avoid the fire. So, contrary to the principles, my mistake is useful to me because my engine is still running, the propeller cuts the fresh snow accumulated by the winds at the bottom of this ravine. It is very cold, but I was very hot!

The retreating infantrymen who pass up there, on what one believes to be the road, come with great difficulty to my rescue and it is with their help, ropes, the mules, and the engine that has been willing to continue to turn that I get out of this bad situation. Going up the slope sideways, I find myself at my starting point. Considering that the cause of my failure is due to a too big admission of icy air in the

Above: A manhunt is organized to capture the crew hiding in the swamps... These French sailors and soldiers have captured the standard. On the left is a 100 hp FBA of the CAM Salonica.
Right: Captured by the French gendarmerie, one of the crew members of the Zeppelin commanded by Oberleutnant Ernst Scherzer.

carburetors, I decide to plug them partially; as the material is lacking, I take my socks that I split in two at the low part and pass them on the air admission pipes with ligatures on the sides. As my feet are going to suffer even more with this cold, I pour some hot motor oil in my shoes and I put my icy feet in it.

New test and departure, always alone since the mechanic will join the base by the road. The take-off is quite good and the climb is normal, direction Guevgueli. I continue to climb progressively but I will have to pull on the stick because the Demir-Kapou mountain range to be crossed must be around 1200 meters; look at the onboard instruments, everything is normal and I continue, the sun shines on the snowy mountain: here is 1500 meters, again look at the instruments. Stupor: the oil indicator is at zero and we have to pass the mountain where the comitadjis are on duty but I don't believe it given the temperature. I am still climbing but with more and more difficulty; nevertheless I have to

Above: The carcass of the LZ 85 is carefully dismantled and reassembled on the place of the White Tower of Salonica.
Facing Page, Above: The carcass of the Zeppelin becomes a tourist attraction of choice.
Facing Page, Below: General Sarrail reviews troops in front of the carcass of the Zeppelin.

finish and I pass near the top, without comitadjis (irregular soldiers in the Balkans), without any hitch and with a big sigh of relief. If the side of the mountain that I have just left represents the great winter, the slope that I am approaching reflects for me the spring; the oil sight is still blocked at zero, the engine stalls but this does not bother me at all anymore; my altimeter indicates 1250 meters, I am above the ground and I begin the descent in spirals. The Serbians have filled in a trench in the middle of the runway that I can see perfectly, but that I think is at ground level; it is not the case and the bump makes me bounce, fortunately without harm. I stopped a little further, engine burned out but plane intact.

Hardly on the ground, my comrades are near me and look at me with amazement. I must say that I am in a good state, black from the gasoline fires under the tent, not shaved since more than a week and very tired by the bad moments spent in the frost and the snow. So I'm going to make a good toilet with the means of the edge and to restore me copiously: I needed it very much. That evening, still under the tent, I tell in great detail my peregrinations on the other side of the mountain, happy to have come out and to have saved my plane.

I spent a few days in Guevgueli, then left for Salonica where our squadron was reformed and moved to Samli-Topsin. There we have a very large and flat area; it is usually used as a pasture for transhumant sheep, because the sun roasts the grass in the middle of summer. The sheep, the shepherds, and their families migrate to the mountains to find cooler weather for some and grass for others.

The Carriage

Our camp was in Samli, a large suburb of Salonica. It is from there that we left to carry out bombardments. Often a large number of aircraft took part, accompanied by Nieuport fighters, in charge of the protection.

Sometimes we took advantage of our days off to go to Salonica for lunch or dinner in awful restaurants where the smell of mutton grease, added to the temperature, destroyed our appetite. Since we had left, we went to the end and when, sometimes, our caterer took us for richer people than we were, the bill was settled a little brutally. (...)

To get home, if no squadron car was available, we took a carriage. The coachman, as well as the horse, wore a superb hat decorated with lace to protect himself, during the day, from the burning sun and the flies, and in the evening from

the mosquitoes which lived in myriads and whose bites we feared a lot because they propagated malaria which caused great ravages in our ranks.

In spite of the compulsory intake of quinine at the morning assembly, the cheaters let us believe that they had taken this medicine. We controlled them by forcing them to stick out their tongues, where the quinine would leave a white trace. In spite of this preventive, we had very violent attacks of fever which exhausted us and left us in a state of extreme weakness. When, for our convalescence, we received the order from the doctor to leave for the surrounding mountains, our satisfaction was great; we had, above all, the hope of returning to France.

How many friends did we leave there? Indeed, those who, before the crisis, were suffering from certain diseases such as albumin, etc., could not resist malaria and remained forever in this foreign and distant land. Some of our dead were repatriated.

The natives of the country, the young people in particular, came almost every day to ask us: "Quinine, Moussié!" What did they do with it? Born in this village, I think they must have been immune. Like any self-respecting Greek, they had to sell the quinine that we gave them. So, from there, we forced them to swallow the pill in front of us. Then they would disappear.

But let's talk about our carriage: we used to travel in this vehicle, usually with four people, because our security was quite relative, especially at the beginning of our arrival here, when some shootings made us think about it; French soldiers were killed in some streets of Salonica: the reaction, as much on the military side as on ours, was quite brutal. Faced with this attitude, calm returned rather quickly. Our trip by carriage ended about 5 kilometers from Salonica, at the Gallico, a river that is dry in summer and overflowing in the bad season, located one kilometer from our airfield. It was there that we left the carriage, which at certain times could not cross the river, but we did not want it to go any further anyway.

We didn't always agree on the price of the trip; this led to a lively altercation between the transporter and the passengers. We must admit that we did not like the Greeks very much, always on the lookout for a bad trick. Nevertheless, I often thought that our way of acting was not correct. Indeed, on some days, as a rule, we took the rear axle of the carriage and turned it over, with the unfortunate horse on the ground and the wheels of the vehicle in the air. The poor coachman would yell loudly in the countryside "Bravo!" which, it seems, meant "thief" in his language.

Our second occupation, on days off, was hunting. Fabre had bought a rifle in Salonica. The one I had at the retreat of Négotin had remained in the fight. I had enough to think about saving my plane and myself, without worrying about small objects that had only been useful to survive in these places and that would have further loaded my unfortunate plane, because, at a few kilograms, I did not pass Demir-Kapou.

In this more than solitary place where a road to Salonica passed, we were accommodated in tents and in our aircraft boxes; these were fitted out by us and despite the great heat of the middle of the day, life was possible there. The insignia of the aviation had been drawn by my care; its dimension was approximately 5 meters length; the wings were represented by small pebbles of various colors: red, blue, white, etc... And the central propeller was in flowers of the country as well as the oval containing the totality of the drawing. The whole raised of about fifty centimeters, gave to the visitor the illusion not to be in this Greek desert.

Scenes of Everyday Life

If our bombardments followed one another, we sometimes received orders from the Grand Quartier Général via Denain to take on board, during a mission, certain passengers whose names were not unknown. One morning, as we were leaving, we saw an officer, Mr. B...., a well-known playwright, get out of a car on the edge of our camp. He was to take a place with one of us for the bombing of Xanti. As we were not unaware of the purpose of this story, we decided to entrust this passenger to be decorated to the most daring pilot among us: Cabanne was designated almost unanimously. The mission was not long and after about an hour and a half of flight we were back. Five planes landed and the sixth one, Cabanne's, started a crazy saraband in the sky, above the squadron. Mr. B... must have remembered this flight because our pilot did with his Voisin what was most daring, at the risk of breaking everything and this for more than an hour, that is to say until the absolute emptiness of the gasoline tank; then he went down in a glide. I don't know if Mr. B. is still alive at the time of writing, but I can't describe to you the state he was in, having evacuated at all ends. The mechanics took him off the plane and after putting him back in his car, we wished him a good trip. His driver was smiling. He never came back, but he was decorated, because it seems he had just won the *croix de guerre*!

We lived there for several months and the few events I will try to tell you about remain in my memory the memory of this small lost corner, Topsin, very far from our homeland. We had hardly arrived two days before a pilot, our friend Carrier, who was probably feeling blue, drowned his boredom by drinking too much alcohol. He was absolutely raging and we had a hard time calming him down in the middle of the night after we tied him to his bed.

Life here was monotonous, the bombings quite frequent; we would leave on the Canton-Unne Voisin with a load of 8 x 90 bomb shells and one or two incendiary bombs. I bombed some cities: Petric, Monastir, etc... with my observer Captain Hug, then batteries on the front.

When I was not on duty, I often went with my shotgun to fetch meat supplies. During the rainy season, the plain on the edge of the Vardar was frequently flooded and, on the un-submerged strips of land, I made quite beautiful pictures; impressive flights of lapwings were looking for their food in these places. Their approach was very delicate. Also I had trained a mule for this approach; the results were very good. Often, the meals were improved by our hunts. But General Sarail got wind, one day, of our expeditions. At the request of certain staff officers, he ordered us to immediately lay down our hunting weapons. This we did, reluctantly, because the only pleasure we had left in the open air was thus taken away. We accepted this punishment with great difficulty.

One morning, we were alerted by the lookout of the outposts: he signaled a group of wild boars heading towards the squadron and soon, indeed, we saw them heading towards us. We did not know the color of the boars and one thing surprised us at first: they were black. The crackling of an airplane's machine gun and the heavy fire that the spahis opened on them, put down almost the entire troop, the survivors going to perish at the next squadron. But the matter soon became complicated, because our cook, a butcher in civilian life, realized while cutting up one of these animals that it was castrated! And the Greek owner of these "pigs" came to see the captain to demand reimbursement for the loss incurred. The matter was settled and we ate pigs for several days.

I don't want to list all the sorties made by the squadron but, in all honesty, the air service was much less hard than on the French front. Unfortunately there was the spleen and the climate. The heat, in this season, was overwhelming; it was forbidden to go out without a haxst. During the afternoon, I sometimes got sunstroke while trying to sunbathe. I didn't know where I was or what I was doing. I found myself in my bed, under my tent, being cared for by my orderly who, very patiently, poured cool water over my head with the help of a quart. The night was already well advanced when I regained consciousness.

One morning, after having received my mission order, I was about to leave when a Nieuport landed a little long and came to throw itself on a Voisin of which it made jump, in the shock, the barrel of its attachment. There was no great harm done and the mechanics, a little quickly I thought, made the repair. The plane was piloted by Girard, passenger Misrachi. We all left but Girard did not come back: I always thought that the cannon, on recoil, had mowed down the two men in the cabin... Very sad the evening meal, without our poor comrade. The Austro-Germans, in their communiqué of the next day, reported a shot down plane. A mass was celebrated in memory of the two missing. The parish priest used the front of a Voisin as an altar and the whole group attended the ceremony. According to some information gathered later, I learned that Misrachi had probably been killed by the recoil of the gun during the shooting, but that Girard had managed to land in the countryside. He was shot with a machine gun by the Bulgarian gunners.

Let's talk a little about my mechanic Lerrano, who had joined the squadron on foot with the infantrymen. He was devoted and an excellent worker; moreover, he was the perfect type of hunter-poacher and often indulged in his pleasure. One day of rest, he went fishing on the banks of

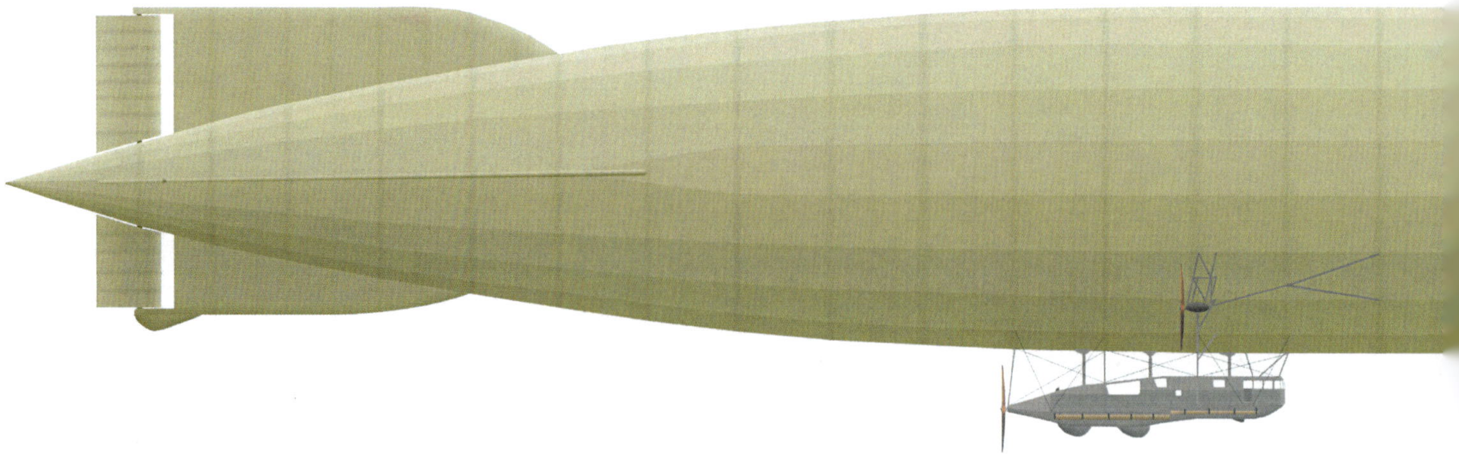

the Vardar. At lunch time, he came to present me with some pieces of fried eel, very well prepared. I enjoyed it. After the meal he offered me a rod covered with a superb snake skin: I understood that the delicious eel was a snake. He also made me taste some turtle in stew (like rabbit) and I must say that I found this dish delicious. Later he caught a few skunks, but no matter how they were arranged, we never managed to eat them.

One evening, while returning from a bombing mission, I was taken by a new attack of fever complicated by cardiac troubles. The squadron doctor gave me some injections and put me to rest. The commander Denain called me to Salonica. He asked me to take charge of a flying school for observers, which I did for some time. But my students (Lucien, Hug, etc...) were not very good at flying and the training ended. (...)

May 5, 1916: The Zeppelin

It was already late, about 10 pm. We are gathered in our dining room, the discussion is in full swing when, suddenly, the alarm sounds. What was going on? A Zeppelin was reported in the direction of Salonica. Two pilots, Deloupy and his partner, immediately took to the air. We go out to look, the sky is illuminated by the ground searchlights and suddenly the lighting intensifies: the navy has just turned on its headlights. The gun thunders, the bursts crackle, the land batteries and the navy are in action, it is a splendid fireworks display. The Zeppelin, hardly visible, passes over Salonica. Will it bomb? No explosion on the ground suggests so. In spite of our two planes and the great quantity of bursts which always crackle and which must dangerously disturb our two pilots, it disappears towards the lands.

The cannonade calmed down, the searchlights swept the sky, and peace returned: the aircraft had crossed the bombed area without being hit and had disappeared. We were about to go to bed when, suddenly, a huge light illuminated the marshes and the sky: the commander had sabotaged his aircraft! All available ground personnel were immediately directed to the spot. The Navy arrived with the patrol boats and was the first to arrive on the scene. But the German crew hid in the reeds on the edge of the sea and at daybreak we continued the search in the Vardar marshes.

A large number of men of all arms were needed to locate and capture the crew. The commander, a man of average height, well-muscled, with a completely shaved head, was caught in the morning. He carried the standard of his Zeppelin. Taken to the police station, he let himself be photographed, unlike many of his compatriots. It will take two days to take all the crew, at least we believe it, because some of them came out of the swamp after having accomplished a rather long journey inland. The prisoners were certainly transferred to France, we never saw them again.

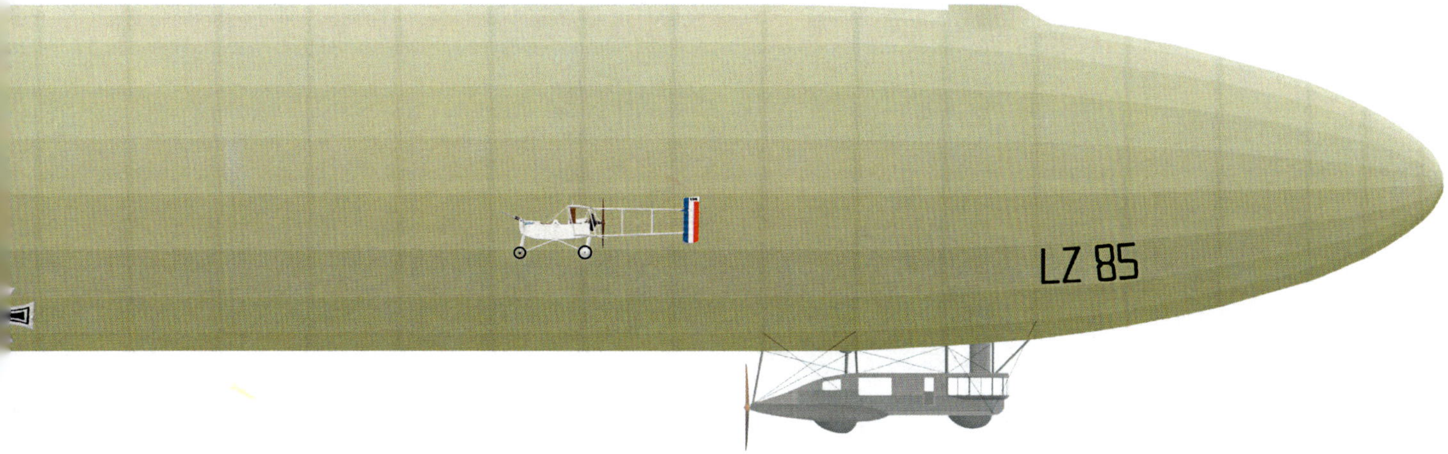

Testimonies of Serbian Aviation by Adjudant Alexandar Deroko and Adjudant Janko Markicevic

Above: Adjudant Aleksandar Deroko in front of a Voisin-Canon of the V 90 S squadron. (Coll Boris Ciglic)

Above: Adjudant Janko Markićević (DR)

Following the Sarajevo bombing on June 28, 1914, Serbia was the scene of the first clashes of the First World War when the powerful Austro-Hungarian Empire declared war on it July 28, 1914. The same night, three river gunboats went down the Danube to shell the Serbian capital Belgrade. But the small Serbian David was going to prove extremely tough for the Austro-Hungarian Goliath, who twice was going to be expelled from Belgrade by the Serbian army during the year 1914... In 1915 France sent to its Balkan ally a small technical help in the form of an artillery battery, as well as a squadron, MF 99 S, which was going to settle in Banjica, south of Belgrade, directed by the commander Roger Vitrat. This squadron would join the small Serbian air force which existed since the Balkan wars of 1912 and 1913. Serbia resisted the assaults of the Austro-Hungarian Empire throughout 1915, until October when Germany decided to take matters into its own hands and sent much more efficient troops to those of Austria-Hungary. Bulgaria entered the war and the combined assault of German, Austro-Hungarian, and Bulgarian troops defeated the Serbian army, which retreated to Kossovo and then evacuated the country through the snow-covered mountains of Albania, where its survivors were taken in by the French and Italian navies and settled on the island of Corfu. The French navy would then transfer all these soldiers to Salonica to reform a Serbian army that would be assigned to the center of the Eastern Front, in a sector of the mountains.

The Serbian army had its own airmen, who were scattered in the French squadrons of the Orient Army in the spring of 1916, and then concentrated in the five squadrons that the French army assigned to the Serbian army, the N 387, MF 382, 384, 398, and 399 – the latter being the former MF 99 S of the 1915 campaign recreated in France, and whose veterans occupied key positions in this new Serbian air force, starting with Major Vitrat, who was in command. The majority of the pilots were French, but the number of Serbian pilots was to increase throughout the war, as the volunteers sent to the flying schools returned from France.

As a young corporal and aspiring pilot in Serbia in 1915, Adjudant Aleksandar Deroko finally received his pilot's license in France on November 7, 1915, and in April 1916

Above: Recognizable by his light kepi and dark uniform, Major Victor Denain poses with the observing officers of the Serbian air force in early 1916. The aircraft in the background is a Farman MF 11bis. (Coll Boris Ciglic).

he was assigned to the V 90 S squadron, one of the three squadrons of the Eastern Bombardment Group (GBO). On May 24, 1916, he participated as an observer in his most daring raid: the bombing of the Xanthi cantonments, in Bulgarian Thrace, where General Mackensen was supposed to be reviewing troops according to the spy service reports. About 180 kilometers separated the Girard-Misrachi camp from the objective, which was at the extreme limit of the Voisin's range of action, and they had to refuel on the way back at the Orfano field, on the Aegean Sea shore. He gives a unique account:

"The planes are slowly gaining altitude, as fast as they can. The pilot Morenas is with me. We each weigh more than 100 kilos in our heavy flight suits... We gain painfully the altitude of 1,200 meters; going higher is impossible because by giving a blow on the stick, the plane stops and falls. Our neighbor is the slowest of the group.

The cabin is open air in this plane, like a boat or a balcony. It is refreshing, pleasant, your nervousness disappears... The sun rises somewhere behind Mount Athos. On the right is the sea, on the left, quite close, mountains, which seem to be holding out their arms to us. Kavala is here like a cape deeply embedded in the sea. The white planes play leapfrog with each other... They keep their speed by trying to remain grouped. Three hours of flight like this, the target is 200 kilometers away, and our speed does not exceed 80 km / hour. In the open-air cabin something runs between my legs. It's small, gray, and runs from right to left: a mouse. As our planes are parked in the open air, without hangars, the bug has found a way to get on board through a small hole to lick the lubricant under the engine. I see it on the ice at my feet, with its little legs, disappearing in a corner.

Finally arrived in the valley, the first signs of welcome appear in front of the leading aircraft. We see the so-called shattering shells with black smoke, accompanied by white shrapnel. With the noise of the engine I hear a "bang". I stand up and see metal shrapnel leaving a set of holes on the starboard wing, before sliding into the void. When a lone plane is fired at like this our instructors advised us to change direction and altitude to deceive the gunners, but when they fire into the pile in a group of aircraft, there is nothing to be done.

The bombs that are carried away are lined up upside down on the sides of the fuselage on a makeshift hitch, held in place by a belt. We have six of them on each side, they are artillery shells on which are adapted fins. First the fuses are operated with a hook, then the belt is detached and we push them by hand, one by one or several at a time. They twist in the air before the fins stabilize them. We see them shrink and disappear in the mist below, then appear clumps of smoke and explosions. Below are houses, a train station, monasteries, what do I know? We were told that Mackensen would be right here. Each plane releases its bombs at a guess (there are no sights), then makes a sharp turn before heading back. Our fighters circle around us. There was no enemy in sight.

Until Morenas slapped me on the back: A Kraut! Where is he? It is far below us. It's a plane with a long fuselage. I yell through the noise of the engine to put full throttle and lean on my shoulders. The Voisin are strange aircraft, if you want to go faster, apart from going full throttle, you have to put it in a dive and need all your strength to push the stick, hence the help given by the observer. The fear is enormous. We are alone, because, being the heaviest, we are the slowest and the lowest... I look around me and I unconsciously feel the presence of something above my head: it is just the moon, pale, always visible in this late morning. Time passes so

Above: Major Roger Vitrat, head of the MF 99 S squadron sent to Serbia in 1915, and head of the Serbian air force on the Orient Front in 1916. (Coll Boris Ciglic)

slowly...

However, we escaped. We breathe better! In sight on our left is the island of Thasos. We were told that there was a small track of fortune. Fortunately, we don't need it. I ask Morenas to check how much fuel we really have left, because we've been flying for three hours, we've got a long way to go and the gauge is dangerously low. He gets up, unbuckles his belt and dips it into the tank right behind him. The belt bends without touching any fuel: dry! This is serious. We have already passed Thasos. We will continue as far as we can. Soon the engine starts to protest. On the indicator, the number of revolutions per minute starts to fall. I look ahead to find a place to land. Suddenly we notice on a strip of sand along the sea one of our fighters that has landed. This means that it is possible and that it seems to be outside the Bulgarian lines. Let's not take any more risks. It is easier to land while the engine is running than when it has stopped. We choose a field, which turns out to be more delicate than it seemed: a swamp. We are too short and our plane hits some reeds or something, and crashes on the nose. We are safe, but the wheels are warped."

Assigned to the Salonica Park, Adjudant Aleksandar Deroko was transferred in December 1916 to the F 384 squadron stationed at Vrbeni, in the mountain sector. In a difficult sector, facing malaria during the summer or a freezing cold during the winter, there is only alcohol to support the morale of the Serbian and French pilots. But alcohol and piloting rarely go together, as our Serbian pilot testifies:

"In the French army, the quart de pinard, that is, a quarter of a liter of wine, was a mandatory allowance at every meal. Everyone was theoretically entitled to it. But one day, some of our pilots were sitting in a tent and were in a French squadron, on a deserted field, in front of Monastir. And Captain Petrašković lamented, "Today is the feast day of my patron saint, and we have to celebrate what the hell, I can't stand to be without wine or mastika!" That's when he had his brilliant idea. He persuaded a mechanic to prepare a Farman for a test flight. He crossed the Gorničevo ridge and followed a road to Sorović, less than twenty minutes away. It was a small Greek village that had a tavern.

His friends waited for him for two or three hours. He should have returned long ago. Nervousness started to reign. The French squadron leader became furious. Then, suddenly on the road, a little further from the field where you usually hear the trucks going and returning for the front, you see someone coming in such a sorry state that you help to walk across the field. Captain Petrašković, drunk and drunk with gasoline, mastika and blood, he dragged himself to the squadron. He had gone to taste the local spirits in a cellar in Sorović until he became drunk, then loaded several bottles as well as smoked ham into his Farman and returned. But on the return flight he lightly clipped the Kajmakčalan mountain (which he didn't understand why it got in the way...) and miraculously survived the crash. He said that he had no sorrow for his destroyed aircraft, because the French state is rich, and it will make others, but was distressed by the loss of his six bottles. In terms of discipline, the wolf eats the donkey, and Captain Petrašković (otherwise an excellent officer and decorated during his past as an infantryman with the Karađorđe star with swords) paid dearly. Not immediately. It was only after the war that his spine betrayed him, and he died of the effects of the wound ten years later."

Another Serbian pilot, Janko Markićević was trained in France and obtained his military pilot's license (n°4196) on August 9, 1916 at the Chartres school. Back on the Eastern Front, he was assigned to the F 398 squadron in January 1917 for a career that would be one of the shortest. A few weeks later, the German twin-engine bombers of Kaghol 1 attacked the Allied airfields and hit several soldiers of the Serbian army with full force. Equipped with outdated equipment, the Serbian squadrons do not have efficient aircraft to oppose their attackers, which will not prevent Adjudant Janko Markićević to go to the attack with his old

Farman taking as observer the French Second Lieutenant Maurice Giraudot. In his memoirs, he recounts his attempt to intercept the bombers on 26 March 1917:

"We see that they have started shelling Sakuljevo. My observer is at this moment clutching the machine gun, ready to open fire. He turns around and, with his usual smile on his face, signals me to start firing at the enemy. At very close range, our machine gun begins to fire and I see our tracers spinning through the enemy group, which has begun to unravel. We make another pass and go at them from the front. Their fighters didn't even pay attention to us, as they were busy with other aircraft coming from all sides. So we were even more encouraged and very close, attacking them front and side. Our machine gun was working perfectly and without jamming. The observer kept turning to me and indicating with satisfaction that we were going to shoot down at least one of them.

During all this time, the flak was firing at the whole sector where the air combat was taking place. The French gunners did not pay much attention to the general mix between our planes and the enemy planes and fired into the heap... I must admit that this general cannonade did disturb the formation of the enemy bombers. But our airmen were panic-stricken and we all started to look for a way out of the danger. All around us, on all sides, shells and shrapnel were exploding with the unpleasant stench of gunpowder. Our plane began to dive and suddenly right itself. The observer signaled me to move away as quickly as possible toward our base and avoid the artillery fire. Just as I managed to turn around, a terrible explosion right next to us shook our plane, flipped it over and threw it on its head. At that moment, I thought we were done for. I reduced the throttle considerably, and the second I did, to my relief, the plane was back in the air and I headed for home with the throttle at idle, getting us out of danger. When we descended below 1000 meters, the weather became very rough and our dove's nest (the Farman plane was so nicknamed by our soldiers at the front because of its unsightly construction without a fuselage, with the engine behind the pilot), started to make jumps and to perform, as always when in bad weather, uncontrolled acrobatics. I then noticed that the ailerons were no longer responding: they were an essential control for this Farman which is very wide and has no fuselage. It is worth mentioning that at that time the pilot was sitting in the front seat, so that the whole plane was behind him; thus, the slightest nod of the head made him feel as if he was at the top of a ship's mast during a gale. So we arrived 200 meters above our field. The weather conditions became more turbulent and our plane once again went into a tailspin and leaned to the left side; I felt we were going to crash because I could not right the plane. I immediately cut the engine to avoid catching fire on impact with the ground. The observer was sitting quietly until then because I didn't want to tell him about the ailerons earlier, so as not to panic him. He turned around and asked, "What happened?" I didn't answer, I just gave him a sign with my hand under his neck, "à la Robespierre," that we were lost. He turned to his seat and bowed in the booth.

Thoughts came to my mind with lightning speed and formed complete visions. I thought of the man who would bring home the news of my death, seeing the village cemetery before me, lamentations and all. I was especially sad that the war was not over. Then I imagined the comrades gathering around our mutilated bodies and speculating on why we had fallen, as usually happens. Suddenly, I woke up, raised my head, and took a look at the surroundings of Monastir; then I looked at where we were going to crash and saw a plowed field. I immediately thought that it would help soften the impact and that we might be able to get away with some serious bruising. That's when our plane nosed over and plunged toward the ground at a very high speed. I thought the whole plane was going to fall on our backs and maim us horribly. Because of the very rapid fall, I couldn't breathe and I felt a terrible pressure on my chest. In this delirium of thought, we violently crashed to the ground. After the impact, I looked with incredible pleasure, as if after a dream, at the remains of the plane around me, feeling no pain at first. Doubting my situation, I wondered if I was still alive or in a dying man's dream. I had heard that you don't feel any pain at this critical moment... And really I didn't feel anything at the moment, even though I was looking around.

Then I heard poor Giraudot wailing, but I couldn't see where he was. My right leg was completely wrapped around my waist. I felt an abnormal heat, but no pain. I thought it was warm blood, I took off my gloves and helmet and tried to untie myself and get out, but I couldn't move. It looked like my legs and arms were broken. The entire gas tank was on top of me and, being punctured, was leaking on my back. Several soldiers who were nearby and hiding from the bombs quickly went to the scene. They pulled us out of the wreckage. I asked what had happened to the observer and they said he was in better shape than me. Immediately, my squadron mates and those from nearby camps arrived."

Contrary to what the soldiers who rescued the two airmen claimed, Second Lieutenant Maurice Giraudot was seriously wounded and would succumb to his injuries. Janko Markićević survived the crash and, after eleven months of convalescence and a shorter right leg, returned to combat in April 1918 in Squadron 521 (which became the 1st Serbian Squadron), and took part in the final offensive in September 1918. Like all the soldiers of the Serbian army, he would return to his liberated homeland after having spent some 34 months fighting at its gates with the French army.

Pierre Dupon, Hero in Spite of Himself

Above: Corporal Pierre Dupon, pictured at the Chartres flying school in 1916. (DR)

Another pilot of the same squadron of Captain Petrašković would experience problems with alcohol bottles. It was a French pilot of the F 384, named Pierre Dupon. Born on 5 March 1882 in Bordeaux, he decided at 19 to join the army and served for three years in the 144th infantry regiment in Gironde. After being discharged in 1904, he got married and started a family while becoming a journalist, following with interest the first exploits of the aviation pioneers, including the work of Captain Ferber. When the war broke out, he was 32 years old and no longer young: he was mobilized in a reserve regiment, the 220th RI, with which he went to fight in the trenches. Like many men, he wanted to get out and volunteered for the air force. His request was accepted on February 21, 1916 and he obtained his military pilot's license (n°3982) in Chartres on July 10, 1916, as well as his promotion to corporal. Just barely: he was 34 years old and the age limit to be accepted in flight school was set at 30 during 1916. Nevertheless, he was more or less considered as a second-rate pilot and was assigned to the Eastern Air Force. He was sent to the Lyon base, from where he went to Marseille to embark for Salonica, arriving at the F 384 squadron in Vrbeni on December 3, 1916.

It would be difficult to find worse living conditions for the pilots: it was bitterly cold and the weather reports of the Eastern Army reported very heavy rain and wind in December 1916, so much so that a Farman of the neighboring F 383 squadron, flying over a valley, was caught up in the eddies and sent westward without being able to turn back until the wind died down... Since December 17, snow and fog have covered the mountains.

The following comes from the memories of Serbian observer Sava Mikić and is open to interpretation.

We can probably think that Corporal Pierre Dupon was not asked to do anything on December 25, 1916, when his squadron leader gave him a special mission: to go to Salonica in a Farman aircraft to bring back food so that the French and Serbian pilots could celebrate New Year's Eve with dignity. He took off and landed at the Serbian park of Mikra, then went shopping in town. He then returned to his squadron with a Farman loaded to the brim with bottles of alcohol, cold cuts, and New Year's Eve festivities. He never arrived in Vrbeni and was reported missing – it is feared that he had hit a mountain in bad weather, the highest peaks in the region exceed 2500 meters, which is about the ceiling of a tired and heavily loaded Farman.

But a miracle happened: a ciphered telegram from the Eastern Army addressed to the Ministry of War and dated February 9, 1917 indicated that the pilot was safe and sound and had been captured by Austrian troops in Elbasan, Albania, on the other side of the Balkan chain and some 120 kilometers from his destination. He had just achieved the feat of being the first pilot to have crossed the mountains of mainland Greece, a raid that would not be officially successful until May 1917 by the famous pre-war pilot Louis Noël, a specialist in long-distance flights, who carried out an exploratory flight on an otherwise more efficient aircraft on orders to organize the delivery of aircraft to Salonica by air from the assembly center in Otranto.

Sava Mikić mentions in his memoirs that Austro-Hungarian soldiers were alerted to the crash of an enemy aircraft in a valley. On their way to capture the pilot, they were stunned to see the wreckage surrounded by a halo of windblown confetti... Dupon was pulled out of his cockpit covered in alcohol, the fuselage filled with broken bottles. Would he have abused the products he was carrying before taking off, to the point of missing the haunts of the great lakes Prespa and Ohrid, and continuing his route for a hundred kilometers by slaloming through the mountains? Or was it caught in a storm? Or perhaps both elements should be taken into account?

A third element, irrefutable, is revealed to us by the short post-war life of the person concerned. Pierre Dupon ended the war in Austria in a prison camp. He was

Above: This Farman F.40 forming a pylon at Vertekop airfield bears the markings of the F.384 squadron: two red stripes on the wings, as well as a white oblique stripe on the fuselage (not visible on the photo). It is an aircraft of this unit that Pierre Dupon was flying on December 25, 1916. (Coll Boris Ciglic)

Right: View of the Albanian mountains (here, Tirana airfield, home of the Austrian *Flik* 6) that Corporal Pierre Dupon flew through on December 25, 1916.

repatriated after the armistice, in January 1919, and had to spend two months in a military hospital to be treated for a disease contracted in the East: malaria. He then returned to Bordeaux and resumed his activity as a journalist, publishing a book on bullfighting in 1922. In 1926 he moved to Rennes where he worked for the newspaper "*Ouest-Eclair*", writing under the pseudonym of Pierre Aël and always showing a passion for aviation. A malaria attack killed him at the age of 45 on September 10, 1927.

We can only underline the feat of the pilot, who, aboard a poorly performing and overloaded aircraft, very weakened by malaria, succeeded in crossing the Balkan chain in the middle of winter in the wind and fog. A feat worthy of that of the pioneer Jean Bielovucic, who just four years earlier succeeded in crossing the Alps for the first time in his Hanriot monoplane!

Above: Lineup of Farman F.40 aircraft of the Serbian squadron F.382 on the field of Vertekop. (Coll Boris Ciglic)

Above: Another view of Farman F.40 of the Serbian squadron F.382 on the field of Vertekop. (Coll Boris Ciglic)

Right: Two French executives of Serbian aviation: Captain Pierre Perrin de Brichambaut (left), and Captain Tulasne, veteran of the MF 99 S squadron in 1915. The two Serbian pilots are Adjudant Aleksandar Deroko and Adjudant Siniša Stefanović (right), both from the F.398 squadron. (Coll Boris Ciglic)

Above: The Serbian Air Park of Mikra bay, near Salonica. (Collection Boris Ciglic)

Below: Aerial view of the Vertekop airfield used by the Serbian air force. In the distance, the plain of Monastir. (Coll Boris Ciglic)

Above: On August 15, 1917, a British BE 12 fighter lands on the airfield of Vertekop. On the left, a Farman F.40 of escadrille 525.

Above: A Nieuport 24 fighter of No. 521 squadron flown by Sergeant Milutin Mihailovic. The aircraft bears a Serbian flag on the fuselage, but its markings are French. (Coll. Boris Ciglic)

Selected Letters by Sergeant Paul Andrillon

Paul Andrillon was born on December 25, 1892 into a Parisian bourgeois family. His father was a an engineer from ***Ecole Centrale*** employed as technical director at the Parisian furrier Revillon. He was the only child of the couple and grew up in the 16th arrondissement of Paris where he was educated by the Jesuits. In early 1914, the family moved to a house in Clamart. The young man began studying law, but he became passionate about aviation in 1910 and attended various aviation meetings in the west of Paris. He also became an accomplished "sportsman" by practicing boxing, running, and especially rugby with the Stade Français. He suffered a double fracture in his leg following an epic match at the beginning of 1912... He left to do his national service in 1913 in the 3rd train squadron and was still there when the war broke out. This assignment did not put him in the front line, but at the back of the front to supply the fighting units.

On October 5, 1914, he served in the region of Reims where he witnessed the first aerial combat between planes, when the Voisin of Sergeant Frantz and Private Quénault *shot down an Aviatik, as indicated in his personal diary: "From the top of the Rosnay plateau, I witnessed the fight between French and German aircraft, German shot down; airmen burned. Their grave is near the Nuizon train station."* This event only strengthened his desire to go to the front and he volunteered to fight in the infantry, which was a very difficult task. This was done at the beginning of November 1914 when he was transferred to the 119th infantry regiment. He wrote to his parents: *"When you are 20 years old, it seems to me that it is a duty to go into battle."*

Several months of life in the trenches began. His level of education allowed him to be appointed corporal on July 1, 1915. But on the following September 25, he was mowed down by a burst of machine gun fire during an assault on enemy positions. Wounded and evacuated, he asked to be transferred to the air force after his convalescence and, from June 1916 onwards, he followed the training course to become a machine gunner (shooting courses in Cazaux and then Pau). In September, he was assigned to the F 36 squadron, on the Farman F.40, in which he served for two months before attending observer courses at Plessis-Belleville in December. He then returned to the F 36 for two months and left on February 15, 1917, joining the flying schools where he obtained his military pilot's license (n° 6889) on June 9, 1917. On August 28, Paul Andrillon was assigned to the Orient, to squadron 507, after a train journey through Italy to Brindisi, then by boat to Salonica. It was there that he met Costes, who took him on as a wingman. He was transferred with Costes to the Franco-Hellenic 531 squadron on March 19, 1918, where he scored two certified and three probable air victories. The armistice of November 11 did not mean a return to France for him, as he was transferred to squadron 509, which was part of the Allied occupation troops in Constantinople.

Above: Sergeant Paul Andrillon. (Benoit Henriet collection)

He was released from his military obligations on August 27, 1919 and was reunited with his parents in Clamart. His mother, frightened by the idea of losing her only son in an air accident, made him swear never to touch the controls of an airplane again... A promise that he kept within the limits of his military obligations, since he did several periods of reserve duty during the interwar period (he was mobilized in September 1939 in a non-aircraft position at the Châteauroux base). Very friendly with Costes who would make him the godfather of his son, Paul got married and became a bookseller. His strong appetite quickly made him acquire an impressive belly... Which he was to lose entirely during the occupation by making his own deliveries on his bicycle dragging a trailer! After the death of his mother in 1941, he obtained from his father the authorization to resume flying and founded at the liberation the aeroclub of the South-West of Paris in Guyancourt. Piloting a Stampe and a Norécrin, he contributed to the training of many pilots, both civilian and military. Paul Andrillon died in 1959. The Guyancourt Flying Club now bears his name; the airfield having been closed in 1989, it has now been moved to Étampes.

Above: Paul Andrillon seated against the mast of his squadron's flag, during the fall of 1917. He ended the war with two confirmed and three probable victories on the Orient Front. (Benoit Henriet collection)

From the long correspondence he sent to his parents, we learn about his daily life as a fighter pilot on the Orient Front. Here is a selection of his letters.

Undated Letter - Circa October 1917

Under the hot sun which in the middle of July makes the atmosphere of Marseille suffocating, among the disparate crowd which at any hour swarms on the port as on the Cannebière, the pilots wandered through the city, waiting for their departure for the Orient Front. Certainly they were melancholic; so close to their last leave spent with their families, the sadness of the farewell taking all their spirit and so many months, so many risks to run, made the hope of the return quite uncertain. After a dozen days of this painful wait, a train took them through Livorno, Rome, Casata, Foggia, as far as Taranto, from where the "*Timgal*", a very comfortable boat crossing the Otranto canal, entered the Gulf of Corinth, then deposited them in the Greek port of Itea. Twenty-four hours of rest in this place, were embellished by a delicious bath in the phosphorescent sea, a spectacle that none of them knew. In a truck they drove to Bralo where they had to wait for the train to Salonica for two days in the scorching heat. Finally, after having completed this last stage at a painful pace, they could finally rest in the capital of Macedonia.

At the Sedès camp they made a few flights to "get the aircraft back in hand", and then a decision by the Orient Aviation Directorate dispersed them to the frontline squadrons. Two of them were assigned to the 507 located at Lembet, one was Maurice Lashermes, the other...myself. A van took me and my luggage on August 28 to the Lembet airfield where the squadron was housed; I remember Salonica crossed under the merciless heat of midday, the road to Zetelnick crisscrossed by a multitude of cars of all kinds and weights which flooded us with burning dust, then the interminable runway of Lembet crossing charred meadows under a too hot sun, the whole was melancholic, but the good reception that I received from the pilots of the 507 came at the right moment to brighten up the situation. Costes, the ace of the squadron, Lefresne, Moine, Maurice (who arrived the day before) were charmingly friendly.

The squadron leader (Lieutenant Callon) having assigned me a Nieuport, I decided the next day to try it... And it was not long; the time to go up to 700 meters and crack, tremendous vibrations, a vague smell of scorch... Having cut the contact I returned to the ground in gliding flight, and extraordinary fact I landed suitably in front of the hangars (I say "extraordinary", because at that time I had the glance and the safety of piloting of a pilot of 25 hours of flight, that

Above: Paul Andrillon at the controls of his new Nieuport 24 in January 1918. (Benoit Henriet collection)

Above: Bad winds in the Vardar Valley destroyed some the tents of escadrille 507. This did not dampen the good mood of Paul Andrillon, sitting on the ground next to a comrade who appears to be Dieudonné Costes.

Above: Nieuport 24 n°4408 of Sergeant Paul Andrillon, escadrille 507, Lambet (N. Salonica) March 1918. (Benoit Henriet collection)

is to say very little talent). My engine being burned out, it had to be replaced; thus I was condemned to rest for two days. On September 1st, the weather being very nice, I flew early for the test of my new engine; this one running to my satisfaction, I could carry out in the afternoon a patrol in company of Costes.

On the lines, Costes saw a Boche gunner passing by me and signaling me to dive vertically on the enemy; yes, but I was very preoccupied by the driving of the engine, I was watching my tachometer or the functioning of the oil bell more than what was happening in the surroundings, so I did not see the signal of my comrade; I simply noticed that there was no more Nieuport in my vicinity; I started to turn on the spot seeing some white flakes far below me, but I was too preoccupied with the disappearance of my comrade, to think that these white flakes could be in the wake of a Boche (Ah, these young pilots!). Having turned around enough, I took the road to the squadron, in a state of great perplexity. Costes landed there a quarter of an hour after me, and as I rushed to meet him to finally know the enigma of the story, "Jerk!" was his only answer. When I think of all the hikes we did together afterwards, the story of our first outing is quite piquant.

On September 8, I was doing a training flight at Sedès, I found myself on a derailed Nieuport which at the first turn did not want to straighten up anymore, I was in a bad situation at 50 meters high, turning on the spot with full engine, I cut by putting the controls to the right (my aircraft was engaged to the left). The Nieuport nosedived, obeyed my direction and I landed in a field after having narrowly jumped a ditch; I had come close to the log that does not forgive. In the afternoon in Salonica wanting to take a tramway, I missed my shot, and made a brutal looping on the ground, while the tramway's handrail brushed my body. The next day a wind of rare violence rose, it was the wind of the Vardar in all its magnitude, all our tents took off, a Bessonneau hangar was half torn off, and for three days there was no question of flying.

Then calm was restored and the patrols started again, without incident except on the 16th when the rod controlling the machine gun was severed and having torn off all the spark plug wires of the rotary, that was worth to me a dry run which occurred without any hitch, without warning, caused me at first a great surprise. But having fiddled with the joysticks and contact without success, I tried to join the squadron by gliding, I succeeded in the enterprise, but I must admit that this time again the share of chance was greater than the share of science. The next day my aircraft being

Above: Maurice Lashermes, Paul Andrillon and Dieudonné Costes pose in front of a SPAD fighter in September 1918 (Benoit Henriet collection)

ready I left on the lines and I found quite natural to make an incursion of a few kilometers at the Bulgarians (poor me, if I had been attacked!).

Everything went well until September 24. That day I was operating at 4,000 meters in the company of Costes on the side of Guevgueli, when suddenly my engine started to sputter losing quite a few revolutions, thanks to the small effort that my engine was willing to give, I was able to drag myself to the field of Gorgop where I landed breaking the end of my leading edge in a wooden horse that I did while trying to avoid Costes who was landing perpendicular to me. Three broken candle wires were put back in place, a makeshift patching ensured the solidity of my leading edge; the whole thing was done in a hurry and I left in the company of Costes who had not wanted to abandon me. It was not long, hardly taken off at about 150 meters high, my engine gave formidable jolts, shaking me roughly in the cabin, I kept it just enough to put me back in front of the ground and cutting the contact I landed again Gorgop; this time it was necessary to give up the idea to join Lembet, my engine was burned out! The cause of this burnout was long to find, and yet it was simple, my mechanic had forgotten to fill up the oil tank before my departure! This inattention could have played a nasty trick on me, but my mechanic having always been very conscientious thereafter I did not hold a grudge against him.

On September 27, a new engine was mounted on the Nieuport and I could finally join Lambet.

Letter Dated April 29, 1918

Dear Dad,
Dear Mum,

There was some sport today at the squadron. Two comrades on patrol this morning very early, got into a fight with a Boche two-seater and forced it to land about 15 kilometers from home. I left so early the news known with Lashermes and Costes in protection of a two-seater which was going to take the picture of the Boche.

We were just about to look for it when five rascals of single-seaters fell on us, at that moment I was flying with Costes over our photo plane next to which was Lashermes. Hearing the crackling of machine guns, I made a quick vertical turn just in time to see Costes grab the 1st Boche at point blank range and the latter fall in a spin.

As four other rascals were coming in single file to get Costes, I stepped on them in full throttle, three of them immediately disengaged, I stuck to the tail of the remaining one and a few meters away, the guy went into a spin and I accompanied him for a moment by pulling him. I had to take the height again because the three other rascals were climbing in spirals on our head, I walked on them side by side with Costes and they put the sails while it can.

Lashermes saw a Boche fall in a spin near him and followed him with his machine gun until 600 meters above the ground, there he saw the Boche still in a spin disappear behind a hill.

We returned quietly with the photo plane, when at the crossing of the lines we saw two single-seaters which followed us from afar.

We went for it but we could not catch them until they were about 20 km away. Costes managed to get close to one of them, stuck in the tail but had a derailment after a few rounds. I followed him at 200 meters, I shot at the Boche, which in front of me was poking so vertically "that I could not approach it effectively". And this time, definitively masters of the battlefield, we returned home as happy fathers.

We are convinced that there was at least one Boche who turned into a pile of matches, but so far we have not had any tips. There was only one observatory that saw a Boche in a spin straighten up at 200 meters from the ground and run away in "low gear" and that's all. If, as is likely, we hit a rascal, we'll find out sooner or later from the prisoners.

That's a busy day, isn't it?

I kiss you with all my heart dear dad dear mom.

Paul

Letter Dated May 1, 1918

Dear Dad,
Dear Mum,

More sport at the squadron!

This morning I was on patrol with my two comrades and my engine was not running properly, I almost wanted to go back to the field, in the end I wanted to hold on anyway and I was right. A Boche two-seater was taking advantage of the cloudy weather to come and take pictures in our lines, we let him go home, and suddenly emerging from a sea of clouds, we fell on his head.

Costes came in the queue, Lashermes on the right, me on the left. Costes, without being seen, shot him at close range under the tail, killing in a few bullets the passenger whose head was bobbing along the cabin. The pilot made a sharp left turn, naturally I fell on him vertically passing him a good squirt of close range (at that moment I had a clutch).

The Boche was then caught by Lashermes who put another pressure on him, and finally the Boche spun a few kilometers away from home. I lost sight of the Boche as he picked himself up, so I was anxious to land at the squadron, but the crowd that was riding in front of us immediately put me on the page. Here is how we knew about the adventure at the squadron. A ground observatory reported:

11:30 a.m. "Enemy aircraft crossing the lines heading south". 11:35 a.m. "the enemy plane attacked by three single-seaters in Greek colors has just crashed north of M..."

We went to the observatory earlier to see the result of our work, and it was with satisfaction that we saw in a spyglass the Boche with his legs in the air, the plans hanging miserably. The guys of the observatory told us that the Boche had made on the ground three or four pirouettes before stopping, undoubtedly is it a cuckoo good for the reform!

This time it was an official Boche for us.

For yesterday's torching, several observatories were unable to follow the fight because of the low clouds, the only one that could follow it, saw a Boche in a spin disappear behind a hill, but did not report it down, not knowing if it had not straightened out at ground level. However, the three of us are convinced that at least one Boche picked up the log yesterday, and we can only hope that a prisoner will come and put the matter to rights.

This evening Mrs. G.... called us to congratulate us, the Commander wrote to her earlier to let her know the success of our two tobacco shots. I forgot! This morning we crossed the lines again in "lowland", the infantrymen jubilated, the helmets were agitated, more behind the artillerymen raised the arms to the sky, and in the squadron the champagne was there: that gassed! This little job of a few minutes will undoubtedly bring us a palm to each one, in any case the proposals left this evening to the Great Headquarters.

I stop my long verbiage by kissing you with all my heart, dear dad, dear mom.

Paul

Letter Dated September 9, 1918.

Dear Dad,
Dear Mum,

Received three letters from you of August 26, 28 and 29. Also received two parcels: milk, chocolate, cakes, candies: thank you very much.

The son Lachet was unlucky to be demolished so quickly, I believe that on the French front many new pilots disappear because the time of school is too shortened, also with the first combat or the first breakdown it is the panic and the log. Now that I have 500 and some hours of flight I realize that it is not from one day to the next that one is able to make the war in a cuckoo, to arrive to be there as at home, it is necessary to try some twirls, and that represents seconds of fright because at the beginning that of times one sticks in spin without wanting it.

A little war story, that will interest you more. Decided with Costes and Lashermes to have the sausage that I missed previously, we did it again yesterday, we drew the order of attack to the Zanzibar game which gave Andrillon, Costes, Lashermes. So last night, this balloon being in the air, we took off in spite of the opinion of Captain Perrin de Brichambaut

Left: The years of war in the Orient forged a solid friendship between Costes and his crewmate Paul Andrillon. The two men are seen here toasting in a restaurant during the 1930s when Costes became an international celebrity. Paul Andrillon, whose wife Suzanne is at the center of the table, has grown somewhat since his years in Salonica... (Coll Benoit Henriet)

(you must know this name; he is an ex-ace of the French front on a two-seater) and of Lieutenant Zannas; both of them came in plane to see the results of the expedition. Departure grouped to the lines, and from there without taking height to the single file full sauce on the sausage. Met the art object at 700 meters from the ground, and as we did not know if it was occupied or not (sometimes the observer is replaced by an explosive) the three SPADs went around the gondola at a respectful distance. Not being able to see if it was inhabited, but as it descended I tried a small attack, oh not very close which gave nothing. My two comrades did as I did, always fearing the thing would blow up. But when the balloon was brought back to the bottom of its ravine, there was no more danger of explosion, we went and shot it copiously; from the surrounding hills, machine guns and revolver cannons were firing, that hampered our shooting, we surely placed some bullets in the balloon but not enough to set it on fire, Maurice had a bullet in his plane, Costes at the moment of shooting the balloon saw two or three luminous shells passing him close by, that made him miss everything, and for my part, being close to the ground, I suddenly saw the temperature of my water (the fixed engines have water cooling) mark 130°, impossible to reduce because I had a hill to jump, that made that I joined our lines cautiously feeling my engine weakening; and it is the damn impression to say to ourselves that we will be obliged to present ourselves to the Boche; damn it, everything but that! Especially since in the evening there was a big dinner at Mrs G.... with Zannas, Costes, Maurice. Perrin and Zannas found that we had been a little strong to push so much the attack, and from there on to Gumendjé where we had a succulent meal.

If we don't leave too early for our move, we'll get the sausage skin, you'll see that; but not in the same conditions because it's so defended on the ground that it's a bit risky. This morning I had lunch at Capt. Perrin's, again a very good meal.

On this I kiss you with all my heart, dear dad, dear mom.

Paul

Facing Page: Paul Andrillon and Maurice Lashermes (right), the two wingmen of ace Dieudonné Costes, with their squadron mascots. (Collection Benoit Henriet)

Fighter Tactics by Dieudonné Costes

Above: Dieudonné Costes, photographed at rest in Barovica in August 1918. He claimed 9 aerial victories in the East, of which 6 would be certified. (Coll. Benoit Henriet)

Dieudonné Costes, before becoming the world famous raid pilot with the first non-stop crossing of the South Atlantic in 1927 and the first Paris / New York in 1930, was a simple corporal pilot who joined the first contingents of the Eastern Air Force at the end of 1915. Very stubborn and resistant to discipline, he nevertheless proved to be a particularly effective fighter by becoming the ace of aces of the Eastern Front in 1918, assigned to several squadrons and ending up at n°531 located at Gorgop. Fifty years after the fact, on March 4, 1968, the journalist Jac Remise interviewed him about his career in the East. His testimony, currently preserved at the Air Museum, is reproduced here in extenso.

Jac Remise: Mr. Costes, during the war you were a fighter pilot. How many aircraft did you shoot down?
Dieudonné Costes: Nine.
J.R.: How did it go? What was your first victory and on which front?
D.C.: It was on the Eastern front, but the first victory, I couldn't place it, I don't remember at all. It's so far away.
J.R.: First, let's talk about your training. In what year did you get the idea to join the air force?
D.C.: Well before the war, in 1912. In 1912 I got my civilian pilot's license, No. 1046 of September 1912, and I was not in the military until September 1913.
J.R.: How was your training at that time? What aircraft were you on?
D.C.: The training, you know, was very rudimentary. To learn to fly in 1912, we had what was called a "rouleur", with which we made straight lines while taking off. It was a Blériot type crossing the English Channel. I passed my license on a Blériot. Afterwards, when we had flown straight lines for 2 kilometers and lost contact with the ground, we did a lap and then, to pass our license, we did a "figure eight" twice and that was it. It was a very rudimentary training. Even when the war came, the pilots who were training usually spent a month, a month and a half at the school. They would come to the front with 15 to 20 hours of flying time.
J.R.: And you think that was enough to make a good pilot?
D.C.: No, certainly not. For some, it was enough: they were acclimatizing. But the greatest losses, whether in the fight, in reconnaissance, or in bombing, occurred during the first three or four months at the front.

After that, if you will, the pilot was ready, if he had made it through those first terrible months.

In aviation jargon, we used to say that the pilot was vaccinated. But for me, the vaccination took much longer. It took a year to a year and a half on the front lines. I had a lot of hard knocks.
J.R.: How much and how long did it take you to be vaccinated, at that time?
D.C.: I don't know, but at least six months. Note that I had already done a period on the French front. From the declaration of war in 1914 until December [of that year] when I was wounded on the side of Reims. Then I returned as an instructor to the Chartres school where I stayed until September 1915. I came back to the front in 1915 where I stayed a few months. At the end of 1915, I left for the army of the East where I stayed until the end.
J.R.: What was your first aviation adventure and your first fight? Do you have any particular memory, such as your injury?
D.C.: No, it was a shrapnel wound. The memory is uneventful.
J.R.: Were you being shot at from the ground?
D.C.: Yes. They were firing shells at us. And it was a shrapnel that hit me in this hand on December 23, 1914.
J.R.: And you were flying over Champagne at that time?
D.C.: It was near the Fort de la Pompelle, in Reims.
J.R.: What plane were you on?
D.C.: On a Voisin.
J.R.: Were you a reconnaissance pilot then?
D.C.: I was a reconnaissance pilot, yes. Besides, at that time, there was practically no such thing as a fighter. Here Frantz shot down a plane, it was in the region of Reims, with a rifle, and on a Voisin. I believe that he must have told you, because if my memories are faithful it was in the neighborhood, at that time.
J.R.: When this feat happened, if we can call it a feat, did it give you the idea, you, the Voisin pilot, to use your machine gun?
D.C.: There was no machine gun at that time.
J.R.: Or to install one?
D.C.: Ah, no. They were installed late. I started getting a machine gun about October or November. Maybe a little earlier.
J.R.: There were even cannons, on those Voisin.
D.C.: Oh yes, but much later. I believe that the first turret-

Above: Dieudonné Costes poses in front of his Nieuport 24 with his teammates, Paul Andrillon and Maurice Lashermes. One can see the Lewis vertical machine gun installed in the cockpit; the ace, according to archive documents, claimed two victories (not confirmed) with this unorthodox system but which prefigures the "Schräge Musik" of the German night fighters of the Second World War. (Coll. Benoit Henriet)

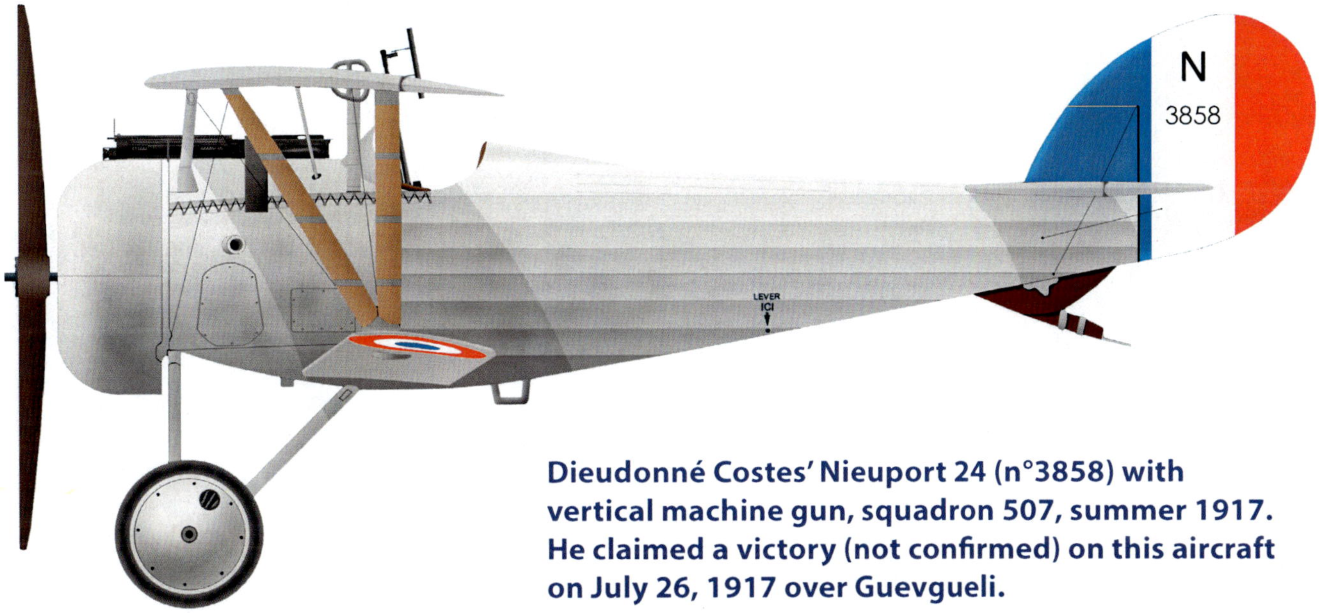

Dieudonné Costes' Nieuport 24 (n°3858) with vertical machine gun, squadron 507, summer 1917. He claimed a victory (not confirmed) on this aircraft on July 26, 1917 over Guevgueli.

Above: Another Costes wingman, Sergeant Maurice Lashermes (1892-1981) aboard a Nieuport 27, decorated with a small plaque taken from a train: "It is dangerous to lean out". He ended the war with 3 certified victories and 4 probable victories. (Coll. Benoit Henriet)

Below: Proof by the image of Costes' account of his taste for hunting large birds aboard his plane: the ace is the second on the left among the pilots and mechanics posing behind the game. (Coll Benoît Henriet)

Above: SPAD VII n°1083 of Dieudonné Costes, equipped in May 1917 by the interested party with two additional Lewis machine guns on the wings. (Coll Benoit Henriet)

mounted machine guns appeared a few months after the declaration of war, which took place in August 1914. The first machine guns appeared around October, that way.
J.R.: When did the idea of becoming a hunter come to you?
D.C.: That, you know, is a pretty long transition.
J.R.: How did it come about? From the exploits of your comrades?
D.C.: No, through a natural inclination. These are things that are hard to explain. I did reconnaissance and night bombing, I was a bit of a jack of all trades. You find hunting interesting.

J.R.: I think you're modest, because I've been told you're a

Photo of the above SPAD VII without its wing guns. (DR)

skilled pilot, and that's one of the main reasons.
D.C.: Yes, I've never been in trouble in air combat, or very rarely. In air combat, with the fighter pilot, there is a personal coefficient that is considerable. I would not want to underestimate the value of many fighter pilots, but it is a very small percentage of them that had a result. If you have seen fighter pilots, they must have told you what I am telling you. Note that most fighter pilots only had results because they were lucky enough to be in a group with a patrol leader who was sort of the matador, the one doing the killing. What happens in a case like that, if you have a patrol of 5 or 6 planes that shoot down an enemy plane, that plane is counted to everybody. So it happened most of the time, it was always the same one in a patrol, that gave the "muerte". It is well known.
J.R.: You mean that all the pilots of this squadron were giving each other a victory?
D.C.: Yes, one victory each. When a plane was shot down by someone in the group, the plane was counted to the group.
J.R.: So there were five victories if there were five French planes?
D.C.: Oh no, it was not five planes. That's not how we counted. But that made five citations, for the pilots of the group, for the plane that came down. Because it was assumed that all of them had participated in the victory. But in truth, it was always the same one, and very rarely a second or third one, who shot down the plane.
J.R.: You tell me that to obtain victories it was necessary to fly in a group, whereas it is well known that some of the great aces of the time flew alone such as Nungesser.
D.C.: No, that's not what I meant. We didn't always fly in groups. But generally speaking, we were in groups of three. Flying alone was quite rare. It happened from time to time, but it was pretty rare. A patrol was made up of a group of three, unless there was a war action such as a ground strafing or a protection mission, where you would form larger groups of 6, 12, 18, 20 or more. Unfortunately, the radius of action at that time was not very large and we did not have the possibility of making a large penetration through the lines. But the isolated fighter is very rare.
J.R.: It was still necessary to know how to defend oneself when flying alone.
D.C.: The question of defending oneself obviously arises, but it is more a question of vigilance than of defense. The vigilant hunter knew how to protect himself, i.e., not to remain always in a flat flight to watch the whole horizon. He always flew on one wing or the other, never in a straight position so as to watch all possible angles where the enemy could appear. It was a merry-go-round, always on one wing or the other. Even as a patrol leader, I sometimes forced my teammates to fly in threes like that, sometimes on one side, sometimes on the other. This has a disadvantage: the light that hits a surface that is more or less inclined to the sun is reflected more to the enemy eye. Whereas if you fly straight, strictly flat, you risk being in angles without light, and therefore the enemy does not see you. But as far as I was concerned, I preferred to be seen and to see, than not to be seen and not to see. Because that's how most of the young pilots were getting killed, so to speak.
J.R.: Was it better to fly high, too?
D.C.: As much as possible, yes. But the ceiling was quite limited. It was better to avoid overloading the aircraft with bullets. For example, we had up to four machine guns on the last aircraft at the time, because I had installed two small Lewis machine guns on the lower plane.
J.R.: On which aircraft?
D.C.: On the SPAD. Even on the Nieuport, previously, I had a machine gun installed right next to the seat that fired vertically, with a sight right next to the head. This was used when we could surprise an enemy plane. We would arrive exactly underneath him, and gradually gain altitude until we could see the strings. At that moment, with the small machine gun, it was a massacre.
J.R.: It was a very French trick. Did you ever postpone a victory like that?
D.C.: I had two with this vertical machine gun. Then I put two Lewis machine guns on a lower plane of SPAD, one on each side. And we had the two Vickers on the hood. All four machine guns were converging at 60–70 yards. The four bullets hit the same point, an area of 0.50 m^2 for example. This gave us a firepower that was quite significant at the time. The Vickers which were synchronized fired at 250–300 rounds/minute. The Lewis fired a little faster at 4 or 500 rounds. That was still 1,600 rounds per minute.
J.R.: General Heurtaux often complained in another interview about the famous machine gun jams. Did this happen to you?
D.C.: Yes, it happened to me very rarely. It was mainly due to the care one took with the machine gun, and the calibration of the bullets. If you do it yourself and are a bit of a technician, I didn't have many jams.
J.R.: Did you always check your equipment?
D.C.: Always, especially the machine guns. And you were talking earlier about the altitude where you should be: as high as possible, to dominate everyone, and not be surprised. On that side, I have been very vigilant, I have never been surprised, while I have surprised many times. The main asset in hunting is the visual factor. I knew Fonck very well, who had an extraordinary visual acuity. He could see the planes in the green devil and signal them to his comrades by swinging his plane from right to left, but nobody saw anything.
J.R.: That's what allowed him to win so many victories.
D.C.: Yes, because he had the advantage of being able to see without being seen, and to position himself. We would load [our weapons] when we had infantry escort missions and ground strafing – 500 rounds for each machine gun. That was heavy. Me, except for ground attack missions, I never carried more than a hundred rounds per machine gun. We never fired them. You know, when you're close to an enemy aircraft, and you have well-tuned sights, you don't need to fire a lot of bullets, two or three are enough. You can achieve a precision with the plane that is hardly imaginable. For example, I'll tell

you about cases where, when I came back from patrol, I often had fun shooting from a training point of view. We had a lot of scavengers in Macedonia: hawks, eagles, vultures, a whole bunch of filth that would get on the dead horses around the squadrons. They were there by tens on the garbage. On the way home, I would often get these critters up as I flew by with my plane. I'd get them up and when they got to an altitude where I could lose speed – these birds fly at 65, 70 km/h – I'd take one of these birds and in one or two shots I'd shoot it down. This happened frequently! I used to shoot geese, too. It was a matter of not getting too close, because you might catch a bird with the propeller. But we shot them like from here to the other side of the street!

J.R.: Do you have any combat memories to tell us about? First on the French front, and then in Macedonia?

D.C.: No, because these battles are uneventful. It is very fast. When you are lucky enough to have this personal coefficient, there is no trouble.

J.R.: Do you have a memory to tell us, against a tough enemy aircraft, for example?

D.C.: Yes, we were on the front where I was the only French pilot who had a few victories in the Eastern Army. We used to spot each other: when someone was doing damage in a sector, we knew from the intelligence services that it was Mr. So-and-so who had moved and was in that sector. We would send someone across the street or vice versa. It happened that I was in the Monastir area in the presence of Fieseler, who was a remarkable boy and a very good pilot. We had two or three clashes with him, but they were clashes that did not amount to anything. Because both of us were maneuvering properly, you maneuver as long as you have altitude. And then we reach the ground, we have to say hello and then separate. And that's it. The first collision I had with Fieseler, it was him who told me about it here when he came to do a meeting on the St-Germain side.

J.R.: Did you see him again after the war?

D.C.: Yes, and we exchanged some common memories. And he told me how he almost got picked up one day when he came with a two-seater to attack a Drachen at our place, a sausage. He was passing inside the lines, he was passing our lines by 30 to 40 km, and he was getting shot at. Then, at that moment, he would come down to his lines, losing altitude, acting as a boy who was hit. I was in the air and I saw a black spot surrounded by gunfire: it was a plane that was diving towards the lines. I thought to myself that it was not going to get past them, that the gunners were going to shoot it down. And since it was 20–25 km inside our lines, I thought it was a two-seater. I was careful not to get hung up on him as he approached, and that's when I saw that he was heading for the tethered balloon. He arrived before me; to prevent him from going down the balloon I fired a burst of machine gun. Then I had a ridiculous idea: I said that as long as that boy was attacking a sausage, it must be a single-seater. At that point, I ventured a little more. I poked a little bit more and I discovered myself. And I got some machine gun fire from behind... It was a two-seater, I went under it and the fight was over. He had good aim because I got a bullet that went a few inches over my head, through the windshield. My career was almost over!

J.R.: Did you have the opportunity to meet other German aces?

D.C.: No, except for him, whom I met two or three times because he had done us a lot of harm.

J.R.: Did he fight the whole war in the East?

D.C.: I don't know, but he was in the region of Monastir.

J.R.: Tell us about the planes you flew. What was your favorite plane? The Nieuport?

D.C.: The Nieuport was the first one we had. There was the 13-meter, then the 15-meter. Then we had the SPAD, the 140 horsepower, and the 200 horsepower with a gearbox. I went through them all in practice.

J.R.: What do you think was the best hunter?

D.C.: There was only the SPAD, which lasted practically the whole war. There were a few British aircraft that weren't bad, but they didn't compare to the SPAD.

J.R.: Compared to the German aircraft, did the SPAD have a place?

D.C.: At times, they had Fokkers that were not bad, the Fokker D.VII. But the SPAD held up well until the end of the war. It is a good aircraft, it would be a tourist plane now, but it was honestly doing its 220 or 230 km/h.

J.R.: What do you think of the evolution of aeronautics from 1914 to 1918?

D.C.: It was considerable. If it hadn't been for the war, aviation would certainly have remained very late. The First World War gave aviation a boost that continued after the war: we had the compressor engine, the retractable landing gear, etc. The big breakthrough came between 1939 and 1945 with the jet engine. Aviation would not be what it is without the two wars, which gave immeasurable credit to development. Perhaps if there had not been the war the airlines would not have grown as they do now.

J.R.: We're going to go back to the war, if you don't mind. You must have some hunting stories to tell us.

D.C.: There is one that is particularly succulent. It is the one I had with an English plane, in 1917, in Macedonia. An English plane was being fired upon by our batteries 15 to 20 kilometers inside our lines. As long as our gunners were firing – they have ways of seeing – I did not try to identify it, thinking it was probably a German. I took up a position to attack him. I get very close, ¾ front. I had my SPAD with 4 machine guns at the time. I send a burst of my 4 machine guns for a fraction of a second when at that moment I see the English roundels. Surprised, I did not even ask myself if it was a German disguised with English roundels. I got out of the way as fast as I could. I did it so hard that the whole leading edge of the upper wing of my SPAD flapped. The entire back cloth of the upper plane was torn off, as if it had been plucked. I was no longer flying, I couldn't glide. The nanny that was in the upper plane was pulled out of her cells and was flapping at the end of a wire spraying me with gasoline. When I tried to put the aircraft back in a more or

Above & Below: Photo of SPAD VII n°1083 on July 8, 1917, destroyed after a forced landing when the canvas of its upper plane was detached in flight. (Coll Benoit Henriet)

SPAD VII n°1083 of Dieudonné Costes, squadron 507, Salonique-Lembet, May 1917.

less suitable flight line, it fell into a spin. I straightened it, raised it a little more, and I realized that as soon as I lost speed it would not hold. If it normally stalled at 110, now it was stalling at 150. I had time to think as I got closer to the ground. I knew the area where we were, it was on the edge of Adzi-Gol lake. I said to myself, I have to go to that corner because there are reeds and the ground is soft. If I pick myself up like I hope to pick myself up, I could hurt myself less. I had time to think about it because it was at 4500–5000 meters. I realized that I had to go fast to get a rudder, I was changing direction with the ailerons. That's how I got to the ground. As I descended to the ground, glancing to the right and left, I saw my Englishman descending in the same way. I had lowered it! I was busy with my descent

Above: Costes was the victim of a second de-entanglement on another SPAD VII on September 29, 1917; however, he managed to crash-land south of Salonica. After this double mishap on SPAD, the ace would fly again on a Nieuport fighter and would fly again on SPAD only at the end of spring 1918 (Gallica)

Above: Dieudonné Costes taking off on alert aboard his Nieuport 24 in March 1918. (Coll Benoit Henriet)

Above: The Greek 531 squadron was created in March 1918 and served in the Vardar sector. Equipped with the best pilots, it won 11 certified victories in a few months. From left to right: Adjudant Basile Sauné (5 victories), Second Lieutenant Maurice Ripoche (2 victories), Maréchal des Logis Maurice Lashermes (3 victories), Second Lieutenant Alexandros Zannas (Greek squadron leader), Adjudant-Chef Dieudonné Costes (6 victories), and Sergeant Paul Andrillon (2 victories) (Coll. Benoit Henriet)

Above: The second ace of the Orient Air Force, adjudant Basile Sauné, born in the Ariège, was shot down by the German ace Gerhard Fieseler on June 20, 1918. Here he poses in front of his Nieuport 24 bis. (Coll. Emilie Durin)

to get out of it as best as I could and he was also descending in a whirlwind; we were about a mile apart from each other. I reached the ground and picked myself up as best I could – naturally, with a rollover. During this time the British landed 500 m away from me. At that moment I was wounded by a fall of motorcycle and I had difficulty to go out of the aircraft, it was necessary to scrape the ground, that took a good moment. During this time the English came to me. And by the time I got out of that SPAD with the upper hut flattened and the upper plane touching the ground, there was not much room left. As I was crawling, my foot caught the Bowden that controlled the machine guns on the lower plane. And as the aircraft was upside down, the lower plane was 1m50 above the ground. And the British were coming in that direction, just as my tracer bullets went towards the British! They thought I was shooting at them! Finally, I came out, waved and raised my hands, and explained to them what had just happened. Everything went very well!

J.R.: You had all come a long way!

D.C.: Yes, that happened a lot.

J.R.: Did you have another story?

D.C.: No, this one was the toughest. This was the second time a SPAD slammed me. It was wood and glue, under the eastern sun this glue broke and cracked. I had another fight like that, during a fight with a German. Of course, the German went home, and I went home – still in a whirl.

J.R.: Did you have an affair with a German, one that you had a hard time shooting down?

D.C.: Oh no. We didn't push it when it was too difficult. Strictly speaking, a good hunter doesn't make a fight. He passes. If he sees that he doesn't have enough assets, he breaks and goes. That's the way it should be. Except with Fieseler where you maneuver from 4,000 yards to the ground, but no one can get hurt if they're alert. On a bad surprise, maybe. But not when there are two good pilots, you can't get down.

J.R.: Do you recognize good pilots in the air?

D.C.: At first sight! From the first turn, you knew who you were dealing with, whether it was a young pigeon or an eagle.

J.R.: What happened when it was a young pigeon?

D.C.: It was uneventful. He would immediately put himself in a bad position. I remember a good boy, whom I caught near Guevgueli, on the bank of the Vardar. He was surprised to hear a machine gun behind him, and he made a turn, he pressed his machine gun without knowing, by reflex, like this. And he went into a spin. And as he spun down, his machine gun was firing all the time. So I just took the trouble to follow him very slowly, sliding down. And when

Above: German ace Gerhard Fieseler, Jasta 25, in front of the wreckage of one of his victims, a Nieuport 23. (DR)

he straightened up... It's terrible, you know.
J.R.: Did you pick it up?
D.C.: Pfff... It's uneventful.
J.R.: And when you saw an eagle? Did you run at it? Or did you avoid the fight?
D.C.: You know, it was mutual. A German, if he didn't catch you by surprise, and if he started the fight, he would see right away who he was dealing with, and he would break off. It was a classic method.
J.R.: You mean if the German saw that he was dealing with an ace, he would break off the fight?
D.C.: Also, yes. Except in a group, where there are 6, 7, 8 isolated fights. The one who is the boss of the group watches the whole fight a bit and comes to the rescue of the one who is in a bad position. There, you can have unpleasant surprises; you fight against a German and don't see the aircraft coming from behind to shoot you. But in individual combat, generally speaking, it was a tactic to break off if you saw that it would get you nowhere.
J.R.: In group flying, did you have your own tactics?
D.C.: Yes, generally speaking. It was always the same person who had to make the approach. Sometimes we would split from the group and fly separately, using the clouds or the sun as protection. The German pilot's eye was drawn to the one in front of him, but not to the one behind. This was a tactic that was sometimes successful. It could not be used all the time, fighter to fighter. For a reconnaissance aircraft, the tactic was different. The aircraft was attacked outright, although sometimes they were good maneuvers.
J.R.: What brand of aircraft?
D.C.: I don't remember the reconnaissance aircraft of that time. I know that once I had a big twin-engine Gotha, in a group, passing by. Always the same thing: they had just done a bombing raid in the Salonica area. I was at a very high altitude: I passed in the middle of the group, I picked one up, and that was it.
J.R.: He was able to defend himself, however.
D.C.: Yes, because they were quite well armed. But when there was a lone recon device, there were always three of us. I had two good comrades, Lashermes and Andrillon, who were not clumsy. They put themselves in different angles from mine. I tried to slip underneath them and get into their landing gear.
J.R.: I see that all the techniques were quite complex and different.
D.C.: It was quite simple, you know. A little bit of thinking...

Above: Gerhard Fieseler of Jasta 25, the German ace of the Orient Front, ended the war with 19 certified victories. Here he is flying a Roland D.II fighter (DR)

Above: A few days of rest before the final offensive. Dieudonné Costes (2nd from the right) is granted with other pilots a few days of relaxation in a beautiful villa of Boresnica by the commander Denain. (Coll Benoit Henriet)

J.R.: Simple when you didn't get shot down. There were many French pilots who were shot down.
D.C.: There may have been a lot of them, but it's hard to shoot somebody, you know. It's hard to shoot someone who is vigilant and doesn't let himself be surprised. I've never been surprised, never, and that's the number one factor.
J.R.: Did you take any bullets in your plane?
D.C.: From the ground, yes. But from other planes, no. Only by Fieseler, in this fight. That may be the only time, or one of the few times. I've had many times bullets come from the ground, including one that went through my broomstick and into my cartridge box and made a fight in it. But nothing exploded.
J.R.: You had a lot of luck!
D.C.: Maybe, yes. There is a luck factor.
J.R.: Who is the French driver you admired the most?
D.C.: It's hard to say. We had on the French front, but I didn't know them, the Foncks, Navarrese, Nungessers, etc. They were really eagles as fighter pilots. They were really eagles as fighter pilots. But they were privileged: they had many planes in front of them. While we, in the East, could go a fortnight without seeing an enemy plane. It is better, perhaps, because if there had been too many, I might not have been there! Because the best ones end up getting shot down... But I didn't know those guys. Guynemer was a brawler, he went for it and was bound to get shot. Fonck was a tactician, he was an extraordinary boy who was just passing through. A shooter with a sure eye. Attacks as fast as they were accurate. He passed like that... It's even questionable whether this boy had any merit, with such extraordinary results. But he did it so simply that we were astonished.
J.R.: In the East, were you kept informed of the exploits that took place in France?
D.C.: More or less, yes. Note that I was rather close to Fonck's tactics, because I hardly picked up any bullets in my plane, except from the ground.
J.R.: You were very prepared for combat.
D.C.: Yes, the advantage was the lightness, except for strafing on the ground where you took your entire load of bullets. That represented about fifty kilos, which for an aircraft of the time had a ceiling of 200 meters.
J.R.: Did you admire some Allied or German pilots?
D.C.: I didn't know them well enough. Our front was a bit far away. You would have had to be in France to see what was going on and the results of each side. But where I was, Fieseler's results forced us to recognize that this boy was a first-rate pilot.
J.R.: After the war, did you have the opportunity to get to know Nungesser, who had the same idea as you about crossing the Atlantic? Did you talk about it?
D.C.: Yes, I knew more or less all the fighter pilots on the French front, I approached them all. Especially Fonck, Nungesser, Haegelen, Sardier, Pinsard.
J.R.: Did you meet Nungesser and Coli before the attempt with the *Oiseau Blanc*?
D.C.: Yes, I saw them several times at the Villacoublay airfield when they were testing their aircraft.
J.R.: Two years later, in 1930, you left for the Atlantic crossing yourself.
D.C.: No, Nungesser did it in '27, I did it three years later. But when Nungesser left for the North Atlantic, we were also preparing for the North Atlantic. But as the weather was not suitable, around September 1927, we left for the South Atlantic. And that's why our aircraft was called the *Nungesser and Coli*.
J.R.: Did your exploits and your experience acquired at the front serve you for your Paris / New York crossing?
D.C.: No. No connection. Either you have "the skill", the airmanship, or you don't. If you have airmanship, which gives you the ability to face the most extraordinary difficulties that may arise, these qualities are applied in commercial aviation as well as in test aviation or war aviation. These are completely different tactics.
J.R.: But didn't you learn during the war to control yourself?
D.C.: Oh yes, certainly! Having started to fly in 1912, one does not fly for 4 or 5 years in the army on a whole type of aircraft without asserting a mastery. But in spite of that, you need personal qualities. I say this without pride, but without a personal factor, a pilot will never achieve anything.
J.R.: When you left to cross the North Atlantic, were you sure you would succeed after Nungesser's failure?
D.C.: No, that's another thing. You can never be sure that you won't have a mechanical failure of some kind. Even if you have prepared your engine, your piping, and checked for all possible failures. There can always be something that happens that you don't anticipate. And even sometimes by doing things too well, you can fall victim to that too well. I'll give you a case in point. For a record attempt I was making in the direction of the great lakes in Central Africa, I was aiming for Lake Tanganyika. I had gone to a manufacturer to build me a special heat-resistant radiator, a larger radiator. I witnessed the welding of this radiator. The worker, thinking he was doing me a favor, probably left the radiator in the welding process for a few moments too long, and it was assembled. But in the process there was a fitting that was not well done. Some tin from the radiator had been left in the molten solder for a few fractions of a second too long, made a nugget inside. And this nugget of tin went for a walk in the engine pipes, and blocked my water pump. I had to land in Aswan after having crossed the Mediterranean and the Libyan desert. I had to dismantle with a wrench all the piping, and I found a nugget of tin this big, which stuck the turbine. I took it out, reassembled everything and off we went. But instead of doing the record, we went back to France. Thirty hours of flight for nothing!
J.R.: Fifty years after your aeronautical exploits, what did you remember about the First World War? What would you do if you were twenty years old and had to do it all over again?
D.C.: I would do the same thing.
J.R.: Were the war years stronger than your raids?
D.C.: Neither of them. The past doesn't leave me with much of a reaction. During the war, we felt like we were doing our

Right: Costes visiting the Parthenon.

duty. Unfortunately, not enough importance was attached to the really useful role we were doing for the country. During the raids, the perspective was different.

Appendix 2: Air Units

Escadrille 501

1916 **1916-1917** **1917-1918**

Created as V 83 on December 1915, renamed F 383 in June 1916 and Escadrille 501 in June 1917

Locations:

- Lyon on December 1, 1915 (creation of the unit)
- Salonica on December 13, 1915.
- Samli on December 19, 1915.
- Ostrovo region on September 1916.
- Kladerop (N. Florina) on November 15, 1916.
- Bach in 1917.
- Koritza on January 15, 1918."

Commanding Officers:

- Cne Gaston de Serre on December 1, 1915.
- Lt René Duvernoy on January 28, 1916.
- Cne Robert Dangelzer on July 1917.
- Lt Charles Fruchard on June 1918.

Equipment:

- Voisin LAS bombers on December 1, 1915.
- Maurice Farman MF 11bis on June 1916, gradually replaced during year 1916 by Farman F.40.
- Dorand AR 1 in 1917, with some Breguet 14 in late 1918.

The unit had a fighter section for escort duties equipped with 2–3 Nieuport fighter (various types) and at last one SPAD VII.

.

Victories: None.

Casualties: 2 pilots lost from enemy action (1 killed, 1 prisoner) + 1 by accident.

History:

Initially one of the three squadrons of the Groupement de Bombardement d'Orient (GBO), the unit participated in several bombing missions behind enemy lines during the first half of 1916. It was then transformed into an army corps unit in June 1916 for the Armée Française d'Orient (AFO) and settled in the Monastir basin. In January 1918, the squadron moved to the extreme west of the front, to Koritza, in a sector held by Albanian troops who had rallied to the French army.

Above: Sergeant observer Vahel near a Dorand AR 1. (Albin Denis collection)

Adjudant Jean Faure, despite being aged 58, flew operationally with escadrille V 83 in 1916.

Voisin LAS belonging to the commanding officer of V 83, Capitaine Gaston de Serre.

Above: View of the airfield of escadrille V 383 in January 1917. We can see a Farman F.40 wearing the second insignia drawn in a shield, at the top of the fuselage.

Below: Pilots and observers of escadrille 501 in 1917, in front of a Nieuport 23. A Farman stands on the foreground. (Albin Denis collection)

Above: Pilot Georges Baudeuf (left) and his mechanic Paul Pedoux on July 15, 1917. Baudeuf changed the unit insignia by putting the word "Zut" ("Heck!") instead of the winged star.

Below: This Nieuport 24bis is quite a mystery, wearing the squadrons markings of both escadrilles 501 and 506. It was captured by the Bulgarians on July 5, 1918, flown by maréchal des logis Marie... A pilot from escadrille 503 !

Voisin LAS (serial unknown) of escadrille V 83 in 1916. The smiley in the star is an unofficial marking.

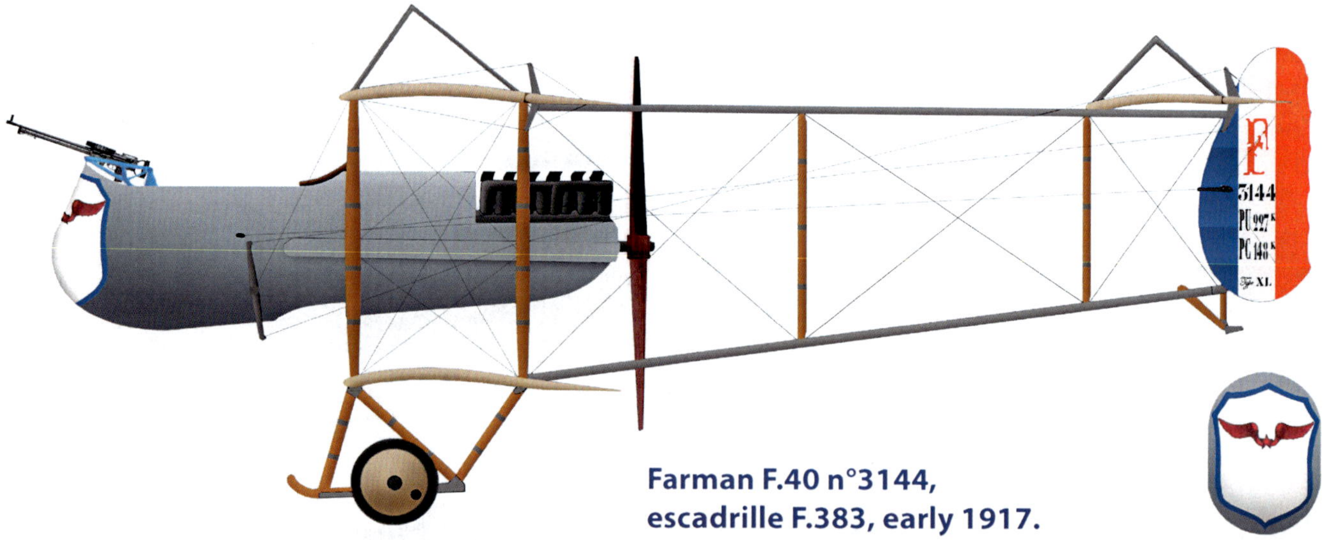

Farman F.40 n°3144, escadrille F.383, early 1917.

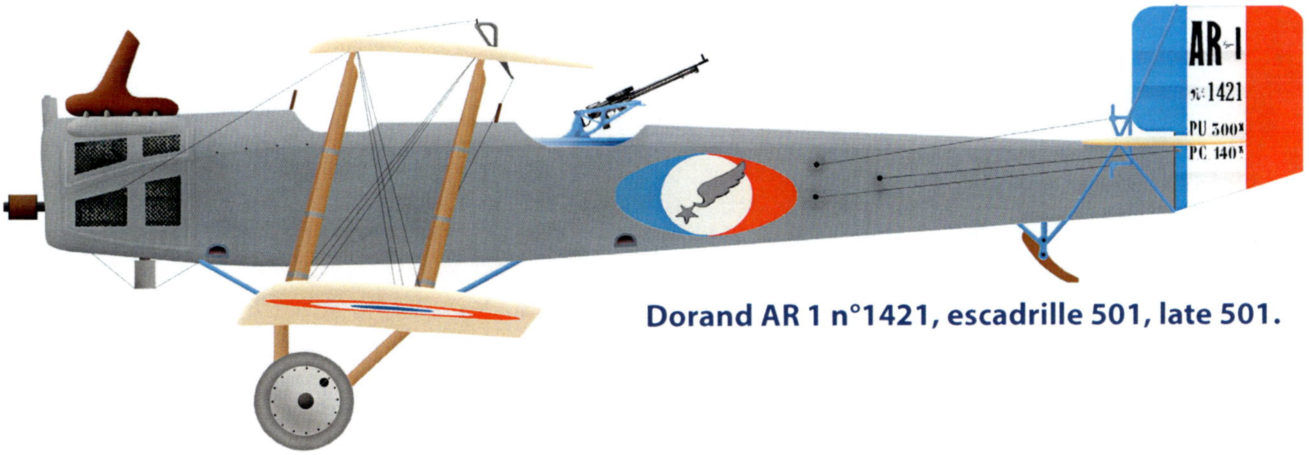

Dorand AR 1 n°1421, escadrille 501, late 501.

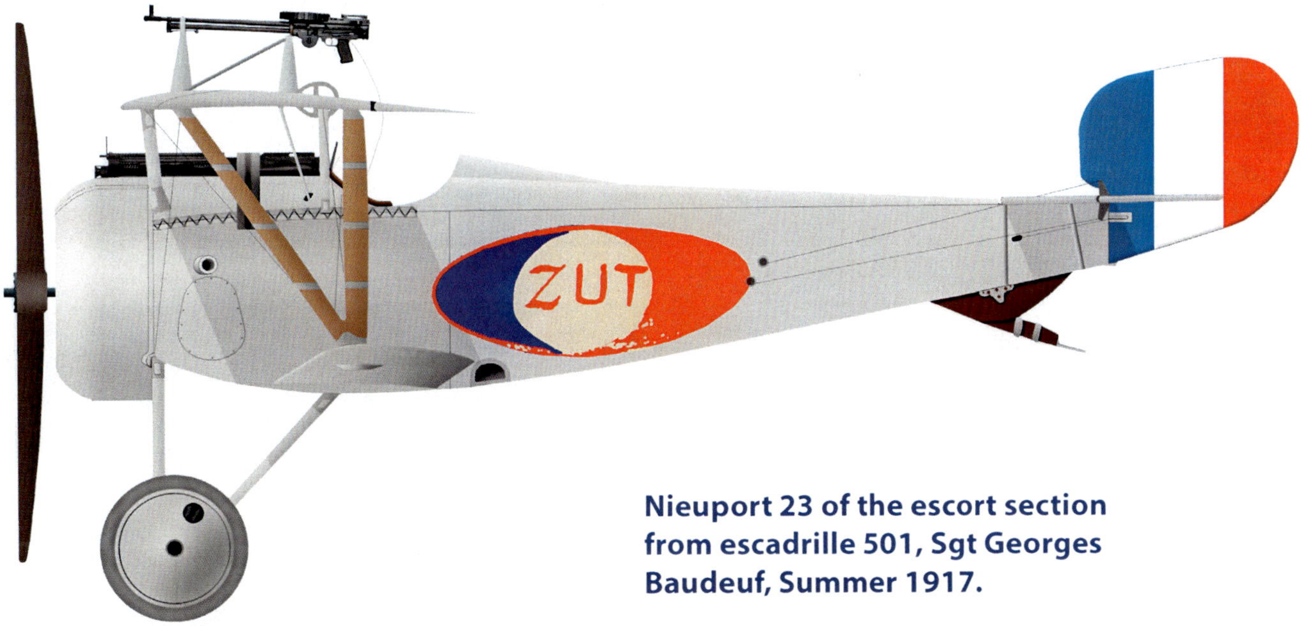

Nieuport 23 of the escort section from escadrille 501, Sgt Georges Baudeuf, Summer 1917.

Escadrille 502

Created as MF 85 on October 1915, renamed F 385 in June 1916 and Escadrille 502 in June 1917.

Locations:

- Lyon on October, 1915 (creation of the unit)
- Salonica on January 16, 1916.
- Samli on January 20, 1916.
- Vardarosi (near Ostrovo) on April 9, 1916.
- Gorgop on August 11, 1916.

Commanding Officers:

- Cne Etienne Cheutin on October 1915.
- Lt Robert de Brun on November 27, 1915.
- Lt Maurice Hamoir on January 16, 1916.
- Cne Pierre Perrin de Brichambault in 1918

Equipment:

- Maurice Farman MF 11bis on October 1915, replaced since mid-1916 by Farman F.40.
- Dorand AR 1 in late 1917, with few Breguet 14 since mid-1918.

The unit had a fighter section for escort duties equipped with 2–3 Nieuport fighters (various types: 21, 23, 24, and 24bis).

Victories: 3 confirmed victories.

Casualties: 2 pilots lost from enemy action (2 killed) + 1 by accident.

History:

Created in Lyon in October 1915 and embarked in the port of Marseille, the unit reached Salonica in January and after accompanying the offensive of the French troops towards the Monastir basin, it settled in Gorgop in August 1916 where it remained for the rest of the war. As a army corps squadron with a fighter patrol, it was not subordinate to the *Aviation Française d'Orient* (AFO) but was placed directly under the orders of Major Denain in the *Commandement des Armées Alliées* (CAA – Allied Army Command). All its aircraft were destroyed on the ground during a raid by Kaghol 1 at the end of February 1917. Immediately reconstituted with aircraft from the Salonica Park, it finished the war at Gorgop while being placed at the disposal of the Serbian army in July 1918, in preparation for the final offensive of September 1918.

Right: Dorand AR 1 of escadrille 502.

Below: Dorand AR 1 of escadrille 502.

Above: Nieuport 24 of the escort section from escadrille 502.

Above: Crashed Sopwith 1A2 from escadrille 502 in 1918.

Farman MF 11bis n°1076 flown by Adjudant Dieudonné Costes, 1916.

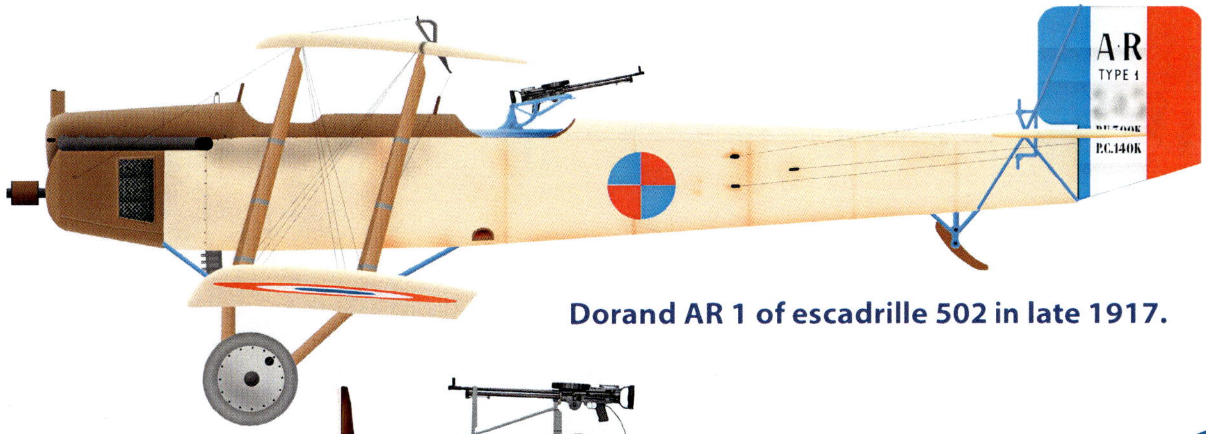

Dorand AR 1 of escadrille 502 in late 1917.

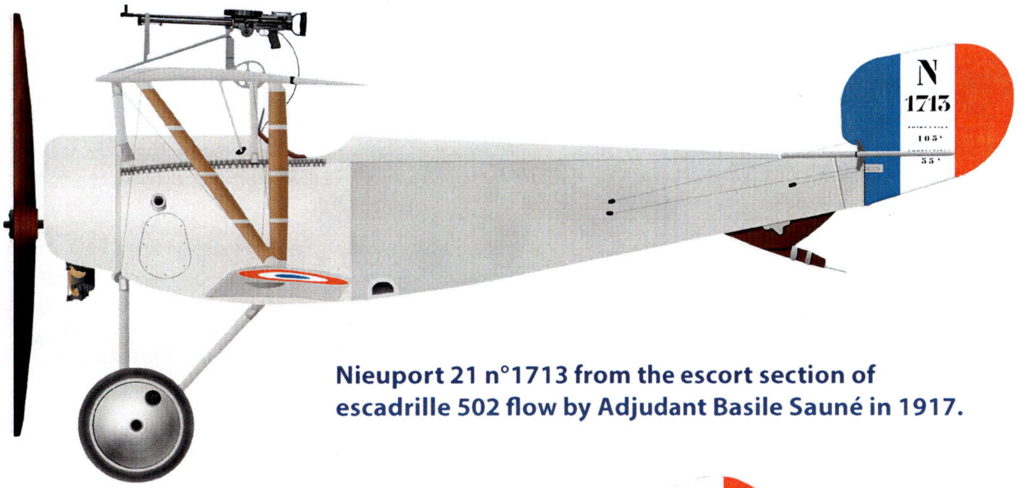

Nieuport 21 n°1713 from the escort section of escadrille 502 flow by Adjudant Basile Sauné in 1917.

Nieuport 21 (serial unknown) from the escort section of escadrille 502 in 1918, nicknamed *"Routchichoune"*.

Escadrille 503

1916-1917

Created as MF 85 S (S=Serbia) on October 1915, then shortly after MF 86, the unit is renamed F 386 in June 1916 and Escadrille 503 in June 1917.

Locations:

- St Priest on October, 1915 (creation of the unit)
- Salonica on November 9, 1916.
- Doiran lake on November 28, 1915.
- Samli on December 14, 1915.
- Mikros on September 8, 1916.
- Kajalar on September 14, 1916.
- Kladerop (N. Florina) on October 1, 1916.
- Bach on January 15, 1918/

Commanding Officers:

- Lt Guy de Kersaint on October 1915.
- Cne Lucien Galloy on May 1916.
- Cne Robert de Brun on June 1916.
- Lt Daniel Jousset on September 3, 1916.
- Cne Gilbert Crussard on December 27, 1916.
- Unknown officer.
- Lt Pierre Chaignon on June 1918.

Equipment:

- Maurice Farman MF 11 on November 1915
- Maurice Farman MF 11bis on March 1916.
- Maurice Farman F.40 on July 1916.
- Dorand AR 1 in May 1917.

1917-1918

The unit had a fighter section for escort duties equipped with 2–3 Nieuport fighters (various types: 21, 23, 24, and 24bis).

Victories: None.

Casualties: 6 pilots lost from enemy action (4 killed, 2 prisoners) + 3 by accident.

Notes:

Created as MF 85 S and destined to come to the rescue of Serbia, the squadron was immediately renamed MF 86 and was one of the first to land at Salonica, on November 9th 1915. After having assembled and tested its planes at the Allatini camp, it left with the French army to go up the Vardar valley and rescue Serbia, setting up in an improvised airfield near the Doiran lake, carrying out its first war mission on November 30th 1915. But the rescue expedition turned out to be short-lived and the French army returned to the city of Salonica, where the MF 86 followed and settled on the Samli field. When the French troops deployed in the Monastir basin in the summer of 1916, F 385 followed the movement and ended the war here as an army corps squadron attached to the *Armée Française d'Orient* (AFO).

Above: Photo taken on November 1915 showing the first operational airfield of escadrille MF 86 near Doiran Lake. The planes appearing are Maurice Farman MF 11.

Below: Farman F.40 of escadrille F.40 in January 1917 on Kladerop airfield.

Above: Farman F.40 n°3335 of escadrille F.386, early 1917. (Collection Roberto Gentili)

Above: Nieuport 17 from the escort section of escadrille 503, 1917.

Below: Pilots of escadrille 503 sharing a drink in their unit's bar. Note the new unit insignia. (Albin Denis collection)

View of escadrille 503's new insignia painted on a Dorand AR 1, after June 1917.

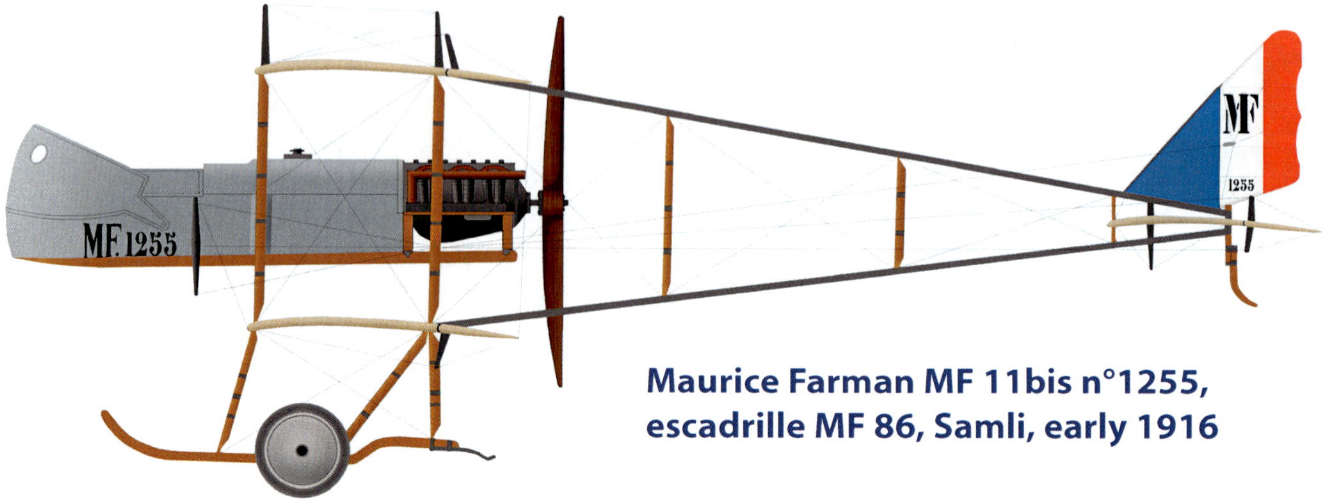

Maurice Farman MF 11bis n°1255, escadrille MF 86, Samli, early 1916

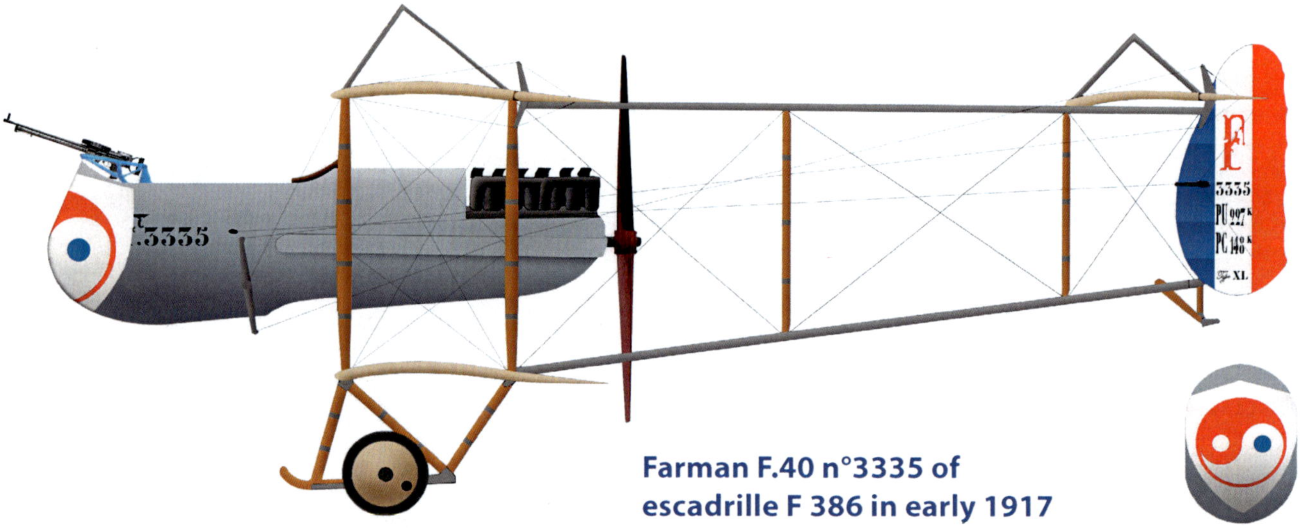

Farman F.40 n°3335 of escadrille F 386 in early 1917

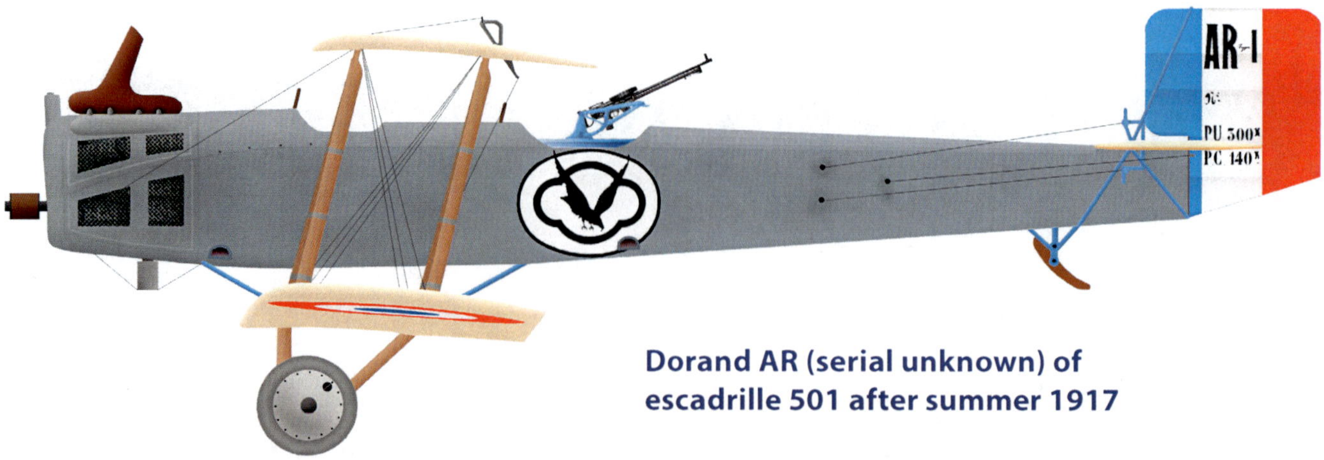

Dorand AR (serial unknown) of escadrille 501 after summer 1917

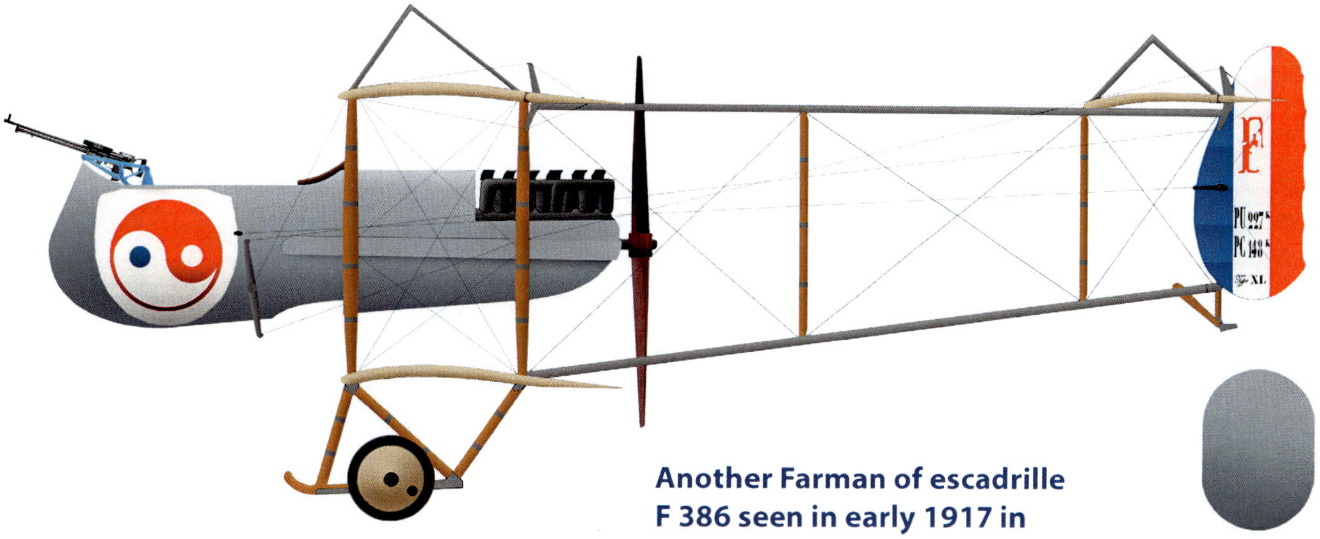

Another Farman of escadrille F 386 seen in early 1917 in Kladerop, showing a different application of the unit insignia.

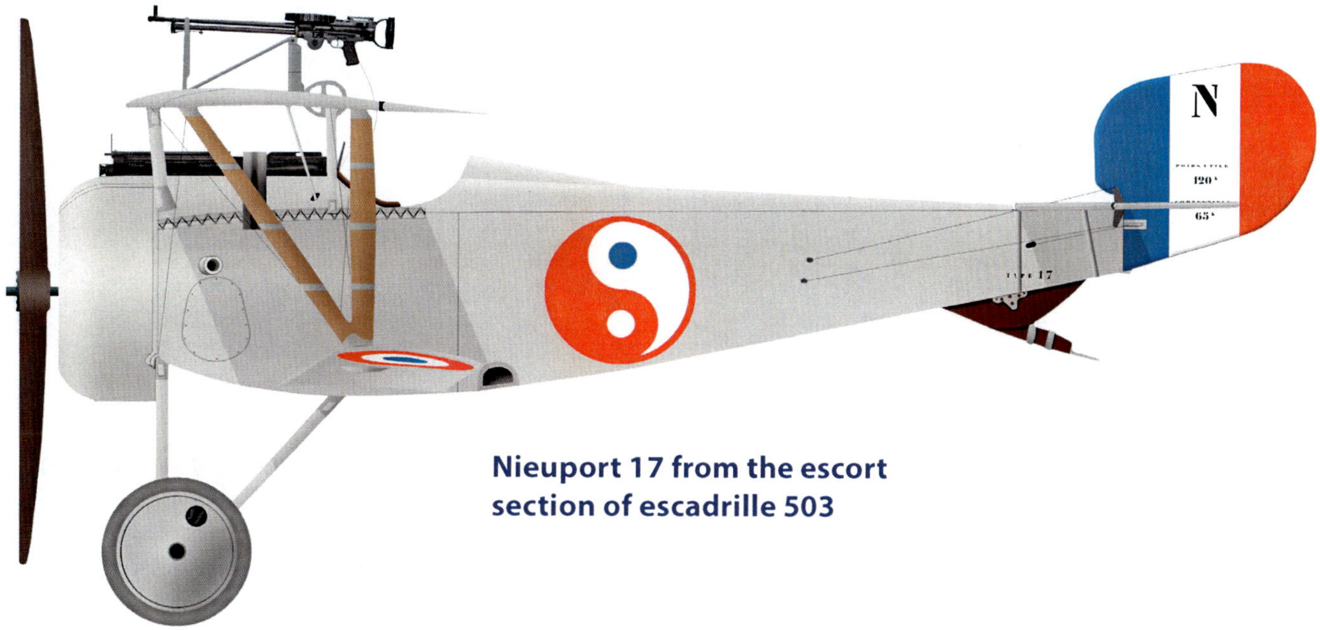

Nieuport 17 from the escort section of escadrille 503

Escadrille 504

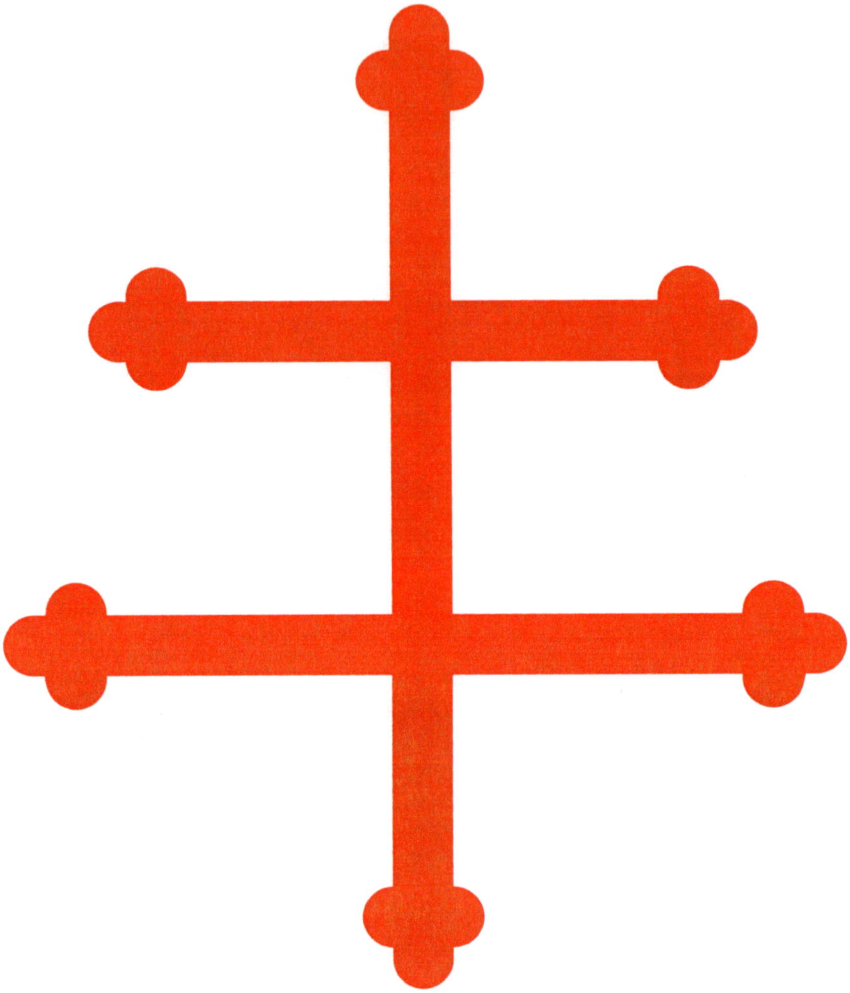

Created as MF 88 on October 1915, renamed F 388 in June 1916 and Escadrille 504 in June 1917.

Locations:

- Salonica on December 1, 1915.
- School of agriculture (Sedes road) on December 12, 1915.
- Kukus on July 23, 1916.
- Snevce on September 9, 1916.
- Florina on November 1, 1916.
- Boresnica on November 3, 1916.
- Negocani on December 1, 1916.
- Odessa (Ukraine) on March 20, 1919.
- Artsyz, (Ukraine) on April 17, 1919.
- Galati (Romania) on July 7, 1919.
- Constantinople on August 1919.

Commanding officers:

- Cne Jean Mortureux on December 1, 1915.
- Cne Louis Maire on June 23, 1916.

Equipment:

- Maurice Farman MF 11bis on December 1915, replaced since July 1916 by Farman F.40.
- Dorand AR 1 in late 1917, with few Sopwith 1A2.
- Breguet 14 since January 1918, slowly replacing the Dorand during first half of 1918.

The unit had since early 1917 a fighter section for escort duties equipped with 2–3 Nieuport fighters (various types: 21, 23, 24, and 24bis).

Victories: No confirmed victories (2 unconfirmed).

Casualties: 2 pilots lost from enemy action (2 killed) + 1 by accident.

History:

Escadrille MF 88 squadron landed in Salonica on December 1st 1915 and stayed for some time at the Allatini camp, where its aircraft were assembled and tested. The first of them, the Farman MF XI n°967, was taken in hand by MdL

Above: Nieuport 10 used for liaison duties by escadrille MF 388 in September 1916. (Collection Boris Ciglic)

Koechlin, taking as a passenger the commander Denain, head of the aviation of the Orient front. On December 12, the MF 88 reached its assigned base, built 5 km from Salonica, on the road to Sedès near the agricultural school. Its first missions consisted in monitoring the movements of Greek troops that General Sarrail had firmly requested to leave the city of Salonica. Adjutant Louis Noël, a famous civilian pilot who had participated in the Aerial Derby in England, led the first surveillance on December 14, 1915. The MF 88 was an army corps unit, in charge of carrying out reconnaissance and observation missions for ground troops. It was initially assigned to the British army in Salonica, which did not yet have an air force. The squadron split on February 1st 1916. The MF 88 bis, which shared its base, was created to carry out observation missions on behalf of the 17th French colonial division, and was dissolved on June 20th.

When the allied troops left the entrenched camp of Salonica to reach the back country, MF 88 reached Kukus on July 23, 1916, east of the Amatovo lake, then Snevce, closer to the front, on September 9, 1916. It was from this airfield that it split again to create a section of three Farmans put at the disposal of the Italian troops in Salonica, entrusted to Captain Lucien Galloy and which lasted until the beginning of 1917. On September 16, 1916, Second Lieutenant Noël (with Lt. Leseur as machine gunner) carried out a raid on Sofia, the Bulgarian capital, with three other Farmans of the Orient Army, bombing it and flying to Romania.

With the French offensive in the Monastir region, MF 388 was assigned to the AFO (Armée Française d'Orient) and migrated to this sector, to the Boresnica airfield on November 3rd and then to Negocani on December 1st 1916, where it remained until the Armistice, carrying out a number of reconnaissance missions for the local troops as well as bombing raids in common with the other escadrilles of the sector (501, 503, 505, 508, and 509). Noteworthy was the splendid exploratory flight led by 2nd Lt Noël, a long-distance flight specialist, who managed in July 1917 to reach Salonica from the Italian town of Otranto, on a Sopwith 1A2, thus inaugurating a route for air deliveries to the Armée d'Orient.

When the armistice of the Orient was signed on November 1st, escadrille 504 was maintained and completed by personnel from disbanded squadrons. It was sent with escadrille 510 squadron to the Danube Army in Romania, where its Breguet 14 and SPAD escort planes were used to fight against the Bolsheviks. Initially based at Arcis, it moved to the Galatz base in July 1919. It was disbanded a few months later on an unknown date, at the same time as the repatriation of French troops from this theater of operations.

Above: Pilots from various units posing in front of a Farman F.40 from escadrille F.388. (Collection Boris Ciglic)

Above: Nieuport 10 used for liaison duties in escadrille 504, during the first half of year 1916. Note the school of agriculture on the right, in the background. (Collection Boris Ciglic)

Possibly the same aircraft after a bad landing. (Collection Albin Denis)

Nieuport 23 n°3660 from the escort section of escadrille 388, crashed on March 8, 1917.

Above: Another accident of a Nieuport 23 from escadrille 504 on July 1917. The plane is nicknamed "*Leo*".
Left: Lieutenant Louis Noël, famous pre-war pilot specialist of long-distance raids, and member of escadrille 504.

Above & Below: A Sopwith 1A2 from escadrille 504.

Above: Dorand AR 1 of escadrille 504 after a bad landing.

Above: Sous-lieutenant Million-Rousseau, observer of escadrille 504, poses in front of a Breguet 13 of his unit.

Above: Breguet 14 from escadrille 504 nicknamed "*Titine*".

Above: Breguet 14 from escadrille 504 nicknamed "*Léon*".

Above: Other Breguet 14 of escadrille 504 with a unreadable nickname. Note the little plush monkey on the left wing.

Below: Three pilots of escadrille 504 bearing the unit flag. Note the Breguet 14 and two Dorand AR 1 in the background.

Maurice Farman MF XI n°967, of MdL Koechlin, escadrille MF 88, Salonica, December 1915

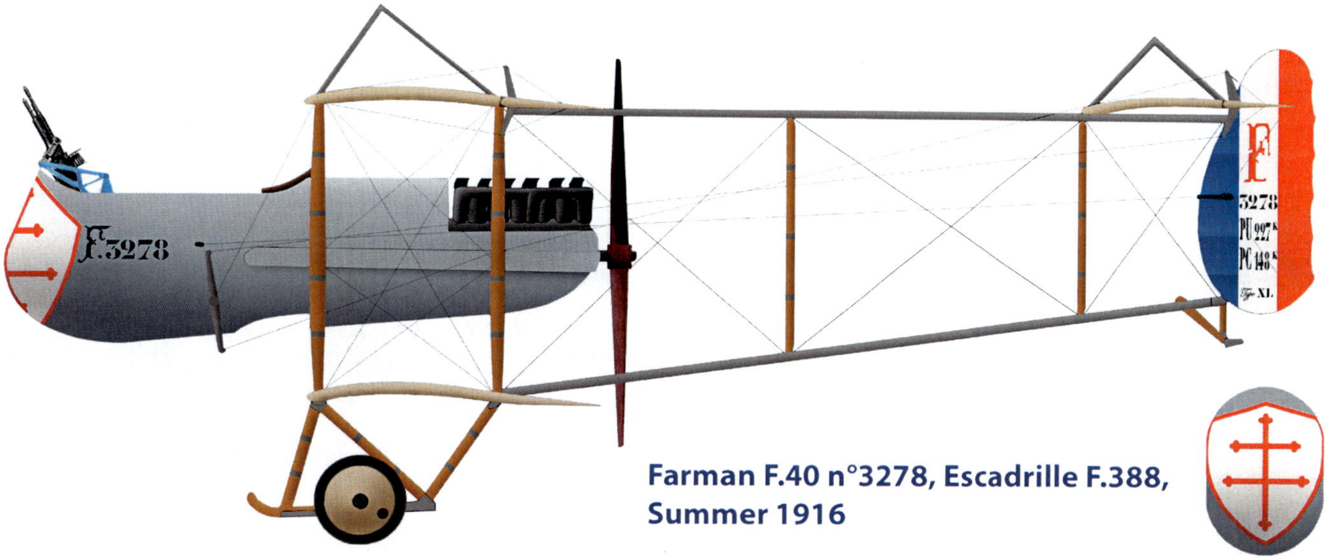

Farman F.40 n°3278, Escadrille F.388, Summer 1916

Nieuport 23 n°3660, Escadrille F.388, March 8, 1917

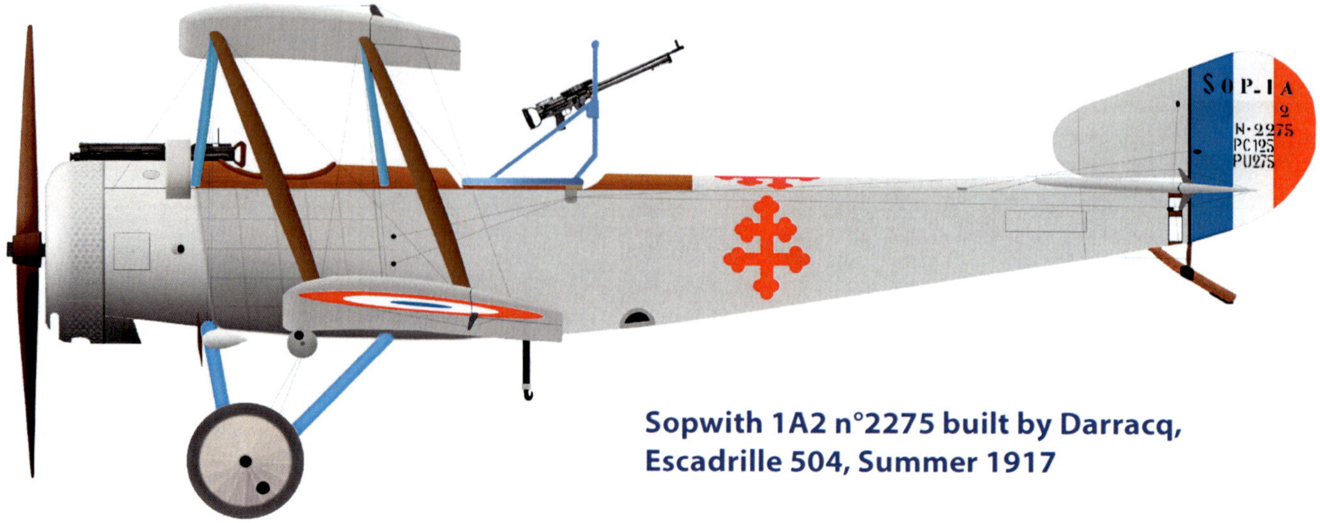

Sopwith 1A2 n°2275 built by Darracq,
Escadrille 504, Summer 1917

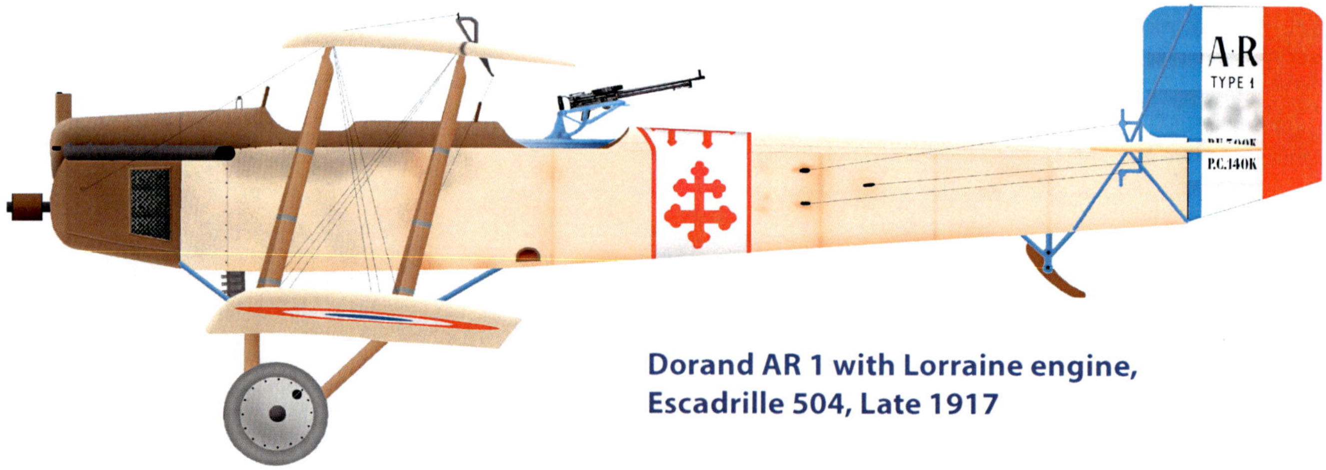

Dorand AR 1 with Lorraine engine,
Escadrille 504, Late 1917

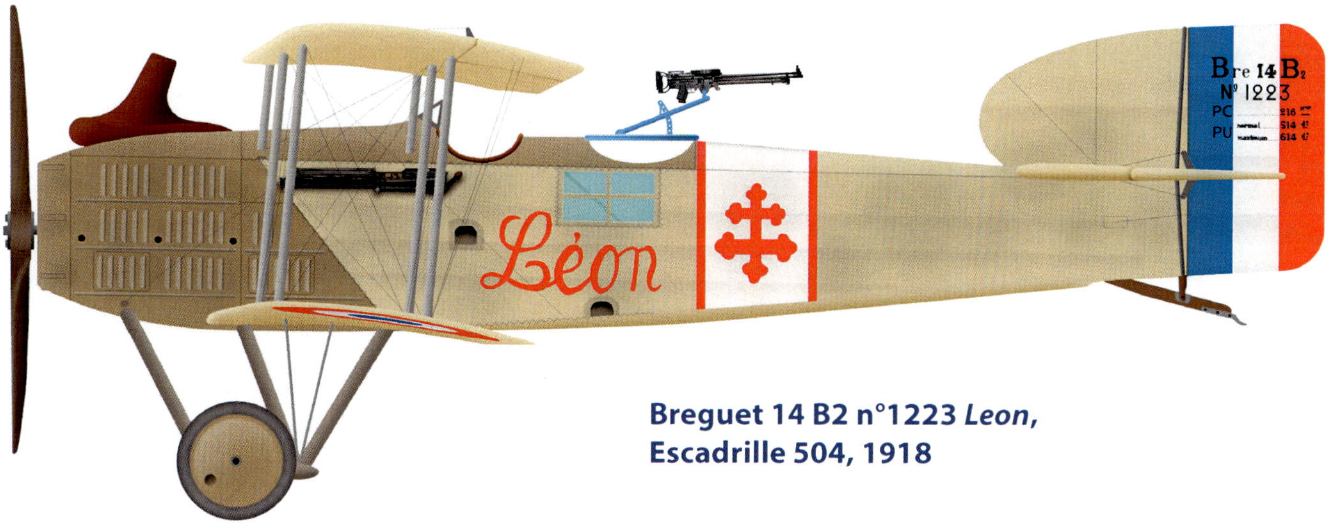

**Breguet 14 B2 n°1223 *Leon*,
Escadrille 504, 1918**

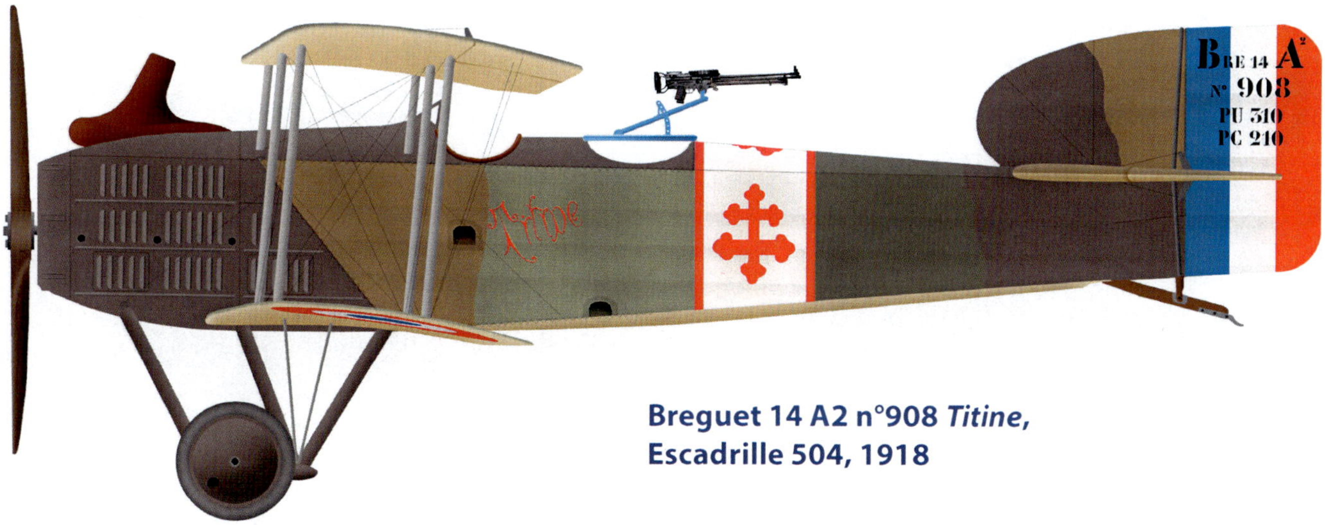

**Breguet 14 A2 n°908 *Titine*,
Escadrille 504, 1918**

Escadrille 505

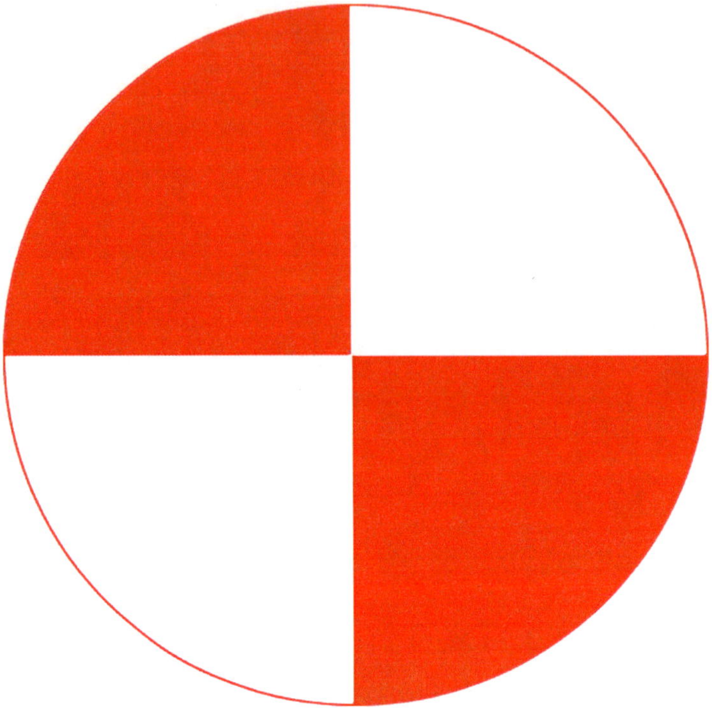

Created as C 89 S (Serbia) on October 1915, renamed C 389 in June 1916 and Escadrille 505 in June 1917.

Locations:

- Lyon on October 15, 1915.
- Salonica on October 19, 1915.
- Guevgueli on October 30, 1915.
- Topcin in December 10, 1915.
- Siret Hissar on May 4, 1916.
- Samli on September 2, 1916.
- Mikros on September 5, 1916.
- Karadjialar on September 12, 1916.
- Sorovicevo (15 km NE Florina) on September 19, 1916.
- Boresnica (E. Florina) on December 15, 1916.
- Larissa (Thessalia) and Lamia (Central Greece) on June 1917.
- Gorgop on June 1917.
- Constantinople on 1919.

Commanding officers:

- Cne Franck Delanney on October 11, 1915.
- Lt Fernand Vigouroux on November 15, 1915.
- Lt Sarda on July 14, 1916 (killed in action on September 16, 1916).
- Cne Bizé on September 22, 1916.
- Lt Picard on June 1918.
- Lt Léon Michaud on unknown date in 1918.

Equipment:

- 6 Caudron G.3 in October 1915.
- Caudron G.4 in February 1916.
- Farman F.60 and Sopwith 1A2 gradually replace the Caudron G.4 since February 1917
- Breguet 14 June 1917, delivered in few numbers.

The unit had since early 1917 a fighter section for escort duties equipped with 2–3 Nieuport fighters (various types: 21, 23, 24, and 24bis).

Victories: 1 official victory (1 unconfirmed).

Casualties: 3 pilots lost from enemy action (3 killed) + 4 by accident.

History:

Assembled in Lyon on October 11, 1915, the C 89 S squadron, made up of 6 Caudron G.3s, was destined to be assigned to Serbia, as attested by the "S" added to its number. It landed in Salonica on October 19, 1915, and took part in the Vardar expedition intended to help Serbia by reaching Guevgueli on October 29, from where its Caudrons launched several reconnaissance and bombing missions against the Bulgarian troops, supporting the French troops going up the Vardar valley. In particular, it carried out missions towards Strumica, locating enemy bivouacs, and even carrying out two bombardments of the camps around this city on November 23 and 24, 1915. However, the Serbia that these troops were supposed to support collapsed under the blows of the combined offensive of German, Austro-Hungarian, and

Above: Caudron G.3 from escadrille C 89 S in late 1915. (Collection Albin Denis)

Bulgarian troops. The French expedition then turned around and returned to Salonica, just like the C 89 S which settled in Topcin, near Salonica, on December 10th 1915.

Re-equipped with Caudron G.4 in February 1916 (only one unit of the Eastern Army to be so equipped), the C 89 (which lost its "S" in the official documents) specialized in long-range reconnaissance of enemy positions, during which its aircraft launched a few bombs on the enemy positions. When the Entente troops left the entrenched camp of Salonica in May 1916, the unit settled in Avret Hissar, east of Lake Amatovo. Renamed C 389 in June, it faced for the first time the enemy hunt and in September migrated westwards to support the Franco-Serbian offensive towards Monastir.

At the beginning of 1917, the squadron received Farman aircraft, which gradually replaced its Caudron G.4s, and was also equipped with a section of Nieuport fighters to escort its aircraft. In June 1917, the C 389, renamed 505 squadron, left the Monastir sector to take part in the Athens expedition, where French troops carried out a show of force in central Greece to force King Constantine I, who was pro-German, to abdicate. If central Greece was secured without too many clashes, short but violent fights took place in Athens between the French soldiers and the King's supporters. This one abdicated on June 12 in favor of his 2nd son Alexander - Greece from now on entirely directed by the prime minister Eleftérios Venizélos enters then in war at the sides of the Entente against the Central Empires. The 505 squadron is stationed for its part for half in Larissa (Thessaly) and for half in Lamia (central Greece). One of its pilots, Maréchal des Logis Pierre Chichet, remembers this expedition: "I was designated to go and kick out Constantine. At the time I was young and proud... When I arrived in Larissa, I went into a bar and saw a painting of King Constantine. I called the owner over and said, "Take it down". The boss complied and put in its place the portrait of Venizelos, which was ready!" He participated in the victory parade with his plane over Athens, and then evaded the local commander who ordered him to stay and fly straight back to his base. The occupation was not to last forever: on June 20th 1917, Major Denain ordered the transfer of his two Nieuport fighters stationed in Larissa to the Monastir sector to reinforce the air force of the Armée Française d'Orient which was still fighting hard. The rest of the squadron followed and continued to participate in the fighting in this sector during the summer of 1917, receiving AR1 bombers in the autumn, four of which carried out a raid on Prilep on October 16, 1917.

At an undetermined date, probably at the end of 1917, escadrille 505 returned to the Vardar sector by joining the Gorgop airfield, which it shared with escadrille 502 and then the Franco-Greek 531 fighter squadron created in April 1918; it was now directly dependent on the Allied Command (CAA). From there it participated in the final offensive which broke the German-Bulgarian front in September 1918, being attached to the 2nd Serbian army, attacking numerous ground targets on the mountains in the center of the front. The war ended quickly after this blitzkrieg, but the airmen had to fight a new enemy: the Spanish flu epidemic, which hit the Armée d'Orient very hard and killed two pilots of the 505, including its commander, Captain Léon Michaud, who died on the day of the armistice in the East, i.e. on November 1, 1918.

However, escadrille 505 was not disbanded immediately, as it left with a French detachment to occupy the city of Constantinople. It was still there in September 1919, where it lost a plane and its crew in an accident – its dissolution probably took place shortly afterwards, when the French occupation troops were evacuated.

Above: Caudron G.4 taking off from the air park of Salonica.

Left: Caudron G.4 from escadrille C 389 ready to go for a reconnaissance mission in 1916.

Above: This Caudron G.4 n°9106 from escadrille C 389 during Fall 1916 belongs to a lot built in Britain, with Anzani engines.

Right: Sergent Gustave Gauthier in front of his Nieuport 21 from the escort section of C.389, Boresnica February 1917.

Above: Caudron G.4 of escadrille C.389 wearing the unit's insignia.

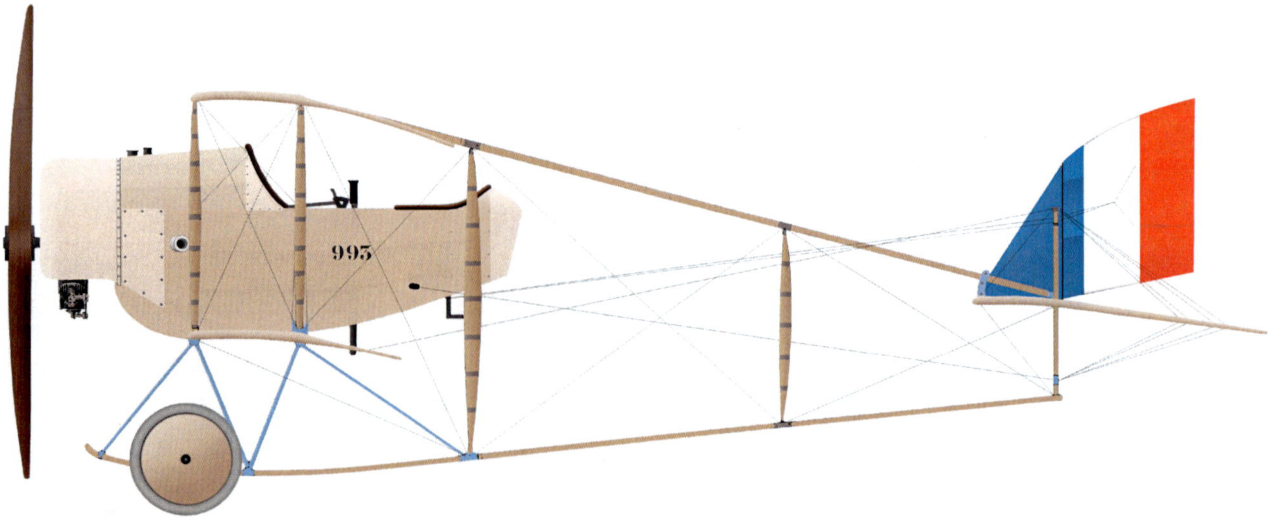

Caudron G.3 n°993 who made the first flight of C 89 S in Salonica on October 25, 1915, under the command of Sous-Lieutenant Potin.

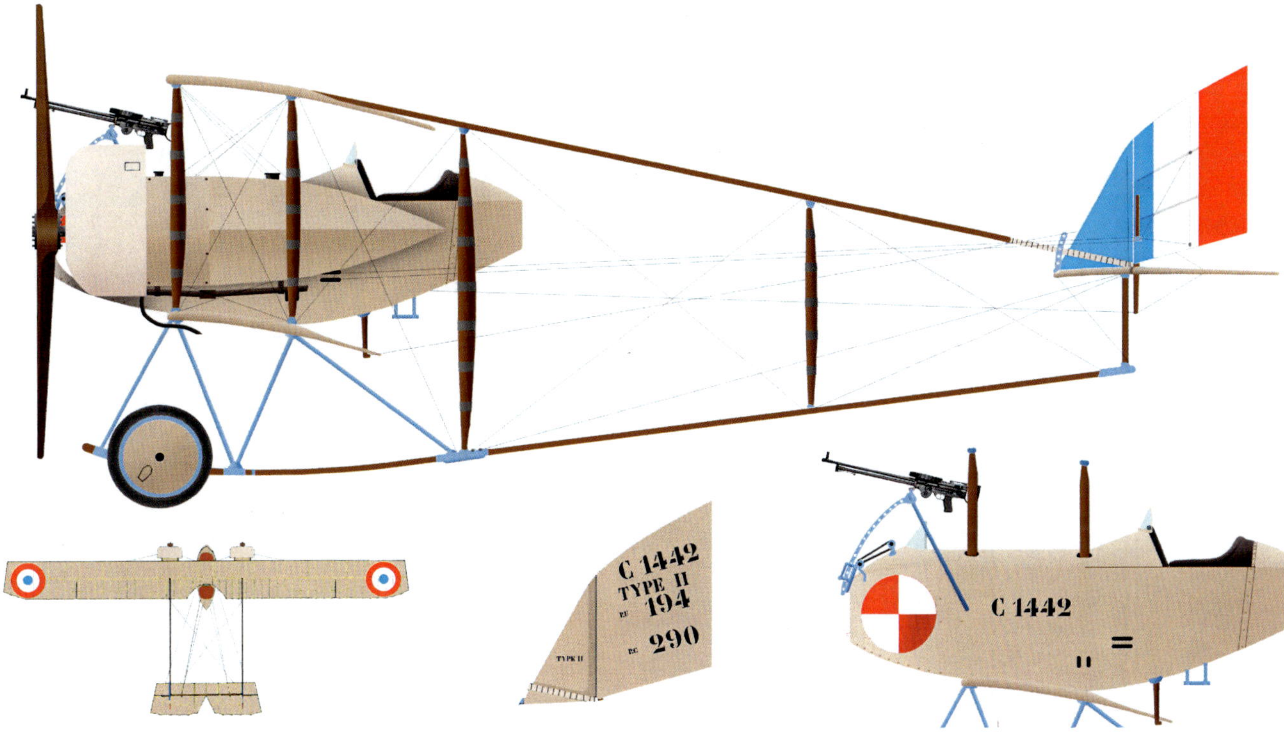

Caudron G.4 n°1442 of Sergent Jean Jaquelin, escadrille C 389, 1916.

Nieuport 21 (serial unknown), escadrille C.389, Sgt Gustave Gauthier, Boresnica February 1917.

Escadrille 506

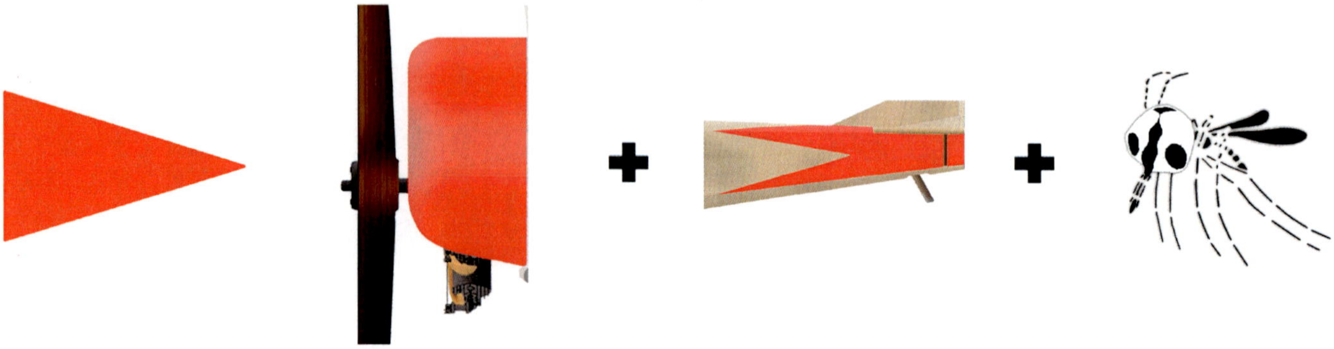

1916-01 1916-06 1917-06 1918

Created as V 90 on September 1915, renamed N 90 in June 5, 1916 and N 390 on July 8, 1916 then Escadrille 501 in June 1917.

Locations:

- Lyon on September 25, 1915 (creation of the unit)
- Salonica on October 31, 1915.
- Krivolak on November 11, 1915.
- Samli-Topsin on December 10, 1915.
- Mikros on September 6, 1916.
- Sorovicevo on October 1, 1916.
- Florina on October 6, 1916.
- Kladerop (N. Florina) on November 15, 1916.
- Kravari on December 1, 1916.
- Cote 619 (2 km S. Negocani) on December 11, 1916.
- Borechnitsa on January 15, 1918.

Commanding Officers:

- Lt Jacques de Thézy on September 25, 1915.
- Lt Jacques Lebouc on February 1918.
- Lt Louis Bonnetête on July 28, 1918.

Equipment:

- Voisin LAS bombers on September 25, 1915.
- A section of Voisin-Canon was affected to the unit in early 1916.
- Nieuport 10 on June 1916, then Nieuport 11 and 21.
- Nieuport 17 and 23 in January 1917.
- A Sopwith 1A2 was affected in March 1917 for strategic photo missions.
- SPAD VII are delivered since May 1917 in few numbers (n°1072, 1083, 1099, 1236)
- Breguet 14 A2 in June 1918 for strategic photo missions.

Victories: 10.5 confirmed victories +5.5 unconfirmed

Casualties: 8 pilots lost from enemy action (7 killed, 1 prisoner) + 1 by accident.

History:

The creation of the V 90 S squadron dates back to September 25, 1915, the date of an order from the French government's under-secretary of state for aeronautics requesting the dispatch of three squadrons to assist the Serbian government, which had been fighting against Austria-Hungary since the early days of the war. The small Serbia was able to contain two offensives and reconquer its capital Belgrade at each resumption. However, the next attack was likely to be the last one because it was reported that German troops were moving towards the Danube... A squadron, the MF 99 S (S for Serbia) was already on the spot since spring but its manpower was insufficient to cover the whole front. Three reinforcement squadrons were then created, one for reconnaissance, the C 89 S (Caudron); one for fighter, the N 91 S (Nieuport) and one for bombing, the V 90 S on Voisin LAS, whose command was entrusted to Lieutenant Jacques de Thézy.

The men and aircraft were hastily assembled in Lyon and embarked in Marseille on October 21, 1915 for the Greek port of Salonica. But when they arrived at their destination on October 31, 1915, the situation in Serbia was already desperate. The V 90 S nevertheless took part in the Vardar expedition and settled near the village of Krivolak on November 11, 1915, in very harsh weather conditions. After a few reconnaissance and artillery adjustment missions, the order to withdraw fell and the unit returned to Salonica to settle on the 10th of December 1915 on the Samli field.

The V 90, which loses its "S" in the official documents, would then be grouped with the two other Voisin squadrons of the Armée d'Orient, the V 83 and V 84, to form the Groupe de Bombardement d'Orient (GBO) commanded by Captain de Serre. It still had six Voisin LAS in line, to which was added a section of three Voisin-Canon equipped with a 37 mm gun, intended to escort the bombers. Another section of three Nieuport escorts is also directly attached to the GBO.

The V 90 squadron would participate in the 12 raids initiated by the GBO on the bivouacs of German troops stationed on the other side of the Greek border, the first

Above: Voisin LAS from escadrille V 90 S in early 1916.

ones taking place against the localities of Pazarli (January 28, 1916), Petric (February 1), Strumica (February 15), Strumica station (February 17), Orfano (March 11), Volovec (March 23), Bogdanci (March 24), Bodorodica (April 13 and 16), Petric (April 21). During these raids the Voisin, two of which were shot down by artillery on March 23, faced German fighters whose Pfalz and Fokker Eindecker began to appear in the orient sky and shot down a Voisin on March 24. However, the escort Voisin-Canon squadrons attached to the V 90 squadron won two air victories for the Voisin-Canon piloted by Lieutenant Albert Mézergues (a future ace with 6 victories) – the first on March 23 during the raid against Volovec, the second on March 27 in defense against a German air raid on Salonica. The Voisin-Canon of Mézergues accompanied by the one piloted by Sergeant Guy Deloupy fired several shells against the Zeppelin LZ 85 shot down by the anti-aircraft artillery of Salonica during the night of May 4th to 5th 1916. The GBO then carried out a daring raid against the Bulgarian headquarters in Xanthi on 24 May, then a last raid on Petric on 6 June 1916. There were no other raids because the potential of the Voisin aircraft was exhausted: 10 of the 18 aircraft participating in the last raid had to turn back before reaching the objective for mechanical reasons.

The escadrille V 90 was then completely reorganized and became a fighter squadron by being re-equipped with Nieuports. Most of the pilots were transferred to France on June 8, 1916 and new pilots, certified on Nieuport fighters, were assigned to the squadron – Lieutenant de Thézy, who was confirmed as squadron leader, was the only veteran of this first period. The squadron, renumbered N 390, was equipped with a combination of 110 hp two-seater Nieuports (second-hand Type 10s) and 80 hp single-seaters (Type 11 "BB", then Type 21 from October). All of them had their engine cowling painted red, perhaps inherited from the time of the GBO where at least one of its Voisins had red markings on the wings.

The N 390 was placed at the disposal of the Armée Française d'Orient (AFO) which took position in the mountains of Macedonia north-west of Salonica, while the Bulgarians, who had remained at the border until then, took Fort Rupel on May 27, 1916, which had been evacuated by Greek troops at the request of their sovereign. After several weeks spent in a state of inactivity, the Bulgarian troops launched a general offensive on August 18, 1916, and seized the town of Florina, which was occupied by French troops. The Eastern Front then became a continuous front and the AFO, with its air component of which the N 390 was a part, worked to dislodge the Bulgarians from the mountains of Macedonia with the help of Serbian troops. On August 28, 1916, the latter took the town of Ostrovo and opened the road to the Monastir plain. N 390 then left the camp of Samli and migrated westwards towards the terrain of Mikros on 6 September 1916, at the foot of the mountains of Macedonia, and then quickly sank into the interior of the relief by gaining the terrain of Sorovicevo. It was on this

Above & Facing Page, Above: Voisin-Canon of Sergent Marcel Girard (pilot) and soldier Albert Misrachi (gunner), shot down in Doiran lake on March 23, 1916. Misrachi was a Jew from Salonica who volunteered in French aviation.

occasion that Sergeant Vermonet won the third certified victory of the unit on September 13, 1916, when during a Farman escort flight in the Bogdanci region, he shot down an Albatross which fell vertically from a height of 3,000 m and fell into his lines, although the German archives, although incomplete, do not record any losses on that day. Vermonet did not have the time to take advantage of his war cross anyway, as he was accidentally killed on the following November 2, while the squadron had meanwhile continued its migration to the west, reaching the town of Florina on October 6.

As the French and Serbian troops approached Monastir, N 390 achieved its fourth certified victory on November 6, 1916, when Maréchal des Logis Jacques Mortagne and Henry Du Lac shot down an enemy aircraft that fell into enemy lines at Kravari – it was then destroyed by two Farman F.40s of the MF 386 squadron, escorted by Lt de Thézy who came to see the work of his pilots. On November 13, the first Le Prieur rockets arrived from France and were delivered to the pilots. They were fixed along the wing struts of two Nieuport 21s with instructions to attack the Drachen of Monastir. The attack was carried out without much success, the piloting of the small fighters being particularly difficult. The squadron moved to Kladerop on November 15, while the city of Monastir was captured four days later, although the German-Bulgarian troops were stationed a few kilometers to the north, still threatening it with their artillery... This is precisely what happened on November 25, 1916, when N 390 lost two of its aircraft (n°1408 and 1660) which were stationed on an advanced ground near the city, shelled by the German artillery.

On December 1, 1916, the squadron moved to the Kravari airfield, then on December 11, to the one built on the cote 619, 2 km south of Negocani. But the main event of the end of 1916 was the renewed aggressiveness of the German aviation, equipped with powerful reconnaissance planes as well as fighter planes delivered in number – a first fighter squadron, Jasta 25, was formed at Prilep in December 1916. With its small Nieuport 21s equipped with 80 hp engines, the N 390 had great difficulty in catching up with enemy reconnaissance aircraft, and could not hope to pursue its fighters. And when a Nieuport finally managed to get into firing position, it was still to be hoped that it would manage to shoot down its prey with the 47 rounds of its Lewis machine gun fixed on the wing... when this one did not simply jam! After these 47 rounds, it was necessary to

Above: Voisin LAS nicknamed *Ninocha*, flow by Russian pilot Alexandre Gomberg and Lt Jacques Richepin from V 90 S. This crew was shot down by a German Fokker E on March 24, 1916, but could land in no-man's-land and walk back to French line after putting their plane on fire.

change the magazine, which was impossible to do in cómbat, and then to return to contact – the enemy generally took advantage of this to escape. Thus, from 16 to 31 December 1916, the Nieuports made 22 attempts to intercept enemy aircraft, but were simply lost on 14 occasions. Of the eight successful interceptions, one was short-lived due to a jammed machine gun, six were unsuccessful, and only one saw the enemy hit by MdL Mortagne, on December 17, 1916, on an aircraft that was diving into the Cerna marshes while reloading its Lewis. MdL Du Lac was shot down by a German fighter escort and managed to land with only minor injuries. In the bar of the squadron, "The red cockroach", we complain every evening about this bad equipment which makes them lose many opportunities to win.

A tornado came to end this difficult year by destroying all the planes of the N 390 on the ground on January 1st 1917. They were quickly reconstituted, but the inefficiency of the French fighters was criticized in high places. Captain

Above: Voisin LAS n°1100 from escadrille V 90, one of the 3 units of Groupe de Bombardement d'Orient (GBO) in early 1916.

Hamoir, who commanded the aviation of the AFO, sent a report to Salonica explaining all the difficulties encountered, stating that his men had done the best they could with the material means at their disposal, and proposing to be relieved of his command if it was felt that he had failed in his task. General Sarrail, who was not very familiar with the air, understood the state of his airmen. He was relayed by Major Denain, who reminded him that there were only 24 fighter planes for a sector of 280 kilometers. Reinforcements finally arrived from France: 97-round Lewis magazines, as well as Nieuport 17 fighters, delivered without their Vickers machine gun belts! Sarrail requested the delivery of SPAD fighters, while the N 390 was reinforced in terms of manpower and increased to 10 aircraft in February 1917, including the first Nieuport 110 hp (Nieuport 17).

The German aviation would make an additional step on February 15, 1917 by the arrival on the field of Hudova of the Kaghol 1, a bombardment unit equipped with three squadrons of twin-engine Gotha, Rumpler, AEG and Friedrichshafen. These powerful aircraft operating en masse were to act with impunity on targets that had never been hit before: on 27 February 1917, the city of Salonica was attacked and suffered a lot of damage, as well as the base of Gorgop which saw all the aircraft of the 385 squadron stationed there destroyed on the ground. Other sectors of the front, including that of the AFO, were also attacked: the Serbs lost an entire ammunition depot which vanished...

Although the bad weather of March allowed some periods of respite in the sector of Monastir, it was the wanking of combat in all the aviation of the East and the fighter squadrons are strongly requested. On March 10, 1917, Major Denain ordered the N 390 fighter squadron, which had eight Nieuports available, to join forces with the Franco-Serbian N 387 squadron stationed at Kravari

Above: Like his brother Jacques, Tiarko Richepin, seen here in front of Voisin LAS n°1100, served in escadrille V 90.

Above: Bad winds of Salonica are more deadly than enemy action : here, a tent and a Voisin-Canon destroyed by a tornado, in early 1916.

to intercept the German bombers. The latter temporarily abandoned the Monastir sector, but the German fighter and reconnaissance aircraft were still present with as much bite and were one of the causes of the failure of the French ground attack launched on this sector. The respite – caused by the bad weather – was short-lived and from April onwards the German twin-engine planes appeared in the Cerna and Monastir sectors. The Nieuport 80 hp (21) and 110 hp (17 or 23) are struggling, and, despite a bomber claimed on April 3, 1917 by the N 387 squadron, Denain decided to concentrate the fighter resources by ordering the N 391 squadron, which was stationed in Salonica, to come and establish itself on the terrain of the N 390 at Negocani. Humiliating for the N 390, the command of the fighter group that settled on April 15, 1917 is entrusted to Lt Potin, head of the N 391 squadron...

The French fighter squadron was thus reinforced, and received its first SPAD VII. It was able to attack the bombers, whose route had already been "cleared" by German fighters. On 24 April, the 390–391 group pursued the bombers in the direction of Prilep but could not join them. April 29 saw the first success obtained by the French fighters, during a raid on the village of Skocivir (south bank of the Cerna loop) which was sprinkled with 2300 kg of bombs: one of the bombers was homologated, but to a fighter of N 391. Other engagements will take place but on May 1, 1917 the German bombers would leave the Orient front to be reassigned in Flanders. The N 391 squadron then returned to Salonica, leaving the N 390 to hold the Monastir sector.

The episode of the German twin-engine aircraft would lead to a serious boost in the use of French aviation,

Left: Lieutenant Albert Mézergues, future 6-victory-ace, won his first two successes in Salonica. He is seen here, posing in front of his Voisin-Canon, christened *T.N.T.P.S.R.* A mysterious abbreviation that could mean "You won't kill without reason".

beneficial for the fighters which receives quality equipment, but also for the bombers. The basic aircraft of the corps squadrons was the Farman F.40, which was now completely outdated; its replacement by Sopwith 1A2s and Dorand AR 1s was then initiated. However, even before the arrival of the replacement aircraft, the old Farmans, imitating the German example, were used for mass raids on the enemy rear by grouping the resources of the AFO squadrons, 501, 503, 504, 505, 508, and 509, plus an Italian squadron. The N 390 squadron, now renumbered 506, took part in the raids not by providing an escort but by carrying out pre-sweep operations in the target area. These new combat tactics paid off and the German fighters were now confined to the defensive. With nine Nieuports, two SPAD VIIs, and a Sopwith 1A2 used for reconnaissance missions, 506 Squadron would win five victories during the summer and autumn 1917.

With the coming of winter, the squadron would leave the field of Negocani at an undetermined date to settle in Boresnica, always in the sector of Monastir. If on January 24, 1918, a victory was claimed by Sgt Perrier and Cprl Lefèvre against an enemy fighter seen stinging in its lines (unconfirmed), the flights were reduced in a drastic way, not because of the weather, but because of a shortage of gasoline

Above: Mechanics from escadrille V 90 posing in front of the Voisin of Mézergues.

Above: Wreck of Zeppelin LZ 85 shot down in the marshes around Salonica, during the night from the 4th to 5th of May 1916. Although the Voisin-Canon shot shells at it, the victory was not confirmed to the French pilots Mézergues and Deloupy.

Above: A Voisin-Canon from escadrille V 90.

which reached all the squadrons of the aviation d'orient... The situation was unblocked with the assistance of the navy of Greece which organizes the transport of material from Italy towards Salonica. The 506 squadron participates moreover on February 12, 1918 in a protection barrage intended to secure the visit of the King of the Greeks to the great park of Salonica... If the aviation d'orient gains in power in 1918, the German fighters also, although fighting on aircraft somewhat outdated compared to those of the Western Front, includes in its ranks pilots of the highest order, such as the Feldwebel Gerhard Fieseler, the ace of the front d'orient aces with 19 victories, who is stationed in Prilep with Jasta 25, and that

Above: A Nieuport 10 from escadrille N 390 during summer 1916.

the pilots of the 506 squadron will face on many occasions.

On April 13, 1918, Slt Cochet-Forthuny, Adj Perrier, Cprl Aumaitre claim a victory on a fighter towards coast 1248, during a fight against three single-seaters, but which is not approved. The German fighters will quickly retaliate and demonstrate their virtuosity by shooting down two 506 pilots in quick succession. On April 24, 1918, Lt Fernand Pivain is shot down and killed on the Presba Lake at the controls of his Nieuport and this in spite of the intervention of Ltt Drozières, on his SPAD, who attends the fall of his teammate. The very next day, Corporal Lefèvre fell in combat at Gramsci, also flying a Nieuport, which clearly appeared to be outdated. At the beginning of May, the squadron was equipped with 10 Nieuports, two SPADs and two Sopwith 1A2s. The Sopwiths were replaced by five Breguet 14s, which formed a section of the squadron specialized in long-range reconnaissance and bombing. The Breguets went into action on May 23, 1918, bombing the town of Ohrida on the lake of the same name, at the extreme west of the front, escorted by SPADs. The SPADs progressively replaced the Nieuports, but it was one of the latter that won the next victory (the 10th) of the squadron on June 16, 1918 with Sgt Mathieu who shot down an enemy two-seater north of Monastir. It was precisely at this time that the two "aces" of Oriental aviation, Adjudants Costes and Sauné of 531 Squadron, were seconded to the squadron for a few days of hunting in the Monastir region. The result was not up to expectations, since on June 20, 1918 a tragic patrol took place. Costes, Sauné and Slt Cochet-Forthuny of the 506 squadron had to leave for a patrol on the frontline aboard their SPADs. Costes' plane did not want to start and his two teammates left without him... To come across a patrol of three German fighters that they try to attack by surprise. Only one of the German planes saw them, and voluntarily lost altitude as if to return to land. Sauné and Cochet-Forthuny then attacked the two remaining aircraft in patrol but were surprised by the Albatros fighter which had dodged and which happened to be piloted by Feldwebel Gerhard Fieseler, the future German top ace of the orient front. A furious whirling fight started between Sauné and Fieseler and ended at ground level where the Frenchman's SPAD could no longer use its advantage which was to dive quickly. Fieseler shot it down near Voprecani but the victory was attributed to his patrol leader, Lt. Rose, which left Fieseler bitter as he recounts in his memoirs. As for Cochet-Forthuny, he managed to return in extremis to his base, engine smoking... Costes took off a little later to try in vain to rescue his teammate. He would avenge him four days later, on June 24, 1918, by mercilessly strafing an LVG with three crewmen of the 506, MdL Hebert, Sgt Leman, and Corporal Aumaitre. The LVG supports the combat courageously against the four Nieuports but must land in catastrophe in its lines, 6 km in the northwest of

Above: Nieuport 11 n°1407 named *Jean Lecompte*, a French pilot shot down in Dardanelles on January 6, 1916.

Above & Following Page: On January 1, 1917, a tornado destroyed on the ground all the planes of the N 390 stationed on the field of Negocani. This photo shows the Nieuport 21 n° 2279 and 2397 which lost the left part of their upper wing. These planes have the engine cowling painted in red.

Dobropolje. It would not be certified.

On July 28, 1918, the Breguet 14 section of 506 squadron was out again and led a joint raid with the Breguets of 510 squadron against various targets on the road between Monastir and Prilep. The eight Breguets managed to reach their objectives by bombing through holes in the cloud cover, but they were intercepted by three Albatros D V of Jasta 25. One of them was piloted by Feldwebel Gerhard Fieseler who attacks several times and shoots down the Breguet 14 of the 506 piloted by Ltt Lebouc with Corporal Roger Baville as observer. He then attacked another Breguet of the 506, that of the crew of Sergeants Giron and Garat, but machine gunner Garat managed to hit the attacker who was seen to dive in flames towards his lines – the victory was certified to him although Fieseler had survived the engagement
On August 5, 1918, the Nieuport patrol composed of Slt Cochet-Forthuny, Adj Perreau, and Sgt Leman claimed a new victory against a fighter south of Podmol (between Prilep and Monastir), but this victory was not certified.
In spite of this success, the commander Denain remains dissatisfied with the results of the fighters of the sector of the AFO which cannot limit the losses among the French reconnaissance aircraft, insufficiently protected by fighter planes too dispersed. For the fourth time during the war, he sent a reinforcement of four fighters from the 507 squadron in order to "regain the moral ascendancy over the enemy". The fact remains that, despite the qualitative reinforcements of the SPADs, the situation has hardly changed since the end of 1916: with its escorts to provide to the five reconnaissance squadrons of the sector, its fighter patrols to carry out, plus the army reconnaissances which are requested of it, the 506 squadron would always have insufficient manpower for the sector of operations which it must cover.

After more than two years of war of position, the French forces were going to succeed in a breakthrough carefully prepared which will allow the victory on the orient front. On September 15, after an intense artillery preparation, the French and Serbian divisions seized the Bulgarian fortified

summits, which opened the road to Prilep, objective of the French and Gradsko, objective of the Serbs. Following this success, the French took Prilep on 23 September. Then, the Chasseurs d'Afrique brigade followed by the Colonial Brigade rode through the mountains to create a surprise. On 29 September, they seized Uskub (today Skopje) where the rear theater area of the XIth Army was located. The latter was obliged to withdraw what remained to Kalkandelen (today Tetovo), then to capitulate. The Bulgarian army was thus cut in two. The road to Sofia was opened. Bulgaria then asked, without delay, for an armistice which was ratified on 5 October.

The air force played a major role in this success because of the narrowness of the communication routes, which were attacked relentlessly. The German aviation was always present... During the preparation of the offensive where numerous aerial reconnaissance flight were conducted, the 506 will claim its ultimate victory (unconfirmed) on September 11 in Resna against a two-seater shot down by Ltt Bonnetête and Sgt Mathieu, both on SPADs. As soon as the summits were taken during the first days of the offensive, the German-Bulgarian troops had to withdraw behind the Cerna river, obliging them to concentrate on the rare bridges which are there, among which the one of Razim-Bey which is vigorously attacked by the French-Serbian aviation, the 506 squadron deploying several fighter patrols and its Breguet bombers. Several French aircraft fell in the air battle that took place from 15 to 20 September and the German air force won its last successes. On September 16, 1918, Adjudant-Chef Henri Leman's SPAD did not return from a patrol over the Cerna loop, and two days later, Adjudant-Chef Jean Charles Perreau was shot down by Feldwebel Gerhard Fieseler over the same area, obtaining his 18th and penultimate success. When the German troops evacuated the Kanatlarci air base on September 21, 1918, German aviation disappeared from the traffic and the activity of the SPAD and Breguet 14 of the 506 squadron was limited to a long series of ground attacks on the German-Bulgarian troops fleeing back to the city of Kalkanderen.

While the French armies spread in all the Balkans, from Greece to Hungary and from Montenegro to Romania, General Franchet d'Esperey submits on December 30, 1918 to the minister of war the dissolution of some squadrons. The 506 squadron was part of the batch and its trace is lost after this date, its active personnel being distributed to the other air units. In three years of air campaign without rest, this squadron from the end of the world, neglected by the staff in terms of equipment, will have fought in extreme conditions on a huge territory. Nevertheless, it won 10.5 air victories and 5 probable victories, against 7 killed in action, 1 prisoner, 2 wounded, and 1 killed by accident.

Above: Nieuport 21 with the red cowling of escadrille N 390, showing a personal marking during year 1917. (Collection Boris Ciglic)

Above: During year 1917, escadrille 506 received some Sopwith 1A2 for strategic reconnaissance missions. (Collection Boris Ciglic).

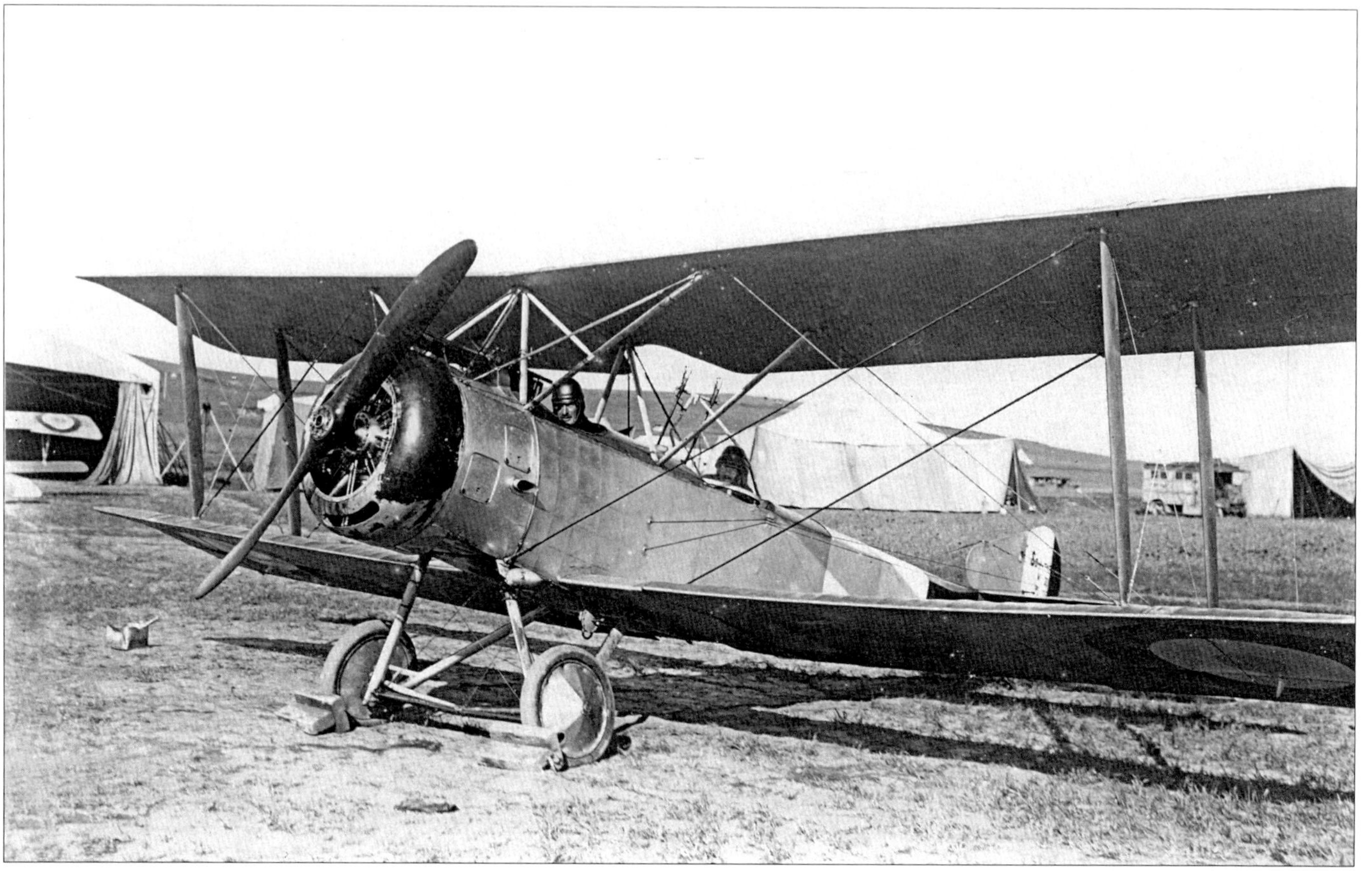

Above: Lieutenant Emile Brüninghaus at the controls of a Sopwith 1A2 in 1917.

Above: Nieuport 23 from escadrille 506 near Florina during year 1917, with Capitaine de Thézy, the commanding officer, on the foreground.

Above: One to the first SPAD VII fighters received by escadrille 506 during summer 1917. The serial number is erased from the tail, but documents show the delivery of SPAD VII n° n°1072, 1083, 1099, and 1236 between April and June 1917.

Below: SPAD VII of escadrille 506 with the mosquito insignia, during the visit of king Alexander of Greece to the Air Park of Salonica on February 07, 1918.

Above: During early 1918, escadrille 506 adopted a unit insignia, a mosquito, a familiar insect more deadly to the French troops than Bulgarian armies.

Above & Below: Fighters of the orient front in Gorgop during June 1918, supporting the offensive of Greek troops against the Serka di egen mountain. The first plane is a SPAD VII from escadrille 506, equipped with two machine guns – a field modification.

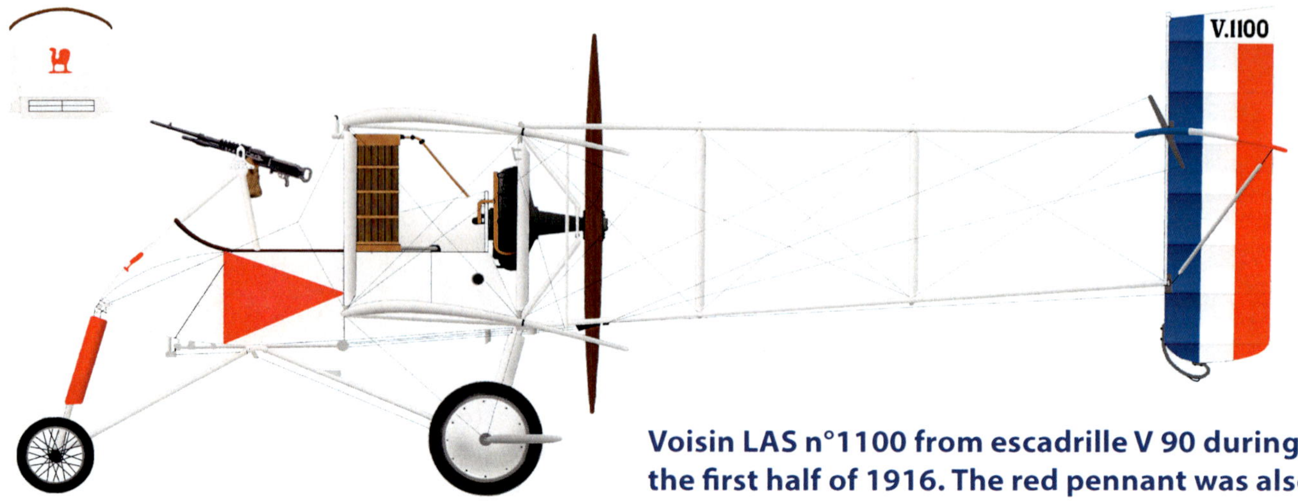

Voisin LAS n°1100 from escadrille V 90 during the first half of 1916. The red pennant was also painted on the top of right wing.

Voisin LB Canon n°985 from escadrille V 90 flown by Lieutenant Mézergues on March 1916.

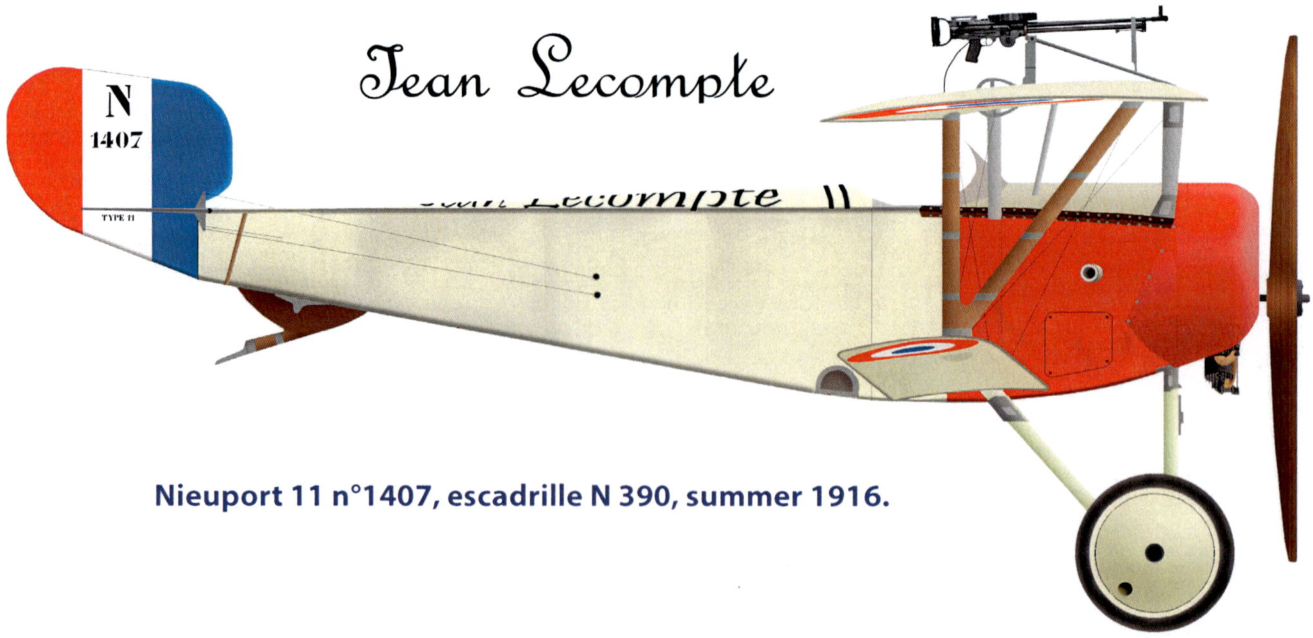

Nieuport 11 n°1407, escadrille N 390, summer 1916.

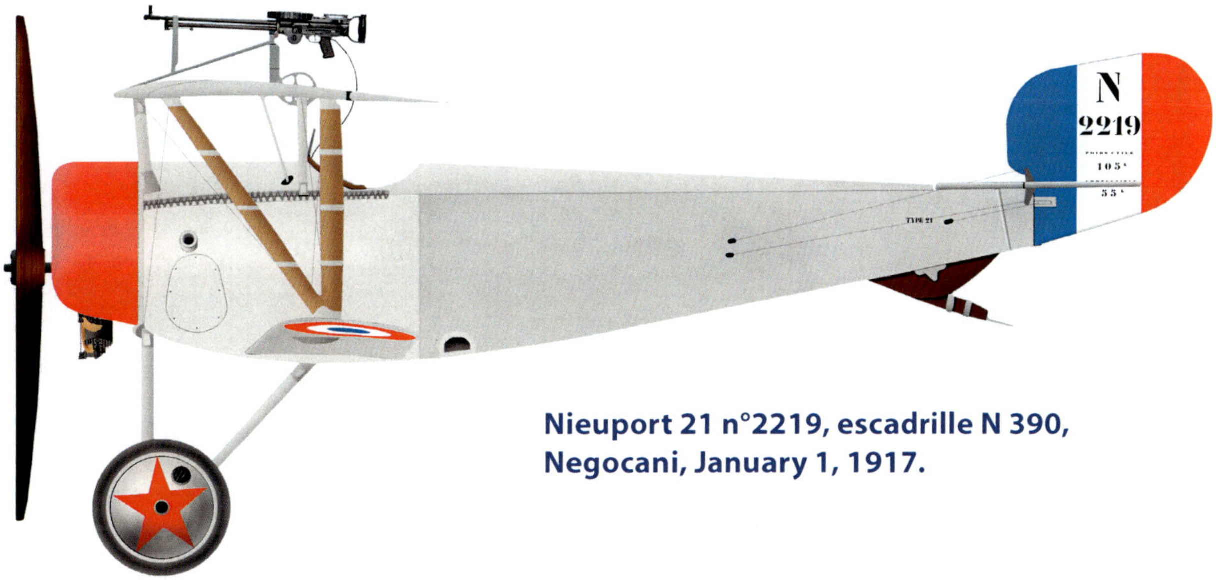

Nieuport 21 n°2219, escadrille N 390, Negocani, January 1, 1917.

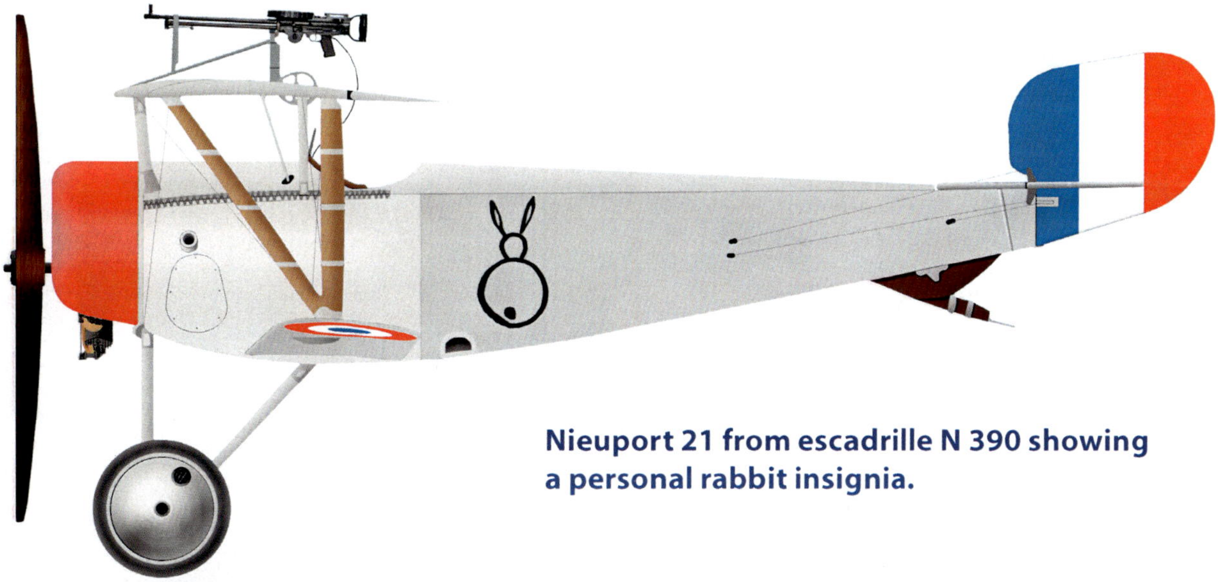

Nieuport 21 from escadrille N 390 showing a personal rabbit insignia.

Nieuport 21 n°2397, Escadrille N 390, January 1, 1917

Sopwith 1A2 from escadrille 506, Summer 1917.

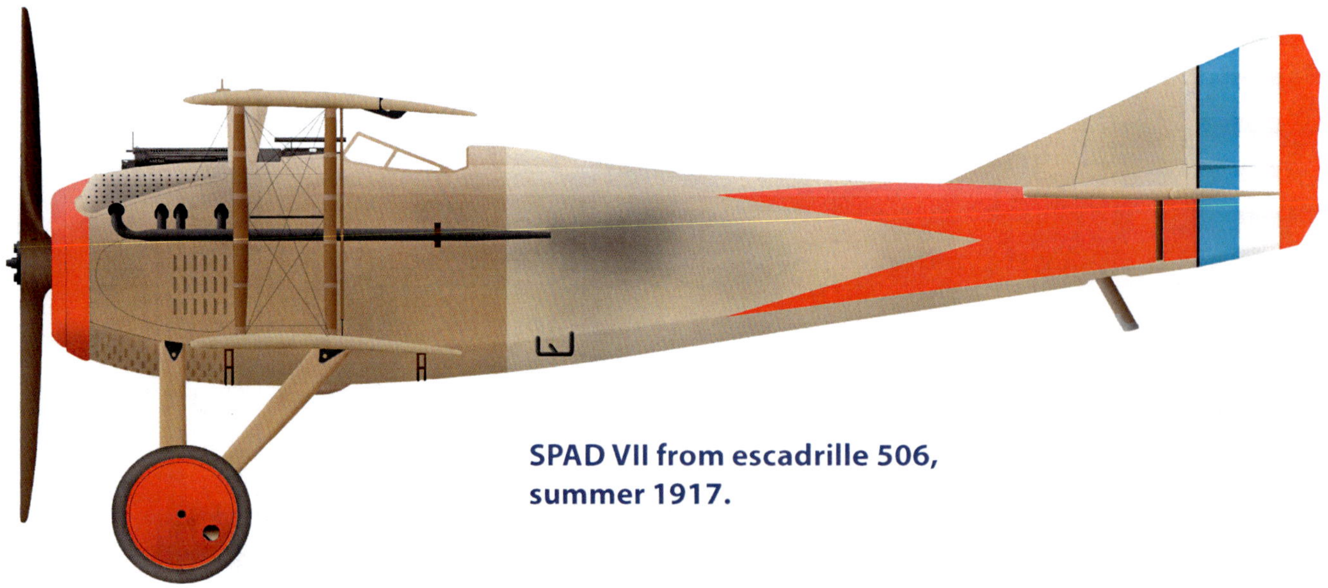

SPAD VII from escadrille 506, summer 1917.

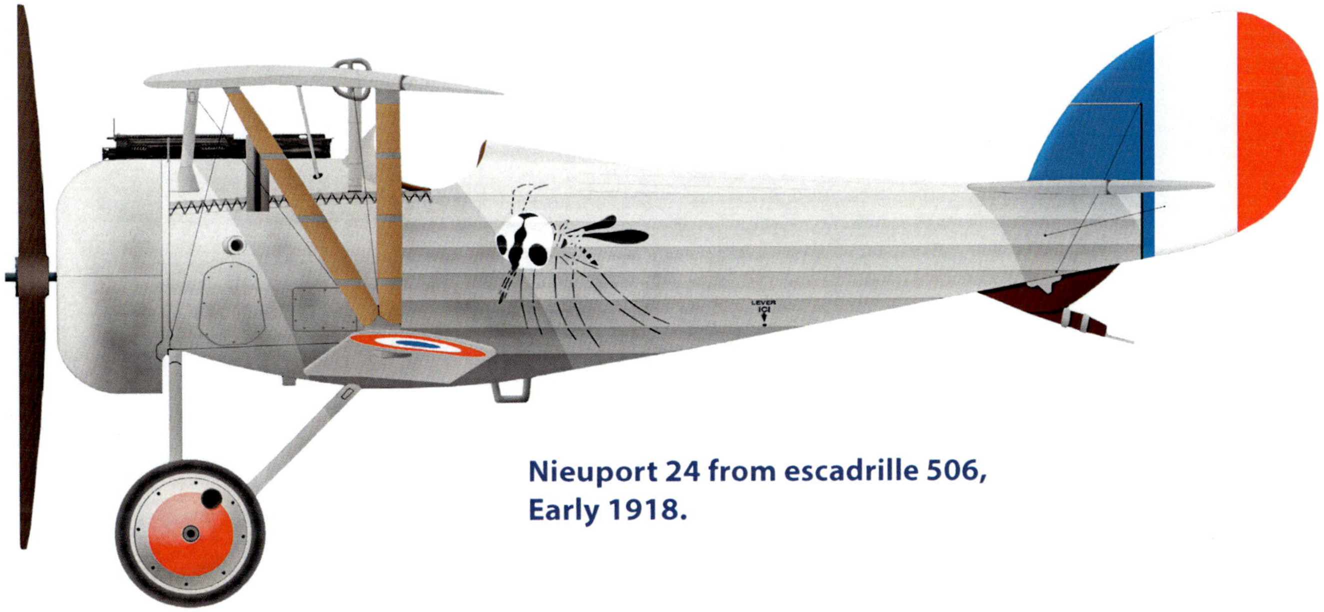

Nieuport 24 from escadrille 506,
Early 1918.

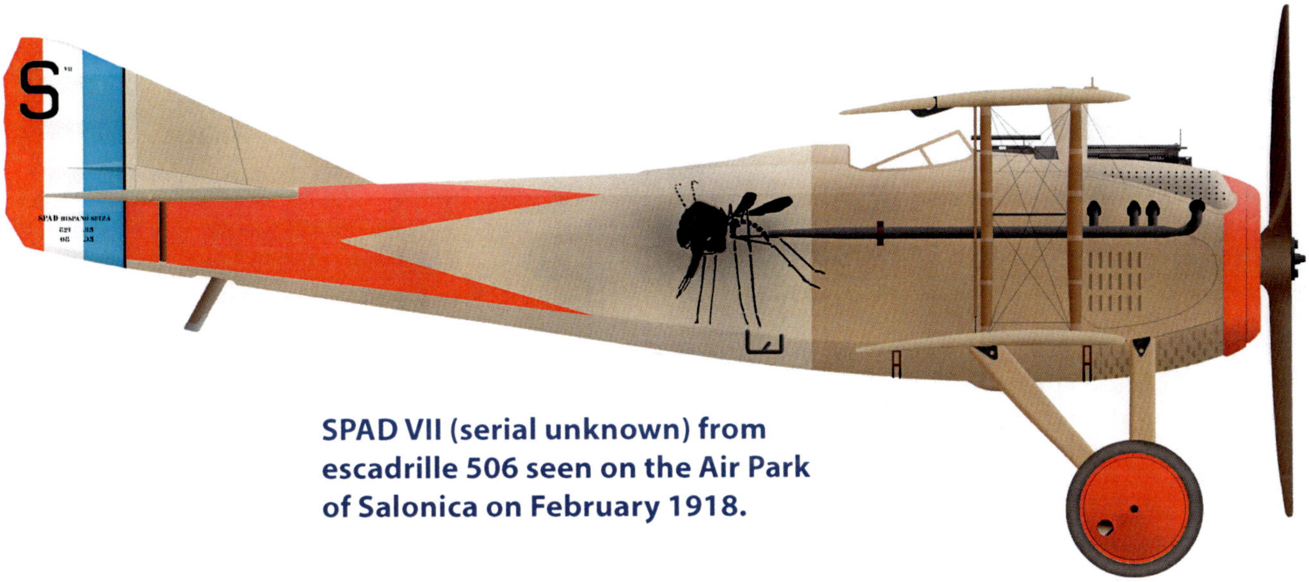

SPAD VII (serial unknown) from
escadrille 506 seen on the Air Park
of Salonica on February 1918.

Escadrille 507

Escadrille N 391 had the cowling of its planes painted in green in late 1916, but this didn't last. In 1917 and 1918, the planes show no insignia at all.

Created as N 91 S on September 1915, renamed N 391 in June 1916, then Escadrille 507 since June 1917.

Locations:

- Lyon on September 29, 1915 (creation of the unit)
- Salonica on November, 1915.
- Cote 262 – Road from Krivolak to Kavadar on November 12, 1915.
- Guevgueli on December 1, 1915.
- Negorci on December 2, 1915.
- Topcin on December 10, 1915.
- Negocani on April 15, 1917
- Lembet (NE Salonica) on July 30, 1917.
- Galatz (Romania) in 1919.

Commanding Officers:

- Cne Pierre Cahuzac on September 29, 1915.
- Cne Jean Potin on Februray 15, 1916.
- Lt Jean Callon on September 1917
- Lt Robert Nast on January, 1918.

Equipment:

- Nieuport 10 on September 25, 1915.
- Nieuport 11 an 21 (80hp) on June 1916.
- Nieuport 23 (110 hp) on March 1917.
- SPAD VII on March 1917.

The unit was also given one or two two-seaters for liaison missions – Caudron G.3, Caudron G.4, Sopwith 1A2, Farman F.40, or Dorand AR1.

Victories: 8.5 confirmed victories + 2 unconfirmed

Casualties: 4 pilots lost from enemy action (3 killed, 1 prisoner) + 1 by accident.

History:

Just like the C 89 S and V 90 S, escadrille N 91 S (S for Serbia) was created in order to help Serbia, which had been fighting bravely against its powerful Austro-Hungarian neighbor since the early days of the war. N 91 S was assembled in Lyon on September 29, 1915. It was a fighter squadron with six pilots flying on two-seater Nieuport 10s, and its command was entrusted to Lieutenant Pierre Cahuzac, a 29-year-old officer from Toulouse who had already fought on the front in the MS 26 squadron. After embarking in the port of Marseille on the steamer "*Amiral Olry*", the unit disembarked in Salonica on November 2, 1915 and immediately participated in the Vardar expedition to try to rescue Serbia, but the country was already collapsing under the blows from central powers armies.

N 91 S received the order on November 8, 1915, jointly with the V 90 S, to go not far from Gradsko to the extreme north of the French positions, to Krivolak where the rolling echelons went by rail. As soon as they left the station, shells fell in the vicinity, fired by the Bulgarians from the northern bank of the Vardar... An airfield was then set up in safety further south, on a plateau called "hill 262" located on the road between Negotin and Kavadar, where the first missions took place on November 13, 1915, under snow and fog. The Nieuports encountered a few German aircraft in the air, but the weather limited their operational possibilities. On 1 December 1915, N 91 S withdrew to Guevgueli, then on 10 December moved to Topcin near Salonica, where it remained for almost a year and a half to ensure the air defense of the city.

While the Entente's ground forces were entrenched in Salonica, which itself was located in a country that had displayed its neutrality in the conflict, the only way for the German-Bulgarian troops, who remained stationed on the Greek border, was to harass them with their air force. Several raids were launched against the city in December 1915, and the Nieuports of escadrille N 91 (whose "S" disappeared from the official documents) distinguished themselves and achieved some successes in the first half of 1916.

After several skirmishes, the first victory of the unit was achieved on February 1, 1916, when Salonica was bombed during the night by Zeppelin LZ 85. A German plane was seen flying over the Topcin airfield at 12:15 am. The Nieuport of Sergeant Terme and Sergeant Astor (machine

Above: The Nieuport 10 n°369 is one of the aircraft from the initial dotation of escadrille N 91 S in September 1915. It underwent two major repairs at Salonica Air Park on February 19 and June 28, 1916.

gunner) took off and climbed to altitude by following it over Salonica, then on the way back towards the west of the city. At 12:45, they are 100 meters behind and below their pursuer, allowing Astor to open fire with his Lewis gun. Good shot, Astor hit the German in the tank and he was forced to land south of the village of Palihor, in the area controlled by the Greek military, which was a buffer between the French and German-Bulgarian troops. The two French airmen landed near their victim and took the two German airmen prisoner. The affair became more complicated when Greek soldiers rushed to the scene and take French and German prisoners. Astor succeeded in negotiating his release, he ran to inform his superiors in Topcin who set in motion a whole detachment of French cavalrymen, who obtained *manu militari* from the Greeks that they were entrusted with planes and pilots. The German aircraft was immediately displayed in front of the white tower of Salonica, coming at the right time to restore the image of the air force, which was reproached for not having been able to oppose the night raid of the Zeppelin which had caused serious damage to warehouses. Terme and Astor became famous among the Army of the East as well as in the French press, where the newspapers were happy to talk about their exploit. They would have others...

On February 17, 1916, it was the turn of the Nieuport piloted by Sergeant Fétu with Slt Minteguiaga as observer to shoot down a German plane in the Allied lines. If the German crew took the key to the field, the aircraft was captured and was exposed for four days in front of the white tower. On March 27, 1916, the Germans launched a mass raid on Salonica at dawn with 12 aircraft from FA 30 and four from FFA 1. The alert was given to escadrille N 91, whose six aircraft took off at 5:15 am and attacked the enemy individually, which they spotted without difficulty by following the black clouds caused by the flak. The other squadrons of armée d'orient also took off planes and several clashes took place; for its part N 91 succeeded in two interceptions. The first one by its squadron leader, Slt Potin, with Private Ponnelle as machine gunner, who confronted a German aircraft which was not confirmed. More skillful was the Nieuport of the crew of Sgt Terme and Sgt Astor, who emptied three and a half rolls of Lewis machine gun onto their prey, which fell into Lake Amatovo where its remains were recovered. The victory (their 2nd) was contested by the Voisin LAS of Ltt Mézergues of the V 90 and was later officially confirmed in common to the two crews, plus one more from GBO (Adjudant Bonnard and S/Lt Prélejean). Astor and Terme did it again on April 25 by shooting down another German two-seater that made a forced landing east of Lake Ostrovo, the crew fleeing. The plane was seized by a company of French fighters and brought back to Salonica where it was displayed on May 1st and was widely photographed, even being published as a postcard... The two airmen were promoted to the rank of second lieutenant.
On 27 May 1916, the Greek army evacuated Fort Rumpel without fighting on the orders of its King, thus opening

Above: Taken from a French magazine of the time, the photo of Nieuport 10 n°577 of Sergeant Alain Terme and his gunner Sergeant Henri Astor, who together won three aerial victories in early 1916.

the door to Thrace to the Bulgarian army which invaded the region and clashed with the Franco-British troops who left their perimeter, the Front d'Orient thus becoming a continuous front. On June 6, 1916, N 91 took part in a long-distance escort of the Voisin bombers of the GBO, which carried out their last sortie by bombing the Bulgarian city of Petric. The French army carried out an offensive towards the west towards the city of Monastir, causing a general commitment of the air force which was reorganized.

Most of the squadrons of Orient would be attached to the French Aviation of Orient (AFO) in the sector of Monastir, while General Sarrail would keep the hand on the air forces of the Allied Command of the Armies (CAA) in Salonica in charge of the defense of the city and the missions of great recognition. Escadrille N 91, guardian of Salonica, was in fact attached to the CAA with the MF 85 squadron and the Air Park detachment. If the enemy incursions on the city were reduced during the summer, the activity of the Nieuports did not weaken with numerous reconnaissance flights launched regularly along the Vardar railroad (up to Gradsko) as well as on Fort Rumpel. In June, N 91, renumbered N 391, received its first single-seat Nieuport 11 fighters, then in autumn the Nieuport 21s which became the standard fighters of the Front d'Orient: small, maneuverable, but without much power or armament. The two-seaters were still used for long-range reconnaissance missions, but the Nieuport 10s were replaced by the more powerful Nieuport 12s. It was precisely during such a two-seater reconnaissance mission along the Vardar valley that on October 23, 1916, the crew of Sergeant Major Emile Sicot and Lieutenant Jean Leune shot down an enemy two-seater that was seen to fall into a spin near Hudova – the victory was certified by a prisoner's account confirming the crew's death. Six days later, on October 29, 1916, Sicot flew with the best machine gunner of the squadron, 2nd Lt Astor, for a new reconnaissance along the Vardar. On the way back, a German plane had the misfortune to cross their path and Astor's Lewis shot it down near the Doiran lake, thus winning his 4th confirmed victory and making him the best shooter of the Armée d'Orient.

On November 19, 1916, a special mission took place for the squadron, consisting of the drop-off in enemy territory of a saboteur charged with blowing up the Vardar railroad in the vicinity of Grasko. Sergeant Lambert took charge of the mission and landed his Nieuport in a field, but his spy, a certain Dimo Christo, tried to leave the Nieuport before it came to a complete stop, and weighed down the aircraft so much that one of the wheels of the landing gear became bent. Lambert could not take off anymore and decided to try his luck on foot after having sabotaged his plane. Spotted by Bulgarian irregular troops, the two men separated on

Above: View in March 1916 of aviators of escadrille N 91 in front of one of their Nieuport 10s. From left to right, standing: Lt Vuillet (observer), Lt Potin (pilot and squadron leader), Lt Astor (observer), Lt de Sant Andrea (observer), Adj Chemet (pilot) and Lt Terme (pilot). Seated: Lt de Menteguiaga (observer), Sgt Brot (pilot), Adj Fétu (pilot), Adj Fenech (pilot). (Guillerez collection)

their own but Lambert was captured and sent to captivity in Germany from where he escaped through Holland at the beginning of 1918 and returned to his squadron... The bad luck was still there for the squadron because the next day, Second Lieutenant Astor was the victim of a jamming of his Lewis machine gun which deprived him of his 5th victory against a German who fled without asking for his rest. The German fighter force was reinforced at this time because the first fighter unit in Macedonia, Jasta 25, was created on November 28, 1916. It was a Nieuport 12 of the 391 squadron that paid the price on December 10, 1916, during a mission over Hudova along the Vardar, where Leutnant Bodo Von Lynker opened the record of his unit by killing Sgt Guillot (pilot) in aerial combat, while the French observer, Slt Soulié, survived the crash seriously wounded and ended the war in a prison camp.

If the winter freezes somewhat the flights and the hostilities, the awakening is brutal for the pilots of escadrille N 391 with the arrival in Hudova of a squadron of heavy bombers, Kaghol 1, which was going to sow the seeds of war on the whole Orient Front as of its arrival around February 15, 1917. Using powerful twin-engine planes of different makes (Gotha, Friedrichafen, Rumpler, AEG), the German unit carried out its raids, which first hit the airfields around Salonica with relative impunity, because of the high altitude at which they flew. It was panic in the Vardar sector: Gorgop's 385 squadron was pulverized on the ground on 27 February 1917 and, in the middle of the afternoon, the German twin-engine planes directly targeted the city of Salonica, on which they dropped their bombs, killing or injuring nearly 330 people. Due to a lack of documents, the activity of the N 391 squadron is not precisely known. The squadron did not claim any victory, but carried out several interceptions against the Gotha: the testimony of a German pilot of the Kaghol 1, the Leutnant der Réserve Braun, mentions a bomber that was heavily damaged during the raid by the fire of four French planes and that broke up on landing at Hudova.

Escadrille N 391 tried to do everything possible to counter the enemy. The efficient equipment that had been requested for so long finally arrived from France: 110 hp Nieuport 23s replaced the very slow Nieuport 21 80 hp, and a first SPAD VII was even delivered to the squadron. However, these reinforcements arrived very slowly and a two-seater Sopwith 1A2 was used as a fighter, even though it was not specifically designed for that role. Astor with his Lewis would be the gunner. A young daredevil just transferred to the squadron, a certain Adjudant Dieudonné Costes, was

Above: The 3rd victim of Terme and Astor, a German two-seater shot down on April 25, 1915, brought back to Salonica by French soldiers to be displayed on the place of the White Tower.

assigned to him as pilot... On March 18, they faced four twin-engine fighters over Mahukovo: Costes' front machine gun jammed and Astor emptied his weapon on four different targets without causing any apparent damage. After a second unsuccessful fight on March 24 over Pardovica, Costes' Sopwith expedition was cut short on April 1 when he and Astor, while leaving to bomb the Hudova bomber base, were intercepted north of Gjevgueli by two Halberstadt fighters. A furious fight started in which Costes' machine gun jammed from the beginning, leaving Astor alone to shoot. The Sopwith, seriously damaged, managed to return to its base, but in such a bad state that it was soon discharged.

Due to the lack of two-seaters to use, Astor was transferred to France, where he went to flying school. His career as a fighter pilot was less brilliant than his exploits as a machine gunner, as he was shot down and captured on December 15, 1917 near Juvincourt, less than a month after his transfer to SPA 80. Remaining in the Orient, Costes was assigned a Nieuport 23 (110 hp) armed with two machine guns, on which he quickly proved himself. Commander Denain, who had obtained some reinforcements from Paris, reorganized the fighter force by trying to concentrate it against enemy bombers and by coordinating it with the British air force. After having attacked targets in the British sector and on the Vardar, Kaghol 1 now attacked targets located to the west of the front in the AFO sector: Denain ordered N 391 to leave to reinforce the N 390 squadron on 15 April 1917 on its field of coast 619, near Negocani, in a fighter group whose command was entrusted to Lieutenant Potin. At Salonique-Topcin, a detachment of five fighters led by Lt. Noël and composed of a SPAD VII, two Nieuport 23, and two Nieuport 21 will remain, but they will hardly see any activity.

The bulk of the squadron left for Negocani and soon saw action against the German bombers and fighters, who did not escort the latter but took care to clear their route of any opposition. On 24 April 1917, the fighter group pursued the bombers in the direction of Prilep but could not join them – another battle took place during the day in which Slt Cousinéry claimed a twin-engine plane that had flown into a cloud but was not confirmed. On 29 April, the French fighters had their first confirmed success in a raid on the village of Skocivir (south bank of the Cerna loop), on which nearly 2300 kg of bombs were dropped. This time, the Nieuports were in a good position to intercept the Gothas, but first had to deal with the Black Cross fighters. MdL Christmas, of N 391, managed to slip his Nieuport 21 between the fighters and join a bomber on which he emptied – without result – the 100-round magazine of his Lewis gun. Adjudant Costes, on his Nieuport 23, was luckier (and also better armed) and shot a bomber which fired a distress flare and continued on its way while losing altitude. Costes was then caught by two fighters and had to let go of his prey. The success of the fighters was, however,

Above: Pilot of escadrille N 391, Lt de Saint-Severin is photographed in front of his Nieuport 21, probably visiting the N 390 squadron in Boresnica in the autumn of 1916.

relative, as on the ground 12 men and 50 horses were killed or wounded... and Slt Cousinéry, who was reported missing with his Nieuport during the fight. May 1st 1917 saw the last battle of the twin-engine Kaghol 1, which bombed the village of Batch where a French squadron was stationed, losing two Farman F.40s on the ground and its fuel depot. The French fighters, led by Costes, succeeded in intercepting the aircraft and engaged in combat, which caused no losses on either side. Kaghol 1, whose aircraft were vulnerable on the ground in Hudova, which started to be attacked by the Franco-British air force, was then transferred by the German staff to Flanders to carry out raids against England, after having led nearly two and a half months of raids that had shaken all the rear of the front.

N 391 would, however, remain some time on the sector of Monastir where its action would allow to regain the air superiority on the local German aviation including Jasta 25 stationing in Prilep. Four fighters flew over the N 391 airfield on May 3, 1917: Adjudant Costes approached them from behind and below, and, with his wing Lewis gun, shot at point blank range one of the fighters which left in a vertical dive, the others escaping into the wild. If the victory would not be confirmed to Costes, he would have on the other hand the rare privilege to be entrusted with the SPAD VII n°1083 with which he will take part in the last missions of the N 391 on the sector of Monastir, consisting in patrols and barrage flights.

N 391, renumbered escadrille 507 in June, returned to its base in Topcin and conducted several battles against the German air force in the Vadar sector, in particular on July 26, 1917, when Adjudant-Chef Costes surprised an enemy aircraft north of Guevgueli and shot it down with the wing gun of his Nieuport. The victory was not certified, although it was mentioned in the bulletin of the Armée d'Orient.
It should be noted that this victory was not obtained on a SPAD: Costes broke his n° 1083 on July 8, victim of the in-flight detachment of the canvas of the upper plane, from which he escaped at the cost of a forced landing.

On July 30, 1917, escadrille 507 would leave the Topcin base to settle not far, in the north-east of Salonica, on the Lembet airfield. This is where it would end the war, in charge of defending the city against the enemy air force, which would hardly ever show its nose there again. In order not to leave the fighters unused, they were frequently used for offensive patrols on the northern front, or sent to reinforce other squadrons, as was the case around August 15, 1917,

Above: The Nieuport 10 n°628 of sergent Lambert, captured in the Bulgarian lines after a special mission on November 19, 1916. The cowling is painted in green.

when the 507 left for Monastir to reinforce escadrille 506 squadron with a contingent of British fighters. On this occasion, a fighter of 507, piloted by MdL Christmas, obtained an unconfirmed aerial victory against an enemy fighter which fell on its field at Kanatlarci. Back in Lembet, new SPADs were delivered to them, but the glue of the canvas covering did not appreciate the climate of Salonica, since Dieudonné Costes was victim of a new accident on August 29, from which he escaped again by crashing on Mount Karakorum near Salonica...

September saw the departure of Lt. Potin, who was replaced at the head of the squadron by Lieutenant Jean Callon. On September 3rd, MdL Moine, during a mission on the front, claimed an enemy two-seater that was seen falling into the British lines. The wreckage was not found and the victory was not certified. It was also during this month that two pilots who were to become Dieudonné Costes' favorite crew members, MdL Maurice Lashermes and Sergeant Paul Andrillon, arrived at the squadron.

The autumn and winter of 1917–1918 passed without any significant events other than the loss in combat of MdL Lefresne who fell near Lake Doiran on November 28, 1917, shot down by the Vicefelwebel Lage of Jasta 38. January 1918 saw gasoline become scarce and flights limited. Lt. Callon gave up his place at the head of the squadron to Lt. Nast, a former 506 pilot who was transferred there with MdL Hébert, a confirmed pilot. On January 27th 1918, the two men won the last success of the squadron against a DFW reconnaissance plane that ventured over Salonica at an altitude of nearly 4,400 meters. A Nieuport patrol already in barrage was able to pass it and fire a few bursts before being outrun. The rest of the squadron took off, but only the two SPAD VIIs, piloted by Nast and Hébert, were able to join him in the company of a British aircraft and accompany him back to his lines by emptying their magazines on him. The enemy, seen stinging strongly towards its base, and having received nearly 1000 rounds of ammunition, was certified only to Nast.

March 22, 1918 saw the creation of a new fighter squadron in which the best pilots were assigned, the 531 squadron or 2nd Hellenic squadron, which settled in Gorgop and took charge of the entire Vardar sector, and which quickly achieved some success. During the attack of the Greek army on the mountain of Serka Di Legen, the French fighters were solicited for missions of ground attack on the Bulgarian fortified positions and the 507 squadron would deploy several aircraft there. Sergeant Gollion's plane had the misfortune to be hit by a direct flak fire – the pilot, with his foot torn off by the enemy shell, found the strength to land in the front lines, from where he was promptly extracted and taken to a field hospital where he was saved. July 17, 1918 saw the last loss of the squadron, Lt. Bordes, who accidentally crashed during a flight at Dimitric, east of the front where the Hellenic 533 squadron had just been installed and for which he was probably to provide an escort.

When the final offensive of the allied forces started on September 15, 1918, escadrille 507 was equipped with 11 SPADs, but they were not used while guarding Salonica against an enemy that had much better things to do. Some of its aircraft left to assist the troops on the front line for ground attack missions. The offensive led by the French cavalry through the mountains previously conquered by the Serbian troops penetrated far into the enemy lines and cut the Bulgarian troops in two, leading this country to ask for an armistice, and allowing the Allied troops to penetrate deeply into the Balkans. We then find traces of the escadrille 507 and its 12 SPADs with the French army of the Danube in Romania under General Berthelot, where it depends on the Galatz park. However, it was not engaged against the Bolshevik troops and then disappeared from official documents, probably disbanded when the French troops returned from this country. Thus ends a campaign of more than three years in which the squadron will have won 8.5 certified victories and 2 probable against three pilots killed in action and one prisoner.

Right: Sergent Paul Pavelka (center), American pilot and veteran from Escadrille Lafayette in France, served with Escadrille 507 in Salonica during 1917. He is seen there in October 1917 with Sgt Paul Andrillon (right). One month later on November 11, 1917, Pavelka died in an accident while riding a horse. (Collection Benoit Henriet)

Above: Escadrille N 391 at Topcin in March 1917, after having received Nieuport 23s with 110 hp engines. We note that the aircraft do not bear any particular marking.

Above: Dieudonné Costes, ace of the aces of the Front d'Orient (left), in front of the Nieuports of his squadron at the end of 1917. (Collection Benoît Henriet)

Above: Take-off on alert of the aircraft of the 507 squadron, early 1918. In the foreground, a Nieuport 23, while a Nieuport 24 can be seen in the background. (Collection Benoît Henriet)

Above: From February 7 to 12, 1918, King Alexander of Greece visited the large Salonica air park and reviewed the aircraft of escadrille 507. He is seen in the center of this photo, accompanied by general Guillaumat (left), commander-in-chief of the Front d'Orient, as well as Cdt Denain (right), chief of the Air Force. (Collection Benoît Henriet)

Above: Camouflaged Nieuport 24 of Sergeant Paul Andrillon, 507 squadron, early 1918. (Collection Benoît Henriet)

Facing Page: Paul Andrillon explains in these two photos the controls of his Nieuport fighter. (Collection Benoit Henriet)

Above: Nieuports of the 507 squadron in spring 1918. (Collection Benoît Henriet)

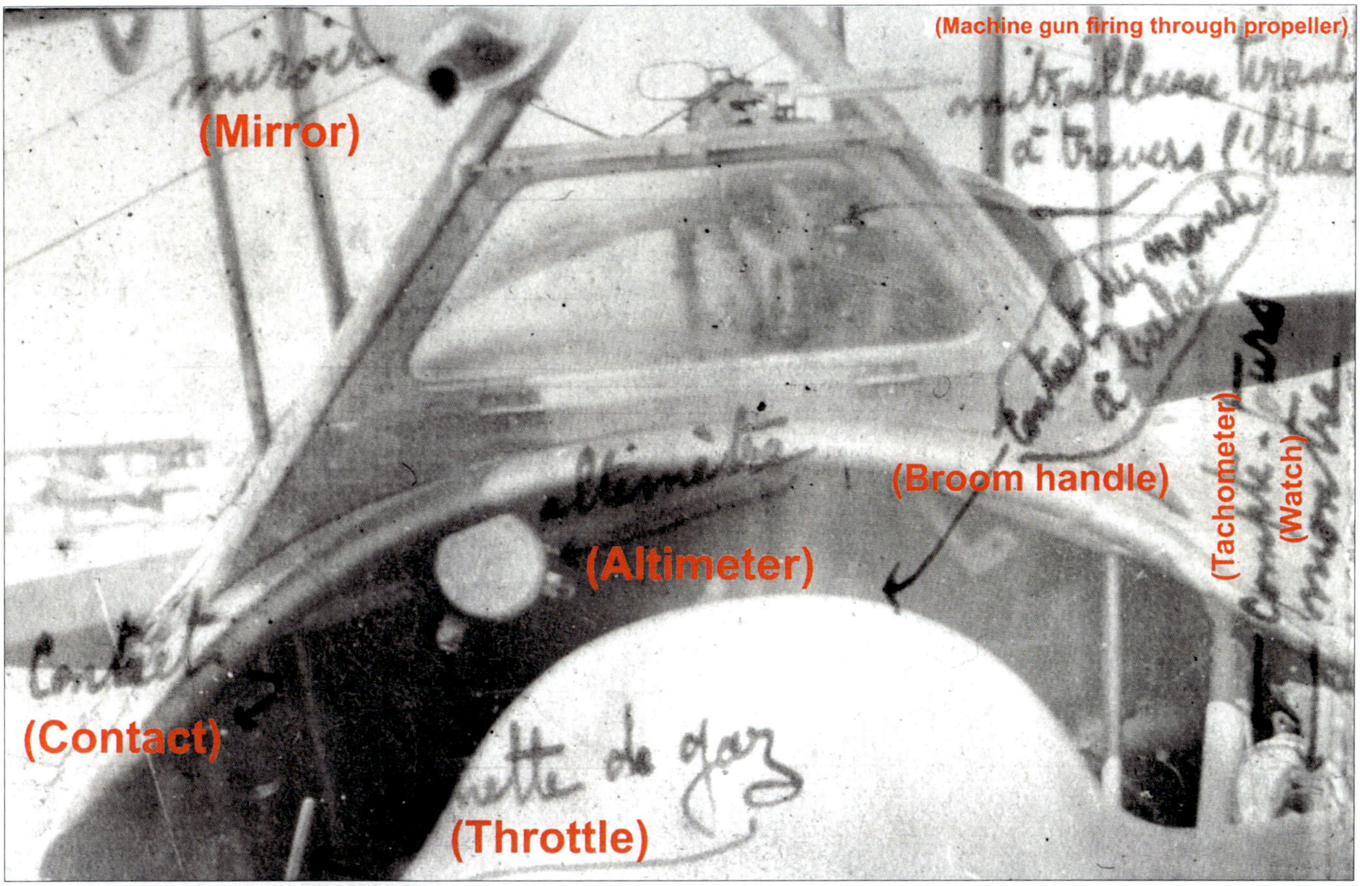
(Machine gun firing through propeller)
(Mirror)
(Broom handle)
(Tachometer)
(Watch)
(Altimeter)
(Contact)
(Throttle)

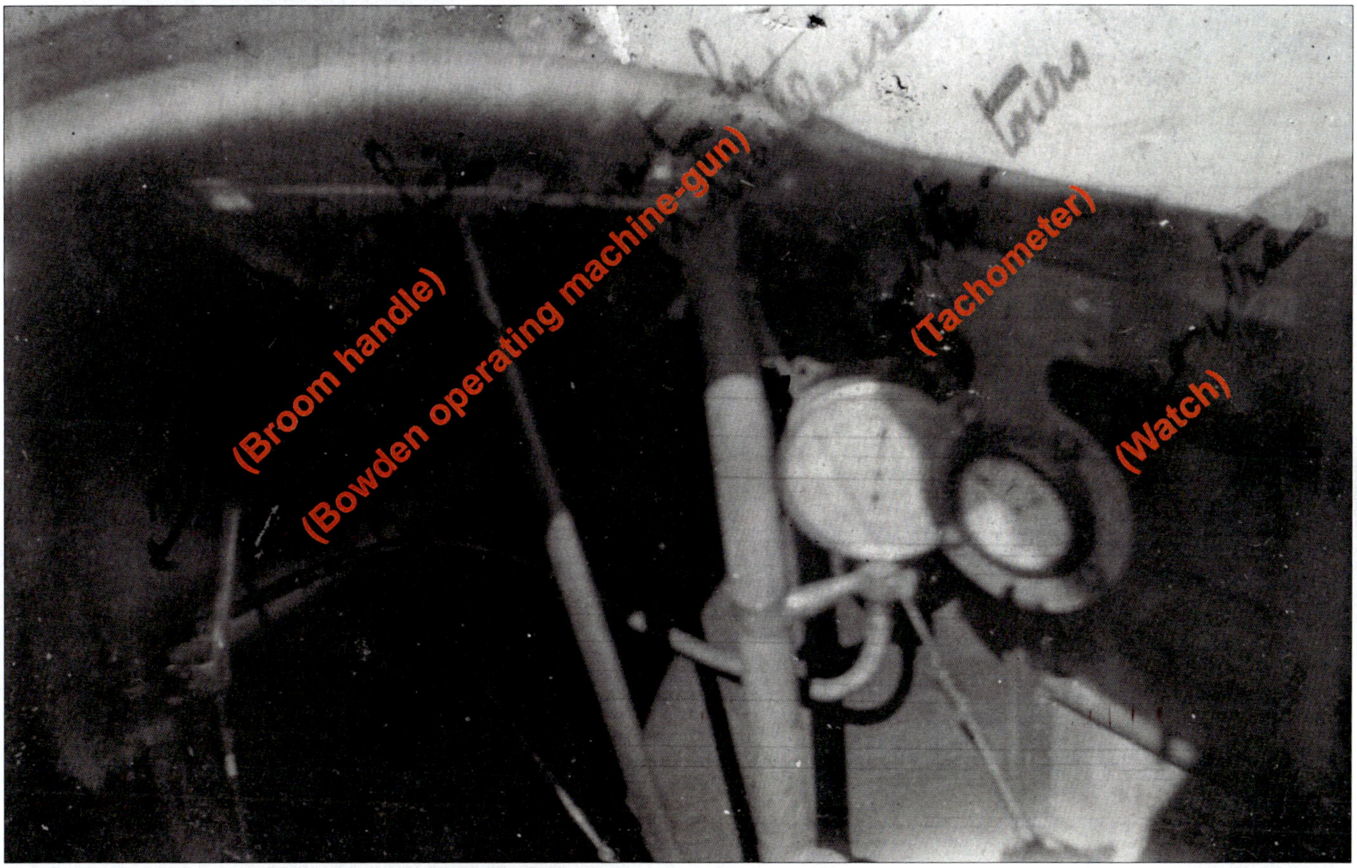
(Broom handle)
(Bowden operating machine-gun)
(Tachometer)
(Watch)

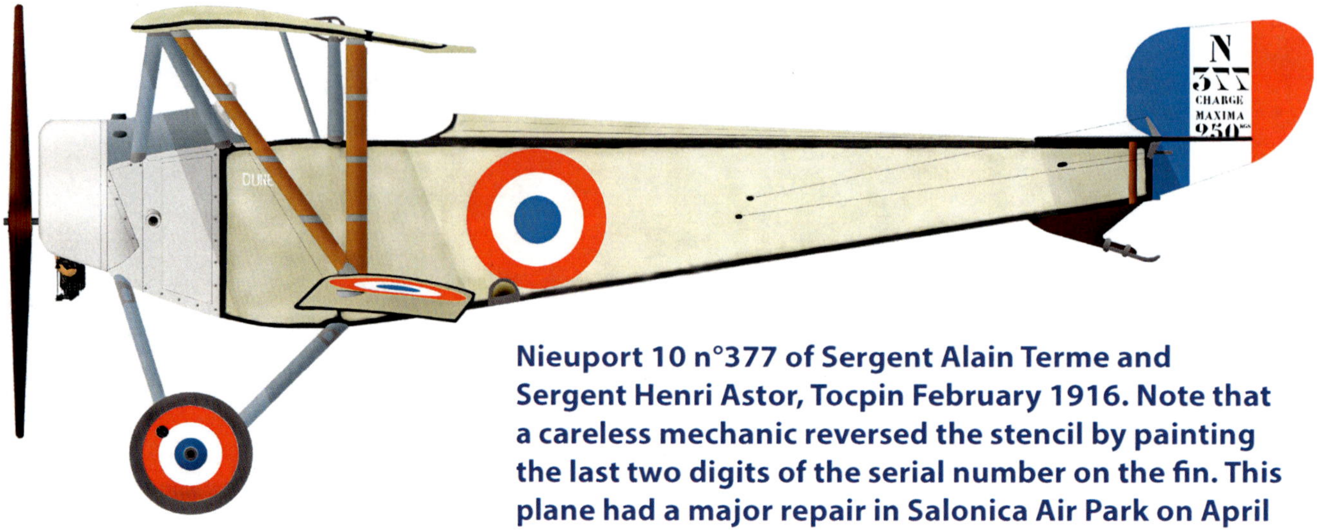

Nieuport 10 n°377 of Sergent Alain Terme and Sergent Henri Astor, Tocpin February 1916. Note that a careless mechanic reversed the stencil by painting the last two digits of the serial number on the fin. This plane had a major repair in Salonica Air Park on April 12, 1916. It was discharged on May 17, 1916.

Nieuport 21 n°2211 of Lt de Saint Severin, escadrille N 391, Automn 1916.

Nieuport 10 n°628 of sergent Lambert, escadrille N 391, November 1916.

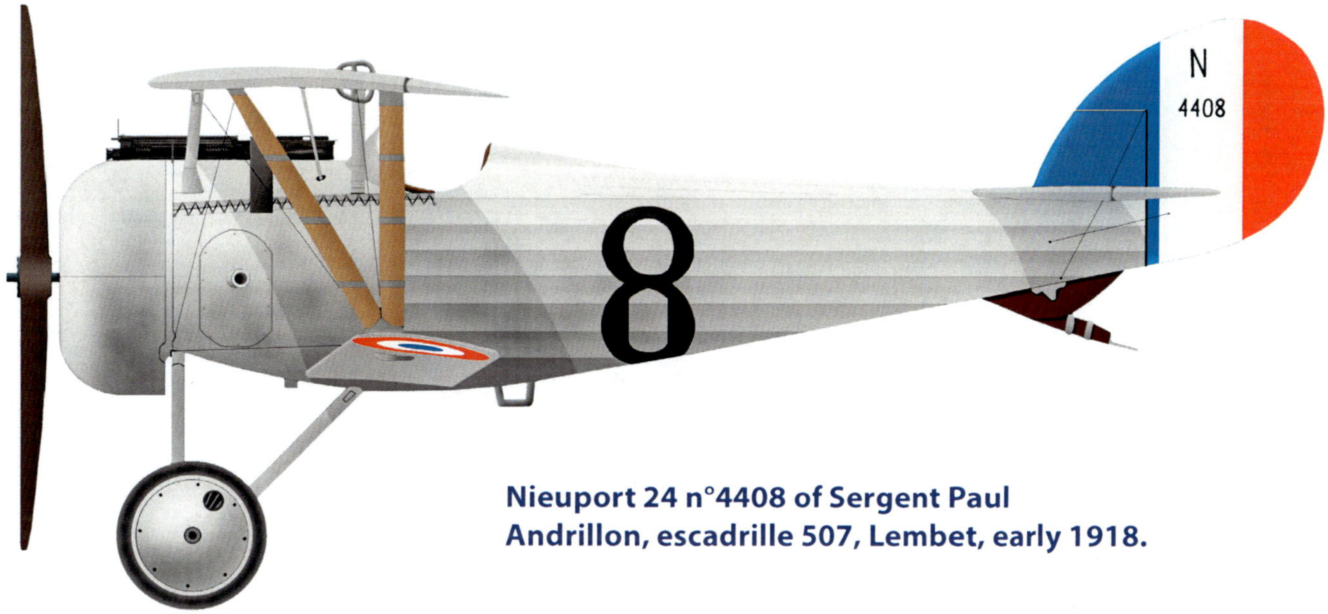

Nieuport 24 n°4408 of Sergent Paul Andrillon, escadrille 507, Lembet, early 1918.

Nieuport 24 (serial unknown) of Sergent Pierre Lefresne, escadrille 507, Lembet November 1917. Lefresne was shot down in aerial combat on November 28, 1917.

Escadrille 508

Created as Escadrille F 383bis in September 1916, then Section d'Artillerie Lourde (heavy artillery section) n°1 in December 1916, then Escadrille 508 in June 1917.

Locations:

- Kladerop on September 1916 (creation of the unit)
- Kravari (S. Monastir) on December 4, 1916.
- Batch in early 1917.
- Novi Sad (Serbia) in 1919.

Commanding Officers:

- Lt de Lavalette on September 1916.
- Cne Frédéric Ruby on June 1918.
- Lt Albert Picot on October 5, 1918.

Equipment:

- Farman F.40 or F 60 on September 1916.
- Dorand AR 1 in June 1917.
- Breguet 14 A2 during year 1918.

Escadrille 508 also had a fighter section of 2–3 Nieuport fighters (type 23, 24, or 24bis).

Victories: 1 confirmed victory + 1 unconfirmed

Casualties: 6 pilots lost from enemy action (4 killed, 2 prisoners)

History:

The creation of this squadron dates back to September 1916, with the detachment of a section of a few Farman F.40 or F.60 from escadrille F 383 (501) intended to specialize in heavy artillery spotting, under the command of Lieutenant de Lavalette. It was then called indifferently in the official documents "Lavalette Detachment", "Farman Army Detachment", or "Escadrille F 383bis". It shared the same airfield as escadrille F 383 at Kladerop (north of Florina), but a document dated December 4, 1916 placed it on the advanced airfield of Kravari, south of Monastir, which was common to all the squadrons of the French Orient Army. Limited to five aircraft and four pilots, the small unit took part in the offensive launched on this front in November 1916. While its aircraft were certified to have had the insignia of the crossed guns as early as November, a first document officializes it on December 22, 1916 as "Section d'Artillerie Lourde n°1", a denomination that will be constantly used afterwards.

One of its aircraft was destroyed by a tornado on January

Above: Farman F.40 n°3070 of the heavy artillery section n°1, destroyed in its hangar following the storm of January 1st 1917.

1, 1917 and a document dated March 13, 1917 mentions three operational Farmans. At an undetermined date in 1917, the squadron migrated with escadrille F 383 to the Batch airfield (still in the Monastir sector) and remained there when F 383 was deployed to Koritza. Renamed Escadrille 508 in June 1917, it effectively reached the size of a full escadrille and received AR 1s, then Breguet 14s, not to mention a handful of Nieuports for escort. It shared its missions between artillery spotting and sector reconnaissance, as well as, more occasionally, joint bombings missions with the other AFO squadrons. During the summer of 1918, the unit suffered heavily from German fighters who shot down three of its aircraft in six months.

The 508 squadron remained active after the armistice, being stationed in Serbia and then participating in the occupation forces of Hungary in 1919. The date of its dissolution is unknown, but most probably in 1920, when the French troops left.

Above: In June 1917, escadrille 508 was equipped with Dorand AR 1s (in the background) with an escort section on Nieuport fighters, such as this type 24 in the foreground.

Below: Nieuport 23 nicknamed "*Médoc*" (a French wine region, near Bordeaux) and decorated with the badge of escadrille 508.

Above: Dorand AR 1 of escadrille 508 in 1917–1918.

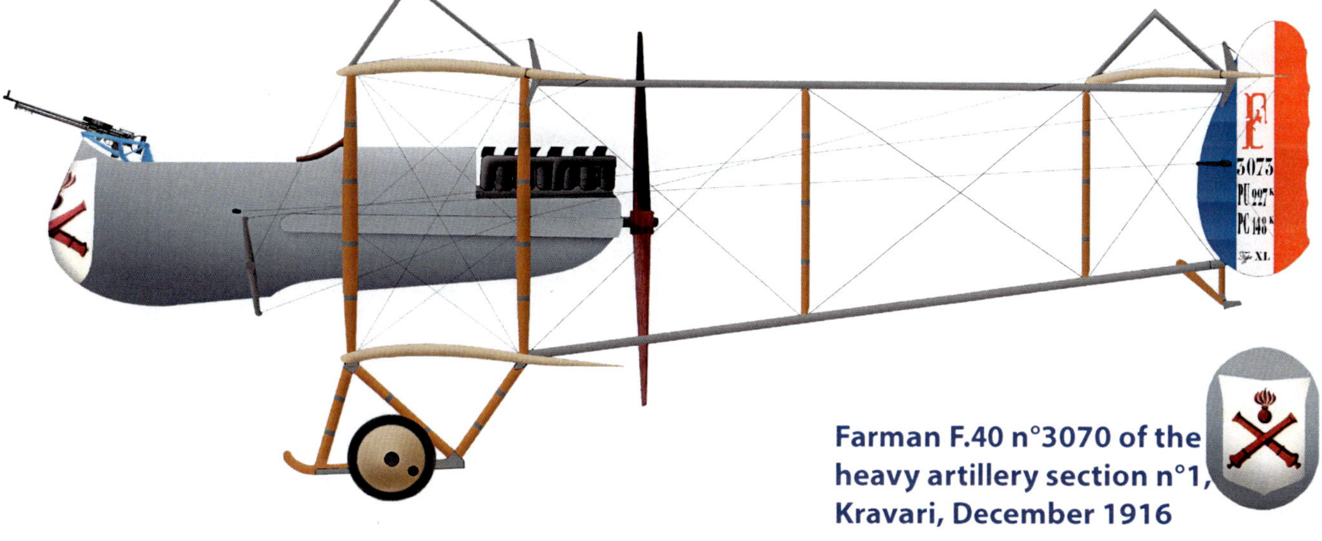

Farman F.40 n°3070 of the heavy artillery section n°1, Kravari, December 1916

Escadrille 509

1917

1918

Created as Section d'Artillerie Lourde (heavy artillery section) n°2 in December 12, 1916, then Escadrille 509 in June 1917.

Locations:

- Negotin on December 12, 1916 (creation of the unit)
- Negocani during year 1917.
- Constantinople in 1919.

Commanding Officers:

- S/Lt Choppin on December 12, 1916.
- Lt Beaune on December 17, 1916.
- Lt Gautier in 1919.

Equipment:

- Farman F 60 on December 1916.
- Dorand AR 1 in June 1917.
- Breguet 14 A2 during year 1918.
- Escadrille 508 also had a fighter section of 2-3
- Nieuport fighters (type 23, 24, or 24bis).

Victories: 1 confirmed victory + 1 unconfirmed

Casualties: 1 pilot lost from enemy action (1 killed).

History:

Escadrille 509 was created on December 12, 1916 as the Section d'Artillerie Lourde n°2 (SAL, or Heavy Artillery Section) of the French Armée d'Orient. It was based at Negotin, in the Monastir sector near the Cerna River, and its commander was Slt Choppin who was replaced five days later by Lieutenant Beaune who remained at this post for the rest of the war.

The SAL 2 was initially equipped with four Farman F.60, one of which was quickly destroyed and another damaged by the storm of January 1st 1917 which caused a lot of damage among the French aircraft stationed in the Monastir sector. On April 12, 1917, an enemy battery, positioned at Dobromir, opened fire on the squadron's installations. The shells followed one another very quickly and were unfortunately precise. Three shells hit the Bessonneau hangar of the unit and several planes, three Farman F.60 and one Nieuport, were destroyed. On April 23, the aircraft allocation was quickly completed by two Farman F.40 replacements (n° 3350 and 3351), to which two Nieuport escort fighters were added.

As its designation indicates, the SAL 2, escadrille 509 in June 1917, would carry out mainly artillery adjustment missions but will occasionally take part in grouped bombardments with the other squadrons of the sector, 501, 503, 504, 505, and 508. No losses in 1917 despite some fights caused by a German fighters more and more biting, the squadron re-equipping itself in Dorand AR 1 and moving at an undetermined date towards the end of the year on the other bank of the Cerna to Negocani. In 1918 the squadron received its first Breguet 14s and suffered one loss, one of its Nieuport escort fighters, shot down by the German ace Gerhard Fieseler on January 30, 1918, with another Nieuport pilot wounded in combat on July 31. Its missions multiplied

Above: A few Sopwith 1A2 were delivered to Aviation d'Orient in 1917 and one was given to escadrille 509, with serial number 2288 meaning it was produced by Darracq.

during the great offensive of September which saw the collapse of the Bulgarian defense.

However, the squadron was not disbanded after the armistice on November 1st 1918 on the Orient Front. It was maintained among the occupation forces of Constantinople (Istanbul), stationed at the San Stefano airfield with escadrille 505. The inactivity and the long isolation far from France caused discipline to be relaxed among the pilots and mechanics.

Most of the activity consisted of training flights, but also of postal links to Bucharest. On August 1st 1920, escadrille 509, under the command of Lt Gautier, was renamed 51st squadron of the 5th RAO. At that time, it had 11 aircraft, Breguet 14s and a handful of SPAD fighters. The French troops remained in Constantinople, observing the terrible war between the Kemalist Turks and the Greek troops in Anatolia... It was in this context that the only special mission of the squadron took place: on September 2, 1920, the Breguet 14 n°1349 piloted by Adjudant Courtet with Ltt Lesourd as observer carried out, in the middle of a fight, a remote photo reconnaissance mission on the town of Eski-Sehir at 2000 m altitude, in central Anatolia. This city is a railway junction where the road coming from Ankara branches off towards Constantinople in the north or towards the center of the country in the south. It was undoubtedly an important supply center for the Kemalist armies, and it is also the location of the first Turkish squadron, the 1nci Tayyare Bölüglü, gathering about ten heterogeneous aircraft of German design. The mission went smoothly and one cannot say if its results were communicated to the Greek government, at that time still led by Venizelos and in excellent terms with France.

The 51st squadron of the 5th RAO continued its occupation activities and its logbook was interrupted on May 5, 1921. Folded back in France, the squadron became the 13th squadron of the 36th aviation group of Pau.

Above: This low quality photo shows the second insignia of escadrille 509, with a dragon around the circle.

Left: Sergent gunner Eugène Jeanperrin of escadrille 509 in his Breguet 14 A2. Note a variation of the dragon, painted in white.

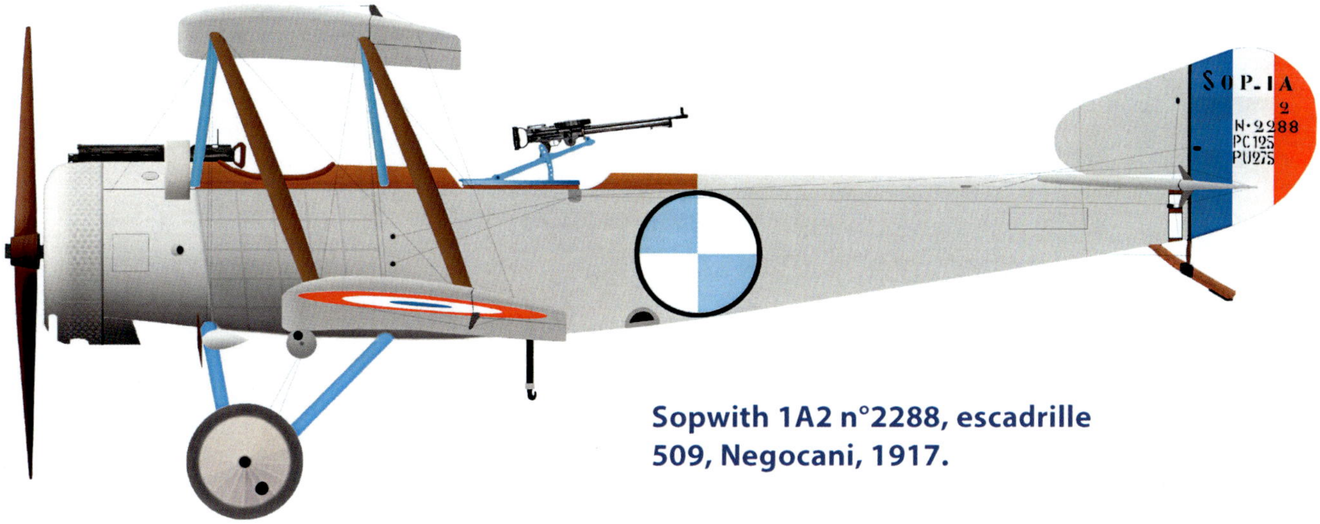

Sopwith 1A2 n°2288, escadrille 509, Negocani, 1917.

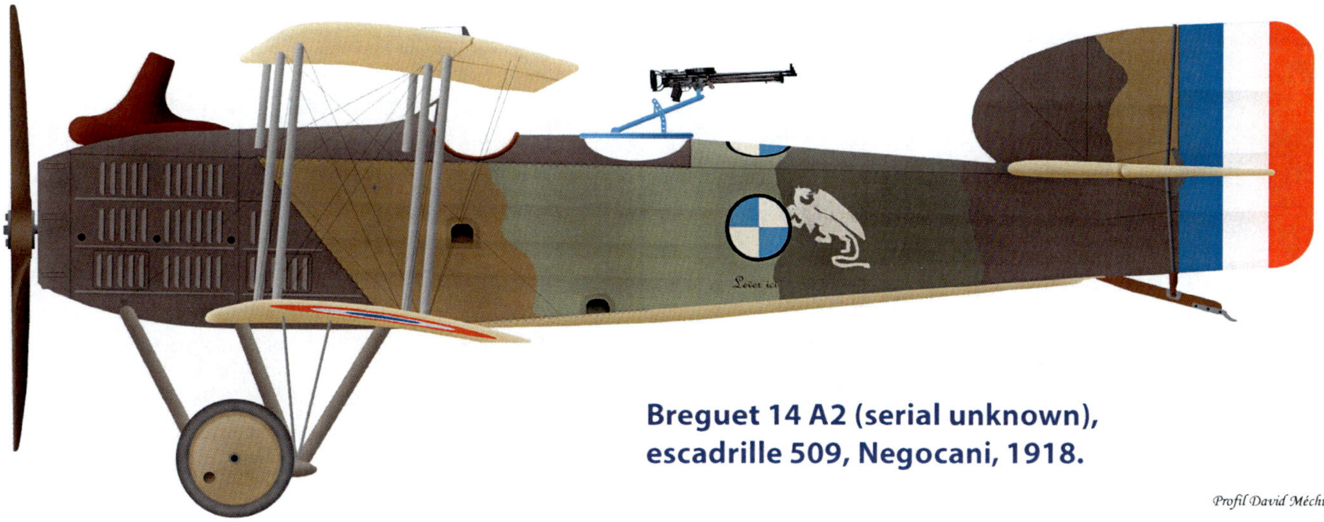

Breguet 14 A2 (serial unknown), escadrille 509, Negocani, 1918.

Profil David Méchin

Escadrille 510

There is no evidence that Escadrille 510 ever had a unit insignia.

Created as BR 510 in France, the unit was called "Escadrille 510" in all documents of Armée d'Orient.

Locations:

- Samli in December 1917
- Litchinew (Chisinau, Moldavia) in Early 1919
- Galatz (Galati, Romania) on July 10, 1919.

Commanding Officers:

- Lt Yvon Rose in September 1917.
- Lt Aimable Lemoine in June 4, 1918.
- Lt Raoul Bergeaux in 1919.

Equipment:

- Breguet 14 B2.
- Escadrille 510 also had one SPAD VII in 1919.

Victories: 2 confirmed victories + 2 unconfirmed

Casualties: 1 pilot lost from enemy action + 1 from friendly fire + 1 by accident.

History:

Since the dissolution of the Orient Bombardment Group (GBO) in June 1916, the Orient Army never had an efficient bombing air force, despite the creation of the short-lived squadron called "section de bombardement" on Farman F.40 in December 1916 and the use of Farman aircraft of the Orient Army squadrons in a grouped manner. The lack was particularly glaring at the beginning of 1917, when the German twin-engine bombers of Kaghol 1 sow terror on the rear of the front: although their base is quickly identified as being in Hudova, all that the French army can send to try to destroy them on the ground are a few Sopwith 1A2 and Nieuport fighters throwing a scattered handful of Brandt grenades without effectiveness...

General Sarrail eventually obtained reinforcements in terms of aircraft and personnel, including a bombing squadron on Breguet 14 B2s, escadrille BR 510, which was formed in Clermont-Ferrand during the summer of 1917. Its personnel left Aulnat in early October 1917. The precise date of its arrival in the Orient is not known, but it was after November 28, 1917 when a document mentions it for the first time.

It set up on Samli airfield, a former GBO base, to begin its missions, which consisted of long-distance army reconnaissance as well as bombing raids: the performance of the Breguet 14s was perfectly suited to this task, and was unmatched by anything else the French Orient Army had ever had at its disposal. The Breguets were sometimes escorted by fighters, including Adjudant-Chef Dieudonné Costes, who was affected there in March 1918. On March 21, 1918, a spin caused the death of the squadron leader, Lieutenant Rose, who was buried with full honors by the aviators of Orient Army. Dieudonné Costes flew over the funeral convoy with his Nieuport. Lieutenant Lemoine took over from him, and the bombing missions carried out by

Above: View of the Breguet 14 flown by Lt Yvon Rose and his gunner Lt Pasteris, before March 21, 1918, when Rose died after a crash. (Collection Benoit Henriet)

the entire squadron on the enemy's rear were to multiply, without losses and with a certain degree of impunity, thanks to the speed of the Breguet 14s, whose crossfire from the rear gunners even enabled them to claim some aerial victories.

At the time of the final offensive launched in September 1918, the squadron was placed at the disposal of the Serbian army and multiplied its missions on enemy communication nodes, causing the only combat losses of the campaign on 16 September 1918, when a Breguet crashed behind the lines during a special mission (crew unhurt), and on 19 September 1918, when a crew was shot down by the German fighter ace of the Balkans, Gerhard Fieseler, who scored his 19th and last victory.

The Bulgarian army collapsed like a house of cards following the offensive and the armies of the entente (French, Serbian, Greek, Italian, British) rushed into the Balkans. Escadrille 510 was not disbanded shortly after the armistice, but was assigned to Romania in the Danube Army, where it was still equipped with Breguet 14s, but reinforced by a patrol of SPAD fighters, and was ordered to fire against Bolshevik troops. But discipline was loosened quite a bit among the French crews, who only aspired to return home. On July 3, 1919, a staff officer went on an inspection tour and discovered the squadron in complete disarray, with neglected equipment and officers living in town, where they only thought of having fun. General Franchet d'Esperey became angry and relieved the squadron leader, Lieutenant Bergeaux, of his command. Escadrille BR 510 squadron did not survive this incident for long: it disappeared from official documents and was probably disbanded shortly after the return of the French occupation troops from Romania on October 11, 1919.

Above: Two Breguet 14 in flight (Collection Boric Ciglic)
Below & Facing Page: Three photos taken during the raid over Gradsko on September 15, 1918, during the final offensive.

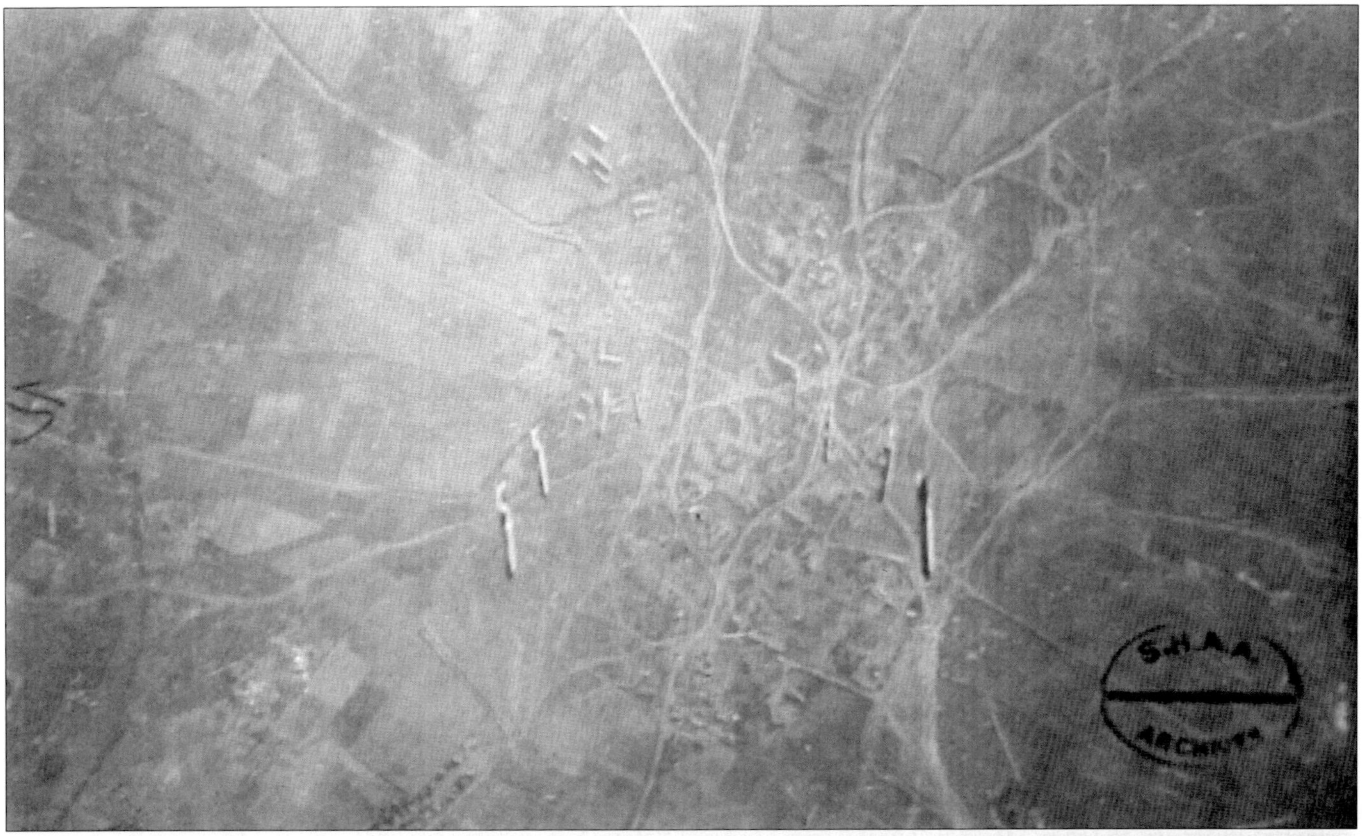
S.H.A.A.
ARCHIVES

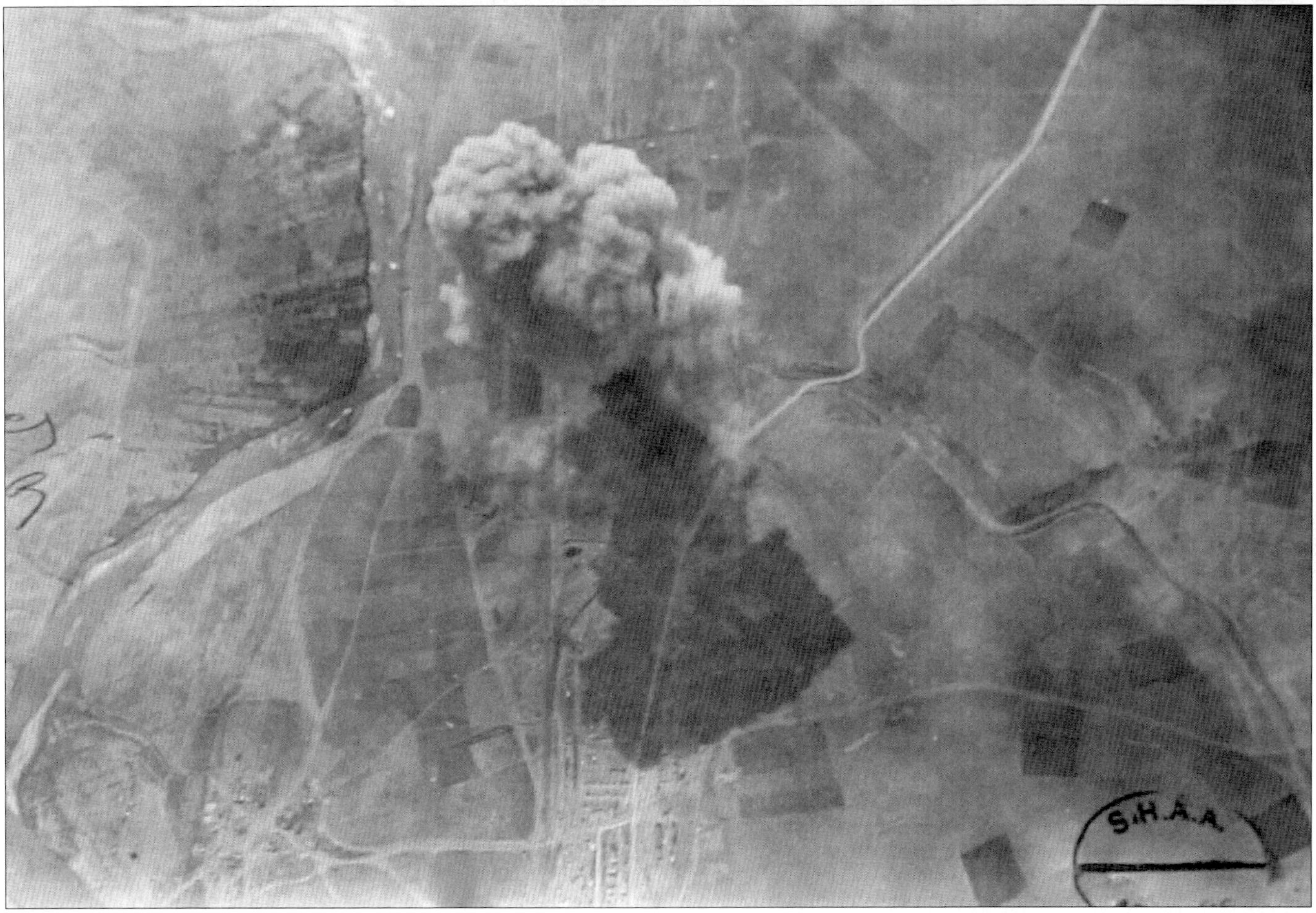
S.H.A.A.

Escadrille 521

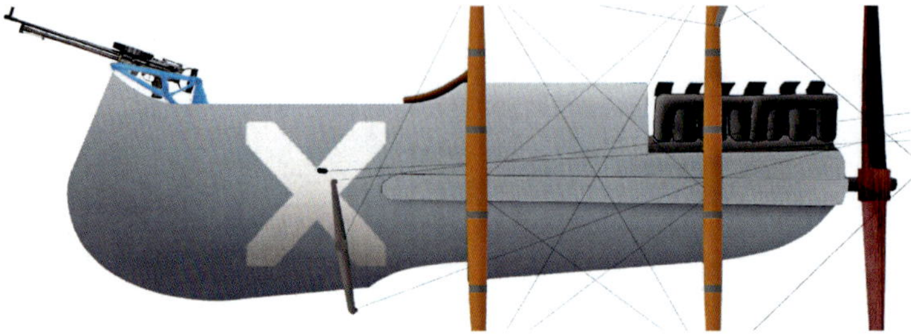

1917-01

LATE 1917

1918-03

Insignia: Escadrille MF 382 adopted in January 1917 specific markings – a white X on fuselage, and a blue chevon on wings. In late 1917 those makings disappeared and were replaced by a black panther on fuselage. After March 1918 this panther was replaced by a Serbian flag.

Created as French Albanian Squadron in January 1916, renamed MF.82 since June 22, 1916 and MF 382 shortly afterwards, then Escadrille 521 in June 1917. It was also called 1ere escadrille Serbe (1st Serbian Squadron) since March 6, 1918.

Locations:

- Potamos (Corfou, Greece) in February 7, 1916.
- Salonica-Sedes on May 30, 1916.
- Vertekop in August 6, 1916.

Commanding Officers:

- Lt Emile Prot in January 1916.
- Lt Marcel Cressole in November 30, 1916.
- Lt Chavannes de Dalmassy in August 19, 1917.
- Lt Michel Marinkovitch in January 16, 1918.
- Cne Nikola Žikić in June 8, 1918.
- Cne Branko Vukosavljević in September 8, 1918.
- Cne Živorad Petrović in Januray 1919.

Equipment:

- Farman MF 11bis in January 1916.
- Farman F.40 in late 1916.
- Dorand AR 1 in June 1917.
- Breguet 14 A2 during year 1918.

Escadrille 521 also had since late 1917 a fighter section of 2–3 Nieuport fighters (type 23, 24, or 24bis).

Victories: 1 confirmed victory + 1 unconfirmed

Casualties: 1 pilot lost from enemy action (1 missing) + 2 by accident.

History:
With the invasion of Serbia by German-Austrian-Bulgarian troops at the end of 1915, most of the Serbian army, accompanied by many civilians, left the country and reached the shores of the Adriatic through the mountains of Albania. The whole of this ragged army was evacuated by the French navy to the island of Corfu to recover and be reformed, before finally being sent to Salonica. It was in this context that the French squadron of Albania was created, offering an air cover to the French troops of Corfu and keeping an eye on the Austro-Hungarian troops occupying the north of Albania.

Commanded by Lieutenant Emile Prot and equipped with Maurice Farman MF XI bis, the "Escadrille Française d'Albanie" (French Albanian squadron) also called "Prot squadron" or "MF Corfu", left Lyon on January 23, 1916 and travelled by train through Italy to Brindisi. It was then transported by ship to Corfu, where its equipment was landed at the Potamos airfield (used a year and a half later by escadrille 562), on February 7, 1916. The Farmans were assembled and patrols and reconnaissance were organized on the Albanian coast. Little is known about the exact operational activity of the squadron, except for the loss of one Farman, that of Second Lieutenant Léon Fabert, which crashed at sea around May 15, 1916. The crew was unharmed and the wreckage of the Farman was recovered, resulting in a photo published in the magazine "*La Guerre Aérienne Illustrée*" (The Illustrated Air War).

At the end of May 1916, the Serbian troops stationed in Corfu were all evacuated from the island and the presence of the squadron was no longer justified. It then embarked for Salonica, leaving Corfu on May 27 and reaching the Salonica-Sedes camp on May 30. At the time when the Corfu squadron landed, the Entente troops left the entrenched camp of Salonica and carried out an overrun westwards on Florina, occupied by the French troops and then taken over by the Bulgarians. In August 1916, these maneuvers turned into a generalized offensive to support Romania, which had just entered the war. The offensive was half a failure: no break in the front was achieved, nevertheless the French and Serbian troops managed to take back Florina and to establish themselves on the town of Monastir even further west, establishing a junction with the Italians in Albania. The Corfu squadron, officially named MF 82 from June 22nd, was put at the disposal of the Serbian army as soon as it arrived in Salonica and conducted reconnaissance missions on its behalf during the offensive, being based in the airfield of Vertekop on August 6th 1916.

It would be stationed here until the end of the war. Equipped with Farman F.40, the MF 82, renumbered MF 382 at the end of 1916, was part of the Serbian aviation under the orders of Major Vitrat with the squadrons MF 384, MF 398, and MF 399 as well as the N 387 fighter squadron. It remained under the command of Lieutenant Prot, who, having become a captain and taken on higher responsibilities in the Serbian air force, left the command to

Above: Lt Emile Prot, a distinguished French pre-war pilot who, in September 1914, made a decisive aerial reconnaissance which allowed to prepare the battle of the Marne which stopped the German advance.

Lieutenant Marcel Cressole. In January 1917, the Farman squadrons of the Serbian army were ordered to wear special markings: its F.40s would henceforth bear a white Saint Andrew's cross on the fuselage and a blue chevron on and under the wings. During the year 1917, when its Farman aircraft will show their obsolescence in front of the enemy aviation, the crews of the MF 382, again renumbered as squadron 521 in June, manage rather well losing only one crew to the enemy but deplore some accidents. The squadron integrated more and more Serbian pilots during their training in France: by January 1st 1917, half of the ten pilots and eight observers were Serbian. The process was completed one year later. Lt. Michel Marinkovitch took command in February 1918 and the squadron officially became the 1st Serbian squadron on March 6, 1918.

Now equipped with two-seater AR 1, the squadron, which included a few Nieuport escort fighters, always faced the mountainous massif of Macedonia, which dominated the center of the Orient Front. It is precisely through this supposedly impassable massif that the victory offensive of September 1918 would be carried out: after a short and violent artillery preparation carried at arm's length

Above: Maurice Farman MF XI bis on March 14, 1916, in Potamos airfield.

on the heights, and regulated by the Serbian planes, the Serbian troops launched themselves to the assault and took the Bulgarian positions. The French cavalry of general Jouinot-Gambetta then launched itself into the breach, and, progressing on mountain roads, cut the Bulgarian front in two, leading Bulgaria to ask for the armistice. The 1st Serbian squadron was then able to return to its liberated country and was then completely under the control of the Serbian government with its aircraft, including a few Breguet 14 A2s assigned in the last months of the conflict.

Above: Maurice Farman MF XI bis n°1225 crashed in Potamos airfield. Luckily, no bombs exploded.

Above: A MF 11 bis of the Corfu squadron fell at sea shortly before May 20, 1916 – it was piloted by S/Lt Léon Fabert, who escaped unharmed, as well as the observer who accompanied him.

Above: Farmans F.40 of escadrille MF 382 in Vertekop, early 1917. (Collection Boris Ciglic)

Above: A rather rare bird, a Farman F.41 (n°1634) of escadrille MF 382 on Januray 18, 1917.

Below: Farman F.40 wearing F 382 markings in Vertekop. Nieuport 21 escort fighters can be seen in the background. (Collection Boris Ciglic)

Above: Farman F.40 of escadrille F 382 wearing a personal decoration. (Collection Boris Ciglic)

Below: Sergent Charles Fugier at the commands of his Nieuport 23 belonging to the escort section of escadrille 521, in late 1917. In front of the propeller is his mechanic, soldier Faux. (Collection Albin Denis)

Above: Probably in March 1918, Lt Michel Marinkovich (French officer of Serbian origins), on the left, in front of his men of escadrille 521 – the unit becomes the 1st Serbian squadron. A Dorand AR 1 is on the background. (Collection Boris Ciglic)

Below: Other view of the same ceremony in Vertekop. Note on the backgroung a Dorand AR 1 wearing the panther insignia. (Collection Boris Ciglic)

Above: In late 1917, the planes abandons the specific markings (white cross on fuselage and blue chevron) and replaces them by a panther, as seen her in this Dorand AR 1. (Collection Boris Ciglic)

Below & Following Page, Top: Nieuport 23 of sergent Maxime Hantz, decorated with the panther insignia. (Collection Albin Denis)

Above: After March 1918, escadrille 521 or 1st Serbian escadrille adopted a Serbian flag as new insignia, as seen here on Dorand AR 1 n°1885 in Vertekop. (Collection Boris Ciglic)

Above: Nieuport 24 n°5515 from the escort section of escadrille 521 in 1918. The tail stand of the aircraft is one of a Nieuport 27. (Collection Boris Ciglic)

Above: Adjudant Dragos Adamovic in front of his Dorand AR 1, wearing the nickname "Mascott". (Collection Boris Ciglic)

Above: Dorand AR 1 n°1871 of escadrille 521. (Collection Boris Ciglic)

Above: Same aircraft after a bad landing in Vertekop on July 23, 1918. (Collection Boris Ciglic)

Above: One of the few Breguet 14 A2 (n°2749) received by escadrille 521, crashed by pilot Petrovaradin on November 1918. The plane carries no specific marking. (Collection Boris Ciglic)

Maurice Farman MF 11 bis n°1256, Escadrille Française d'Albanie, Potamos (Corfou, Greece), March 1916.

Dorand AR 1 n°1871, Escadrille 521,
Vertekop, Late 1917.

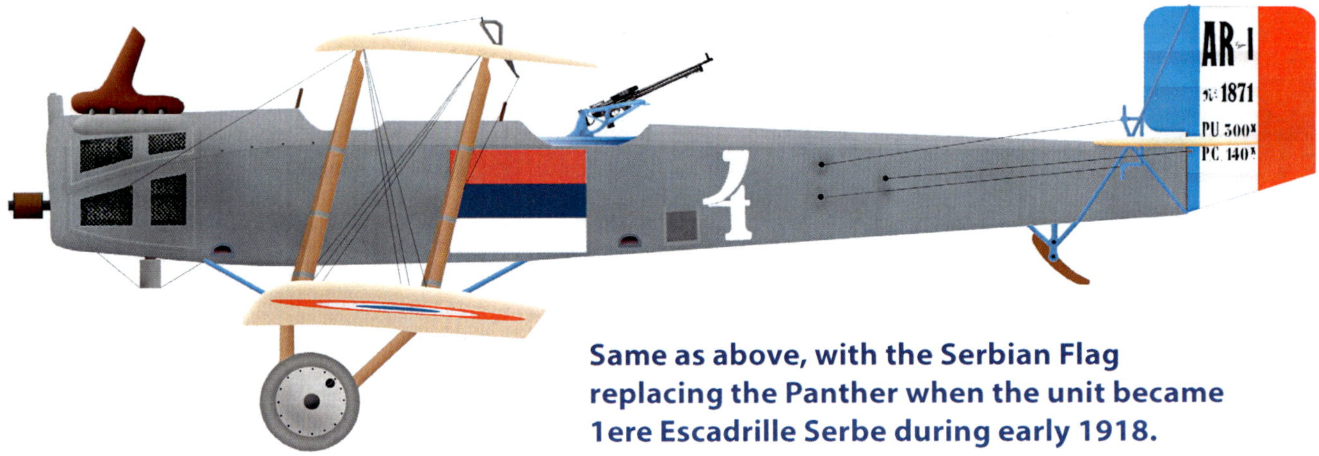

Same as above, with the Serbian Flag
replacing the Panther when the unit became
1ere Escadrille Serbe during early 1918.

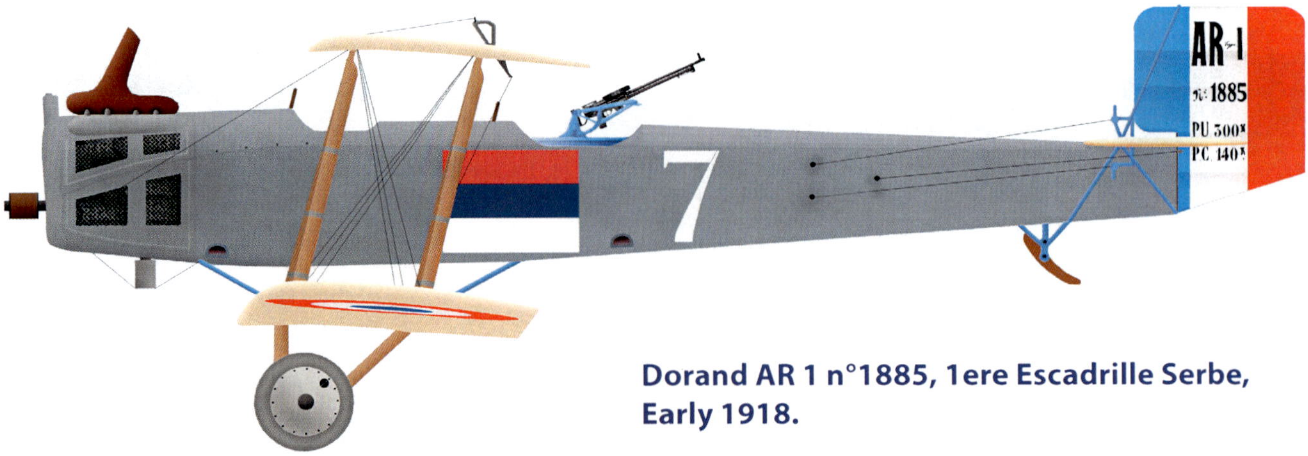

Dorand AR 1 n°1885, 1ere Escadrille Serbe,
Early 1918.

Nieuport 23 (serial unknown), Escadrille 521, Vertekop, Late 1917.

Nieuport 24 n°5515, 1ere Escadrille Serbe, Vertekop, March 1918.

Escadrille 522

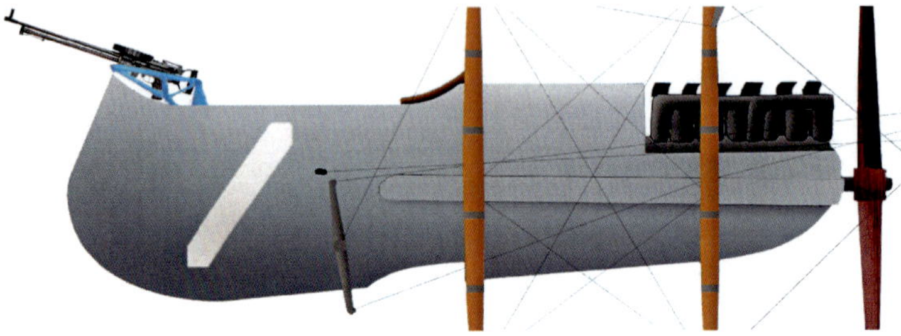

Created as MF 84 in December 1915, renamed MF 384 in July 1916 then Escadrille 522 in June 1917.

Locations:

- Salonica in December 19, 1915.
- Samli in December 20, 1915.
- Sarovitchevo in October 5, 1916.
- Vrebeni (NE Florina) in November 2, 1916.
- Vertekop in May 1917.

Commanding Officers:

- Cne François de Castel in December 1915.
- Cne Charles Hug in July 25, 1916.
- S/Lt Alexandre Guibert in 1917.
- Lt Philippe Chavane de Dalmassy in February 1918.

Equipment:

- Farman F.40 in late 1916.
- Dorand AR 1 in June 1917.
- Escadrille 522 also had since late 1917 a fighter section of 2-3 Nieuport fighters (type 23, 24 or 24bis).

Victories: None.

Casualties: 3 pilots lost by enemy action (1 killed, 2 captured) and 1 by accident.

History:

Escadrille V 84 was formed in France, probably in Lyon, at the beginning of December 1915 under the command of Captain de Castel and consisted of six pilots. Originally destined for the Armée d'Orient, it embarked on December 9, 1915 in Marseille and disembarked in Salonica on December 19, 1915, settling in the Samli camp, 12 km west of Salonica. It was grouped with the two other Voisin squadrons, the V 90 S and V 83, which formed the Groupe de Bombardement d'Orient (GBO).

From then on, the Voisin squadrons of the GBO were used *en masse* against enemy targets, sometimes far from the front line. The first action took place on February 1st 1916 against the Bulgarian camps of Petric, in retaliation to the raid of a Zeppelin on Salonica carried out the day before: 16 Voisins, including six of the V 83, carried out the mission without any other opposition than the flak but also the sometimes defective mechanics: a Voisin of the V 84 had to turn back before the objective. From then on, the missions multiplied on distant targets such as Strumica, Volovec, and Xanthi, sometimes meeting the opposition of the still stammering German fighter pilots on Fokker Eindecker or Pfalz Eindecker. But the main enemy of the Voisin was the mechanics, the Canton-Unné engines not easily withstanding the climate. Command of the GBO reverted to Captain de Castel on March 25, who kept command of the V 84 but delegated certain tasks to his second officer, SLt Charles Boudoux d'Hautefeuille. For the eleventh mission of the GBO on June 6, 1916, the objective of which was once again Petric, the rate of unavailability was so high that barely four (including three from the V 84) of the 14 Voisin that had taken off for the raid reached the objective, the others having turned back due to mechanical problems. Major Denain, head of the Eastern Air Force, took the necessary decision: the dissolution of the GBO in June.

Imitating V 83, V 84 would exchange its tired Voisin for Farman MF 11 bis and F.40, becoming an observation squadron. The pilots also changed, a good part of them for medical reasons due to the malaria that raged in Greece. On July 25, 1916, the command returned to Captain Charles Hug and the F 384 (new name of the squadron) was put at the disposal of the Serbian army, with some Serbian pilots assigned to it.

The F 384 was to participate in the Franco-Serbian

Above: Farman F.40 of escadrille MF 384 after a bad landing, showing its red stripes under the wings. (Collection Boris Ciglic)

offensive on Monastir, west of Salonica, but was also to send crews on bombing missions over Sofia, the Bulgarian capital. It was during one of these missions that the Farman MF 11 bis flown by Corporal Maurice Rouable and carrying the Serbian Slt Maoumovitch as observer was shot down. Both men were slightly wounded and captured, they were shot down by a Fokker Eindecker piloted by a German and a two-seater piloted by a Bulgarian crew.

On October 5, 1916, the squadron left the Samli airfield to migrate westward in the mountains to Sarovitchevo, then to Vrbeni on November 2, 1916. It would remain in the central sector of the front devolved to the Serbian army until the end of the war, migrating one last time to Vertekop at an undetermined date in 1917.

In January 1917, a command directive ordered the painting of distinctive signs on the Farmans of the Serbian army – those of the MF 384 now had red stripes on and under the wings as well as a white slash on the sides of the fuselage. The Farmans were all decorated in this way, as well as the first Dorand AR 1 which replaced them in mid-1917, probably the date when the new squadron leader, Slt Charles Guibert, took command. The squadron, renumbered as 522 in June 1917, was also equipped with Nieuport escort planes, whose decoration is unknown. In November 1917, the German fighters were biting and one observer was killed and another wounded, while a Nieuport escort also fell with its pilot. In January 1918, the Serbian command sought to create entirely Serbian squadrons. The Serbian pilots then left the 522 squadron and it was to survive undermanned for several months under the command of Lieutenant Chavane de Dalmassy. It was dissolved in June 1918, its pilots being distributed to other French squadrons.

Above: View of the hangars of Vertekop and Farmans of MF 384. (Collection Boris Ciglic)

Above: A Dorand AR 1 of escadrille 522 in September 1917.

Above: Sergent Emile Grillet (marked with cross) with other French and Serbian pilots in front of a Dorand AR 1 of Escadrille 522 in late 1917.

Right: Lt Alexandre Guibert in front of a Nieuport fighter from the escort section of escadrille 522.

Above & Below: Two views of Dorand AR 1 n°1870 of escadrille 522 in late 1917. (Collection Boris Ciglic)

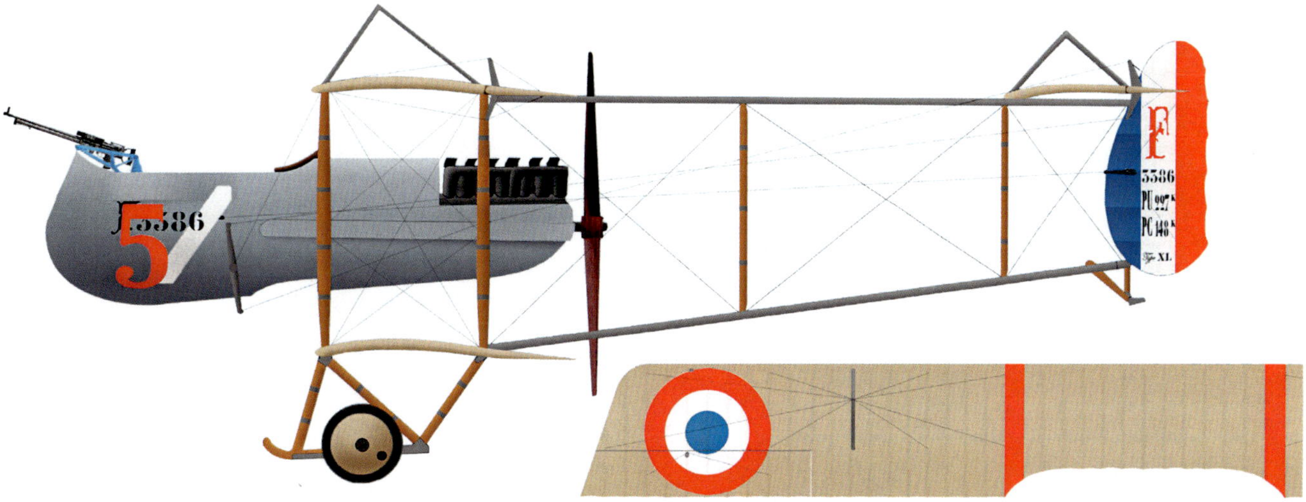

Farman F.40 n°3386, MF 384, Vertekop, 1917.

Dorand AR 1 n°1870, Escadrille 522, Vertekop, 1918.

Escadrille 523

There is no evidence that Escadrille 523 ever had a unit insignia.

Created as N 87 in October 1915, renamed N 387 in June 1916 then Escadrille 523 in June 1917.

Locations:

- Salonica (Allatini camp) in November 20, 1915.
- Sédès in December 1915.
- Amatovo in July 6, 1916
- Vertekop in August 5, 1916.
- Sorovitchevo in October 2, 1916.
- Vrbeni in November 7, 1916.
- Vertekop in April 7, 1917.
- Gradsko in October 1918.
- Nish (Serbia) in Late 1918.
- Novi Sad (Serbia) in March 20, 1919.
- Szeged (Hungary) in May 11, 1919.

Commanding Officers:

- Lt Jean de L'Hermitte in October 13, 1915.
- Lt Louis Delrieu (interim) in July 24, 1916.
- Cne Henri Gallet in August 6, 1916.
- S/Lt Marcel Thirouin in November 1916.
- S/Lt Michel Marinkovitch (interim) in December 21, 1916.
- Cne Henri Gallet in January 1, 1917.
- S/Lt Marcel Thirouin (interim) in January 4, 1917.
- Lt Mihajlo Marinković in August 1917.
- Lt Pierre Boucher in February 1918.
- Lt Léon Sutter after the armistice in late 1918.

Equipment:

- Nieuport 10 in October 1915.
- Nieuport 11 and 21 in Summer 1916.
- Nieuport 23, 24, 24 and 27 in March 1917.
- SPAD VII in May 1917.

Escadrille 522 also had since Summer 1918 a section of Dorand AR 1 for strategic reconnaissance missions. These planes were replaced by Breguet 14 after the armistice.

Victories: 8 confirmed victories + 3 unconfirmed.

Casualties: 1 pilot lost by enemy action (captured) + 4 by accident.

History:

Escadrille N 87 was created at the end of 1915 from the 2nd aviation depot of Lyon-Bron, with the group of squadrons destined to join the Armée d'Orient in Salonica. The N 87, originally intended to be the fighter component, comprised eight Nieuport aircraft (type 10 two-seaters and one or two single-seaters) under the command of Lieutenant de L'Hermitte, an active officer from N 37. He had at his disposal seven other pilots, Slt Delrieu, Adjudant Havet, Maréchal des Logis-Chef Monrouzeau, Sergeants Biot and Fenech, as well as Corporals Chappaz and Cherié. The squadron, which also included 40 mules, 12 horses, 22 cars, 3 tractors, 1 trailer and 2 light cars, embarked from Marseille on November 10, 1915 on the liner "*Sinaï*" and disembarked in Salonica ten days later, on November 20, 1915, at camp Allatini.

After a few days, it moved to the Sedès airfield, 12 km ESE of Salonica, where its first operational mission

Above: Nieuport 10 fighters of squadron N 87 at Salonica-Sedes airfield, early 1916. Note the different colors of the two aircraft in the foreground: the combat reports of the Armée d'Orient mention "blue" and "yellow" Nieuports. (Collection Boris Ciglic)

took place on December 24, for a reconnaissance mission led by Adjudant Havet on a single-seater Nieuport. At that time, the Franco-British army was entrenched in the city of Salonica and the Bulgarian army was still holding out on the Greek border. Between the two armies, a no-man's-land occupied by Greek troops... The airmen were the only ones to see action, although N 87, because of its position at Sedès, was not in a better position to intercept German planes coming from the North to bomb the Allied warehouses in Salonica. However, there was no lack of encounters: on January 7, 1916, Adjudant Havet, piloting one of the squadron's single-seaters, intercepted an enemy two-seater over Salonica. Getting into firing position, his Lewis machine gun fixed on the wing fell on his knees... He nevertheless continued the fight by firing his Mauser pistol at the enemy who escaped. Havet did not admit defeat: he landed at Samli on the MF 86 field, changed his machine gun, and went back to the fight but could not find his prey. He spotted another enemy and fired a magazine at him without success. On March 27, 1916, during a mass raid conducted by the German air force on Salonica, the N 87 sent two aircraft into the air, Nieuport No. 416 and the one piloted by Warrant Officer Havet, who emptied the magazines of his Lewis machine gun on two different targets without success. He witnessed the aerial combats carried out by the Nieuports of the N 91 squadron, which shot down a German aircraft.

While the French army was entrenched in the city of Salonica since January 1916, an event changed the situation in April 1916: the arrival of the Serbian army reconstituted on the island of Corfu after the December 1915 debacle that saw the invasion of Serbia. With these reinforcements, the French army moved towards the hinterland and came into contact with the German-Bulgarian army that crossed the border. The orient front then became a continuous front.

In May and June 1916, the N 87, renumbered N 387, carried out several interceptions and escorts on the front, which led it to intervene on the Doiran lake or up to the village of Guevgueli where the front line of the Vardar sector was fixed. Enemy aircraft were confronted on several occasions, notably by Adjudant Havet and Second Lieutenant Thirouin. The squadron moved to the Amatovo airfield, near the lake of the same name on the banks of the Vardar, getting closer to the combat zone. It was there that it officially became available to the Serbian army, along with four other reconnaissance squadrons on Maurice Farman aircraft, and for several months it had been integrating several Serbian pilots and mechanics among its personnel. The Serbian squadrons formed an autonomous air force under the command of Major Roger Vitrat.

On August 2, 1916, N 387 established itself in the mountains of Macedonia at Vertekop, in the sector which would be its place of assignment for the remainder of the war. It was the central point of the front where the Serbian troops would be stationed; in the immediate future, a generalized offensive to support the entry of Romania into the war is launched on the Vardar and triggers a violent Bulgarian counter-attack in the plain of Monastir. On August 6, its new commander, Captain Henri Gallet, arrived, sent to replace Lieutenant de L'Hermitte who had been

Above: Sergeant Alexandre Cherrié, pilot of escadrille N 87, at the controls of his two-seater Nieuport 10 in early 1916. A rare practice in military aviation, some Nieuport 10s had a roundel on the fuselage (Collection Albin Denis)

sent back to France, after an interim period carried out by Lt. Delrieu distinguished himself on August 11, 1916 by carrying out a special mission on a two-seater Nieuport, dropping a Serbian spy behind enemy lines. August 25 was another eventful day: taking Captain Mavet as an observer, Lieutenant Delrieu saw his two-seater Nieuport attacked by three enemy fighters. The same day, Second Lieutenant Marcel Thirouin, on a Nieuport 11, shot down an enemy aircraft (the unit's first certified victory) but returned home with his aircraft seriously damaged. Veteran of the Serbia campaign in 1915 with the MF 99 S squadron, where he won a victory, he obtained here his second personal victory. On October 2, 1916, the squadron settled in Sorovitchevo and a few days later, on October 8, one of its pilots, Adjudant Charles Denti, had to be sent back to the rear because he was exhausted. He was indeed a victim of the endemic malaria that was rampant in northern Greece at the time, and he was frightened when he fainted at the controls of his Nieuport during a flight at 4,000 meters – his plane started to sting slightly and he woke up 3,000 meters below, dangerously surrounded by the mountain walls... More generally, malaria was responsible for the poor sanitary state of the entire squadron, as well as that of all the squadrons placed at the disposal of the Serbs, who had to deplore an understaffing until the summer of 1917.

November 1916 saw the stabilization of the front with the capture of the town of Monastir on the 19th by Franco-Serbian troops, putting an end to a bloody battle that had cost the lives of nearly 32,000 Serbian fighters. On 7 November, N 387 was established in Vrbeni under the interim command of S/lt Thirouin. The situation of the front would remain frozen for almost two years until the final offensive of September 1918: the Serbian troops would stand guard in front of the mountains of Macedonia, as well as its 4 corps squadrons (MF 382 and MF 384 in Vertekop and MF 398 and 399 in Vrbeni) and its fighter squadron, N 387. However, the Serbian sector was cut off by a transverse mountain range that separated the Vertekop and Vrbeni sectors. In order to offer a hunting cover to the Vertekop squadrons, the Nieuport detachment of Vertekop was created from N 387 on December 17, 1916 under the orders of the Serbian Captain Branko Vukosavljevic. On December 21, 1916, a Serbian, SLt Marinkovic, took over as

Above: Nieuport 10 n°416 of escadrille N 87 in early 1916. The aircraft is probably used as a single-seater fighter even if the machine gun fixed on the upper wing seems to be mobile. (Collection Boris Ciglic)

Above &Facing Page, Right: NNieuport 11 n°1482 christened "*Le môme*" and whose pilot is probably Second Lieutenant Raymond Havet, killed in action in France at the N 77 squadron on March 16, 1917 (Boris Ciglic Collection)

Above: The Nieuport 21, a hybrid version of the Nieuport 11 with its 80 hp engine and the Nieuport 17 with its wing, became the standard fighter of the French Armée d'Orient in the fall of 1916. The three aircraft in the picture, two of which are wearing personal decorations, were stationed at Vrbeni in November 1916 (Collection Boris Ciglic)

acting commander of the squadron for a short period, and was relieved by Slt Thirouin on January 4. Two days later, MdL Viallet obtained the second certified victory of the unit southeast of Prilep.

On February 15, 1917, Kaghol 1, a German squadron of heavy twin-engine bombers, set up shop in Hudova and soon caused panic in the Vardar sector and over the city of Salonica. They then targeted the French Eastern Army (AFO) sector in the Monastir plain, then the rear of the Serbian sector in the center of the front... Commander Denain, expecting this movement, ordered N 387 to regroup with the AFO fighter squadron (the N 390) to intercept together the bombers and their escort fighters which cleared the bombers' route beforehand. The first raid took place on 12 March 1917 over Vertekop with 18 bombers and six fighters. Only two Nieuports of the Vertekop detachment were able to take off and were unable to counter the raid, which destroyed the ammunition depot of the 2nd Serbian army, while also hitting the nearby hospital. Other bomber raids took place and the Serbian air force had some success. On April 2, 1917, a two-seater Nieuport 12 of the Vertekop detachment (pilot Slt Miletic, machine gunner Cpt Vukosavljevic) won a victory against a reconnaissance plane of the 2nd Bulgarian squadron. The following day, April 3, it was one of the formidable AEG G.III bombers that the Nieuport of Lt Miodrag Miletic claimed (not confirmed), during a bombardment on Vertekop. The complete squadron settled besides in Vertekop on April 7, 1917 but left a detachment in Vrbeni, the detachment of Vertekop being for its part dissolved on May 17, 1917 and amalgamated with N 387 which is soon renamed squadron 523. It would receive some two-seaters because it also had the task to carry out strategic recognitions for the staff of the Serbian army.

Kaghol 1 left the Balkan front but German aviation remained present; the 523 squadron received some SPAD VII fighters. On May 20, 1917, Lt. Marcel Thirouin and MdL Charles Martin shot down a plane in the Cerna Gorge (4th victory of the unit) but the next day the squadron mourned the accidental death of the Serbian Adjutant Friedrich Roth, who did not recover from a spin on his Nieuport 21 near the airfield. On 9 July, MdL Fernand Meignen managed to shoot down a German reconnaissance aircraft near the Serka di Legen mountain in the French lines, where the crew was captured and the wreckage of the aircraft was displayed near the White Tower in Salonica. The same feat was repeated on July 21, 1917 by a patrol composed of Lt. Miodrag Tomic, MdL Charles Martin, and MdL Robert Mollien against a two-seater of the FA 34 whose crew was also captured.

In August 1917, Lt. Marinkovic officially took command of 523 Squadron. The Nieuport detachment of Vrbeni was disbanded on the 25th, but its pilots remained in place, divided between 524 and 525 squadrons (four pilots each), whose two-seater reconnaissance aircraft they escorted. A year passed without any particular event, during which the command passed to Lt Pierre Boucher in February 1918. The month of August 1918 saw the last successes of the squadron when on the 7th, Lt. Miodrag Miletic won his 3rd and last victory at the expense of a two-seater forced to land in his lines. Miletic became the best fighter pilot of the Serbian army. On August 25, 1918, the Sgt-Major Mateya Hodiera closed the table of hunting of the squadron by shooting

Above: Pilots and mechanics of escadrille N 387 at Vrbeni in November 1916. The 7th from the right, with his hands in his pockets, is Lt Marcel Thirouin, veteran of escadrille MF 99 S in Serbia and one of the best pilots of the Armée d'Orient, with 3 confirmed victories and one probable one. (Collection Boris Ciglic)

Below: Nieuport 21 n°1965 of the Vertekop detachment, depending from escadrille 523 then stationed at Vrbeni, in March 1917. This detachment was composed of Serbian pilots, such as Captain Branko Vukosavljevic, leaning here against the propeller of his aircraft which bears no particular marking. (Collection Boris Ciglic)

Above: Nieuport 12 two-seater n°1757 of the Vertekop detachment in March 1917 – Probably the aircraft with which the crew composed of S/Lt Miletic as pilot and Captain Branko Vukosavljevic as gunner shot down a Bulgarian two-seater on April 2, 1917 (Collection Boris Ciglic)

down within sight of the ground observatories a two-seater near Dobropolje.

The Bulgarian army was thus cut in two. The road to Sofia was opened. Bulgaria then asked, without delay, for an armistice which was ratified on October 5. During this period, many reconnaissance missions were carried out by the Breguet 14s of the 523 squadron, as well as escort missions by the fighters. A few aircraft fell due to flak, but without damage to the crews. Shortly after the armistice, 523 Squadron reached Serbia in Nish and was maintained among the air forces of the Eastern Army. It moved to Novi Sad and fired against the troops of the Hungarian communist government of Bel Kun, before being stationed in Hungary with the French occupation troops. Three airmen lost their lives accidentally during this period. In 1920, the Serbian Aeronautics was officially disbanded as well as the 523 squadron, whose equipment was entrusted to the Serbian army.

Above: Nieuport 23 n°2780 of S/Lt Miodrag Tomic decorated with the Serbian flag, seen on April 30, 1917. (Collection Boris Ciglic)

Below: Alignment of Nieuport 23 of escadrille 523 in June 1917, now stationed at Vertekop. Two SPAD VII's can be seen on the right. The Nieuports all bear personal symbols, although on several of them a flower seems to be visible (Collection Albin Denis)

Above: One of the Nieuport 23 of escadrille 523 squadron on its nose. (Collection Albin Denis)

TOP GUNS OF SERBIAN AIR FORCE 1914-1918
Escadrille 523

Lieutenant
Miodrag MILETIC
2 victories + 1 unconfirmed

Capitaine
Branko VUKOSAVLJEVIC
1 victory (as gunner)

Sous-lieutenant
Miodrag TOMIC
1 victory

Sergent-major
Mateya HODIERA
1 victory

Above: Portraits of the victorious Serbian pilots of escadrille 523. All were trained in France.

Right: Corporal Emile Delauzun, mechanic of escadrille 523, poses in front of a Nieuport 23 of the unit at Vertekop on August 13, 1917 (Collection Albin Denis)

Below: Line-up of three Nieuport 23s of escadrille 523 at Vertekop during the summer of 1917. All have a tactical number, but no squadron or personal insignia. (Collection Albin Denis)

Above: Pilots of escadrille 523 Squadron in 1918. From left to right, front row: Sergeant observer unknown, Maréchal des Logis (MdL) Charles Fugier, Sgt unknown, Adjudant (Adj) Pierre Rasting, Lt Pierre Boucher (squadron leader), MdL René Glarner, MdL Maurice Bergès, - Adj Sergnic Marinovitch, Sgt Jean Tallien de Cabarrus. From left to right in the background: Sgt Etienne Ingrain, Sgt André Ravise (observer), unknown. (Collection Albin Denis)

Above: Lt Miodrag Tomic, pilot of escadrille N 387, surrounded by his mechanics, poses next to his Nieuport 24 n° 4408 on the Vertekop field. In the background, the SPAD VII coded "10". (Coll Boris Ciglic)

Above: Sgt Charles Fugier, pilot of escadrille 523, in front of a camouflaged Nieuport 24 at Vertekop airfield in May 1918 (Collection Albin Denis)

Below: Pilots of escadrille 523 in front of a SPAD VII – Lt Pierre Boucher is in the center, in front of the propeller of the aircraft. (Collection Albin Denis)

Above: SPAD VII n°7203 (produced by De Marcay) of escadrille 523 photographed at Vertekop on September 15, 1918. The aircraft is wearing Serbian roundels (Albin Denis Collection)

Above: View of a Nieuport 24 of 523 squadron photographed in Serbia in 1919 – the aircraft is now wearing Serbian roundels (Collection Boris Ciglic)

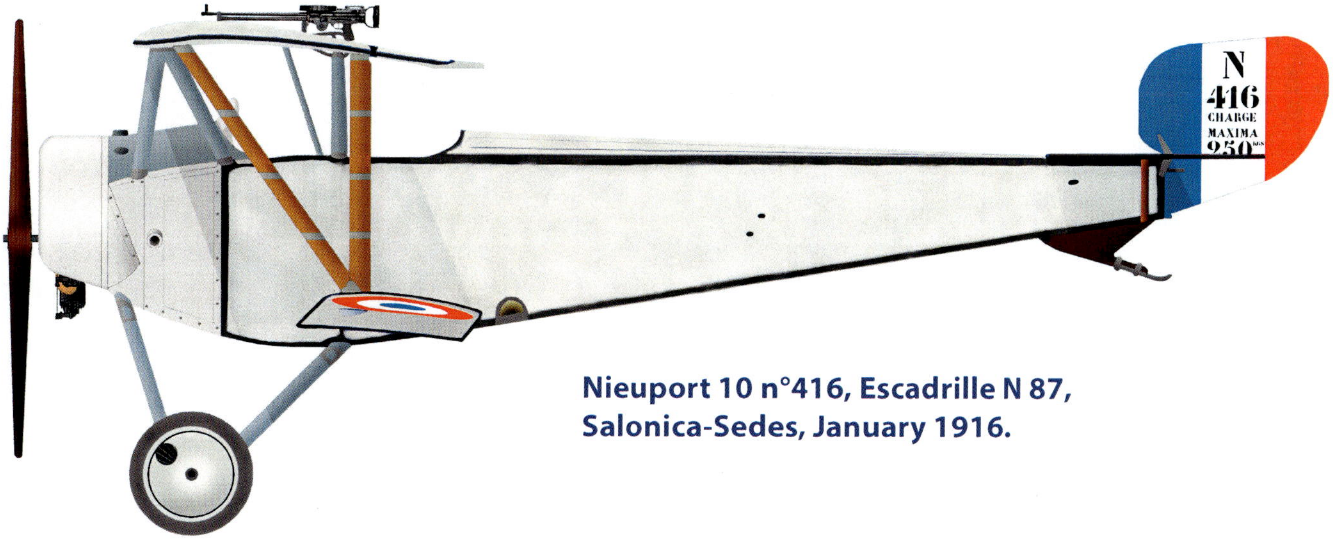

Nieuport 10 n°416, Escadrille N 87, Salonica-Sedes, January 1916.

Morane Parasol n°416, Escadrille N 87, Salonica-Sedes, June 1916. Two Morane Parasols from escadrille MF 98 T joined escadrille N 87 in April 1916, but were soon removed from service, being less performants than Nieuports.

Nieuport 11 n°1482, Escadrille N 87, Salonica-Sedes, Spring 1916.

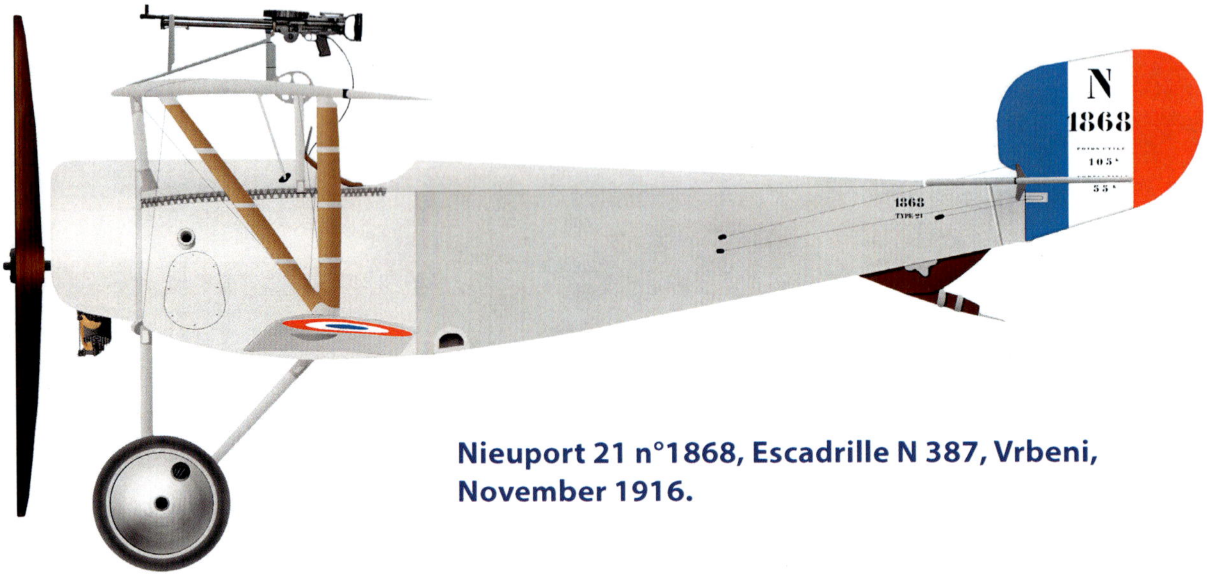

Nieuport 21 n°1868, Escadrille N 387, Vrbeni, November 1916.

Nieuport 17 n°2780, Escadrille N 387, S/Lt Miodrag Tomic (1 victory), Vertekop, April 1917.

Nieuport 24 n°4400, Escadrille 523, S/Lt Miodrag Tomic (1 victory), Vertekop, April 1917.

SPAD VII (serial unknown), Escadrille 523, Vertekop, Summer 1917.

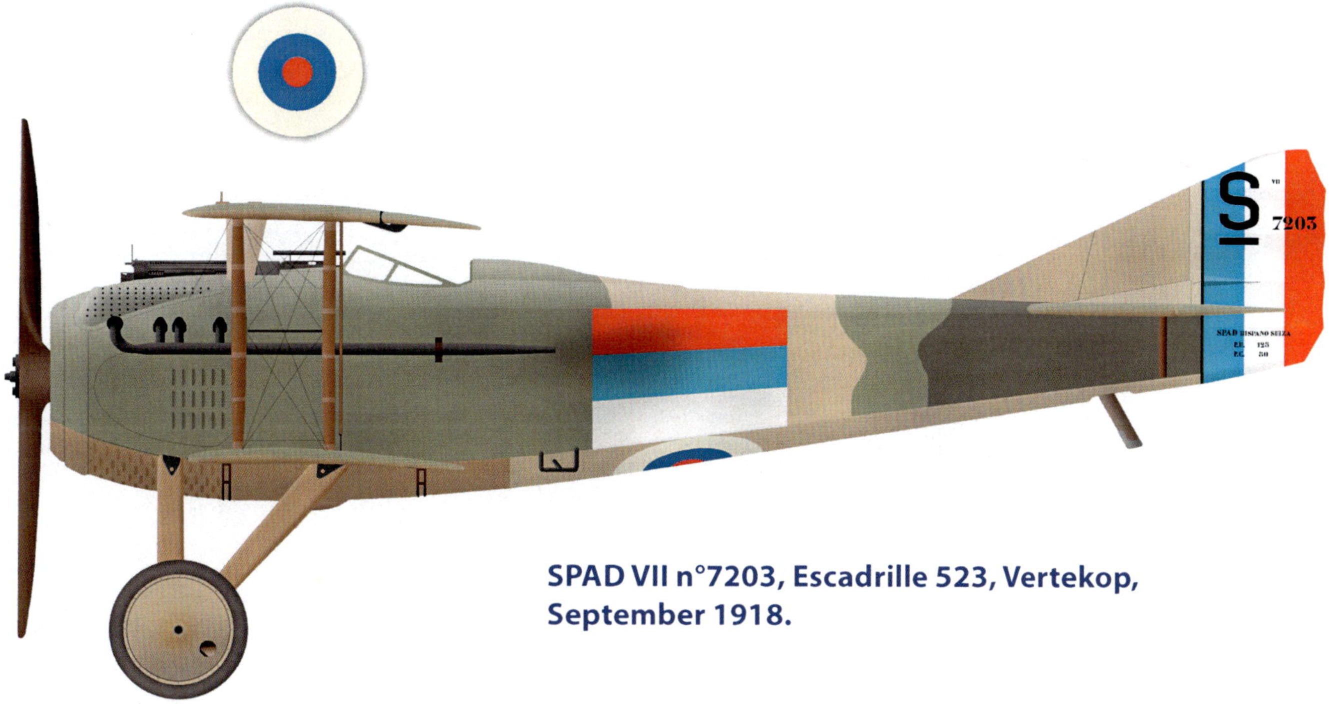

SPAD VII n°7203, Escadrille 523, Vertekop, September 1918.

Escadrille 524

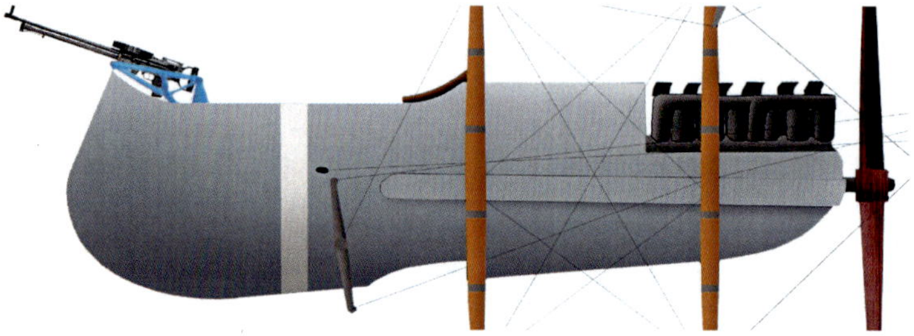

Created as MF 98 T in March 1915, renamed F 398 in August 1916 then Escadrille 524 in June 1917.

Locations:

- Tenedos in April 29, 1915.
- Mytilene (detachment) in January 27, 1916
- Salonica in March 1, 1916.
- Kilkich (South Lake Doiran), unknown date.
- Kukus, unknown date.
- Mikra, in July 22, 1916.
- Gumenze, unknown date.
- Ostrovo, in August 31, 1916.
- Vrbeni, in October 6, 1916.
- Nish (Serbia) in Late 1918.

Commanding Officers:

- Cne Antoine Césari in March 1915.
- Lt Roger Sarda in March 1, 1916.
- Lt Henri Lamasse in June 1, 1916.
- Cne Jacques Sabattier de Vignolle in November 16, 1916.
- Lt Yves Allain-Launay in April 15, 1917.
- Unknown officer in April 31, 1917.
- Lt Maurice Paulin, unknown date.

Equipment:

- Farman MF XI in March 1915.
- Farman MF XIbis during Summer 1916
- Morane L during Summer 1916, for escort duties.
- Farman F.40 during Fall 1916.
- Nieuport 10 during Fall 1916.
- Dorand AR 1 during Summer 1917.
- Breguet 14 durign late 1918.

The unit also had an escort section of 4 Nieuport fighters (21, 23, 24, or 24bis) since June 1917.

Victories: 1 official victory + 1 unconfirmed.

Casualties: 9 pilots killed from enemy action + 6 by accident.

History:
Escadrille MF 98 T, commanded by Capitaine Antoine Césari, was created in Lyon (France) to support the expedition in Gallipoli (Dardanelles for the French), landing in the island of Tenedos in April 29, 1915. The detailed history of this expedition is told in Chapter 3. Escadrille MF 98 T ended the campaign divided in two parts: one detachment was sent in Mytilene under the direct orders of Capitaine Césari (MF 98 M), while the main part of the unit remained in Tenedos (MF 98 T) under the orders of Lt Sarda.

After the withdrawal from Gallipoli, escadrille MF 98 T was affected in Salonica with all French troops from the expedition. The arrival of the Farman aircraft occurred on March 1st 1916, and marked a new era for the squadron. Its leader during the Gallipoli epic, Captain Césari, immediately returned to France to hand over command to Lt Sarda. With eight Farmans, including more powerful MF XI bis, MF 98 T quickly left Salonica for Kilkich and then Kukus, south of Lake Doiran, and carried out reconnaissance missions on behalf of the British army, which was in charge of the eastern part of the entrenched Salonica camp. Meanwhile, the detachment left in Mythilene (MF 98 M) commanded by Lt Lamasse was brought back to Salonica by boat on March 23 and settled on the ground of the big Park in the South of the city. MF 98 M was kept in reserve for the Serbian army which was reformed in Salonica but was finally disbanded on May 30, 1916, while the next day its leader took command of the reunified squadron which was folded on July 22, 1916 in Mikra, on the Serbian air force park, and which would henceforth be definitively at its service.

Above: Capitaine Sabattier de Vignolle, under the fuselage of the Farman F.40, stands with all his pilots in November 1916. It is not known if the insignia on the Farman is the first unit insignia, or a personal insignia of the pilot. (Collection Boris Ciglic)

Receiving Farman F.40s, MF 98 T, renamed F 398, accompanied the Serbian army in its movement towards the West in the battle of Monastir and settled in Ostrovo on August 31, 1916, then Vrbeni on the following October 6 where it remained until the end of the war. Two days later, the squadron came under the command of Captain Jacques Sabattier de Vignolle, who was confirmed in his command on November 16.

It is there that during the two years which followed F 398 would lead missions of recognition on behalf of the Serbian army on the central sector of the front, occupied by the mountains of Macedonia. If the losses were rather limited in 1916 (two observers killed and one wounded in combat), it is different in 1917 when the German fighters would shoot down three crews, at the price of four killed. The Farman F.40s, which were decorated with a vertical white stripe on the fuselage and two blue stripes on the wings since January 1, 1917, were replaced by Dorand AR 1s at the beginning of the summer when the new commander, Lt Yves Allain-Launay, took up his post on April 15, 1917. Renumbered as Escadrille 524 in June, the formation received on August 25th 1917 a section of four Nieuport fighters to escort its two-seaters.

At dawn on 14 September 1918, the Serbian air force redoubled its activity in bombing the enemy lines in preparation for the ground offensive which was to take place in the Serbian sector, through the Dobropolje and Sokol mountains. Escadrille 524 was no exception and targeted the Razimbey bridge, a necessary passage for supplies on the Cerna river. On September 15, 1918, several of its Dorand AR 1s escorted by a few Nieuports attacked the bridge and were intercepted by German fighters led by the ace Fieseler, who hit the Nieuport of Adjudant René Leblond, who, wounded, found the strength to land in his lines where he died a few hours later. One of the Dorands, piloted by Adjudant Raingo Pelouse with Lt Albert Nicot as observer, was also hit by the fighters and had to land in the wilderness. Its crew evacuated it and ran back to their lines while their aircraft exploded under German artillery shells. The next day, during a new attack on the objective, another Nieuport of the squadron, piloted by Brigadier René Nugues, disappeared in its turn, victim of the German fighters. There were other battles in the following days, but with the capture of Kanatlarci by the Serbian troops, the German air force disappeared from the sky and the Bulgarian defense collapsed like a house of cards. The armistice was quickly signed and the 524 squadron, which entered liberated Serbia, won the Nish airfield. It was gradually emptied of its French pilots and handed over to the Serbian government with its aircraft.

Above: Capitaine Sabattier de Vignolle (center, standing in dark uniform) poses in front of a Nieuport 10 with his pilots and mechanics during early 1917. Some of the pilots and the mechanics wear Serbian uniforms. (Collection Boris Ciglic)

Below: Capitaine Sabattier de Vignolle and Serbian pilot Stefanovic in front of a Nieuport 10, used by the unit as a fast reconnaissance plane. (Collection Boris Ciglic)

Another view of the Nieuport 10, showing the unit markings – a vertical white stripe on the fuselage, and blue stripes in and under the wings. (Collection Boris Ciglic)

Below: Closeup of the nacelle of a Farman F.40 from escadrilleF.398, showing the squadron marking – a vertical white stripe.

Above: Serbian observer Sondermayer (left) with his pilot in front of a Farman F.40 in Vertekop, early 1917. (Collection Boris Ciglic)

Above: Serbian pilot Borivoje Popadic (right) and his mechanic in front of his Farman F.40 wearing the squadron markings of escadrille F.398 and a personal insignia in the fuselage nacelle. (Collection Boris Ciglic)

Facing Page, Below: Serbian pilots and mechanics in front of a Farman F.40 wearing the squadron markings of escadrille F.398 in 1917. (Collection Boric Ciglic)

Below: Serbian general Vasic visits escadrille 524 during after summer 1917 – a Dorand AR 1 is in the background. (Collection Boris Ciglic)

Above: Dorand AR 1 of escadrille 524 crashed by Serbian pilot Petar Petrovic in February 1918. (Collection Boris Ciglic)

Below: Dorand AR 1 of escadrille 524 flying above the mountains of Macedonia, during Winter 1917–1918. (Collection Boris Ciglic)

Above: Serbian pilot Sinisa Sefanovic in a Nieuport 21 from escort section of escadrille 524. (Collection Boris Ciglic)

Below: Another view of the same aircraft, wearing a personal insignia – actually the squadron insignia of escadrille 13 in France. (Collection Boris Ciglic)

Above: Breguet 14 A2 with squadron markings of escadrille 524, in late 1918. (Collection Boris Ciglic)

Farman MF 11 n°505, escadrille MF 98 T, Tenedos 1915.

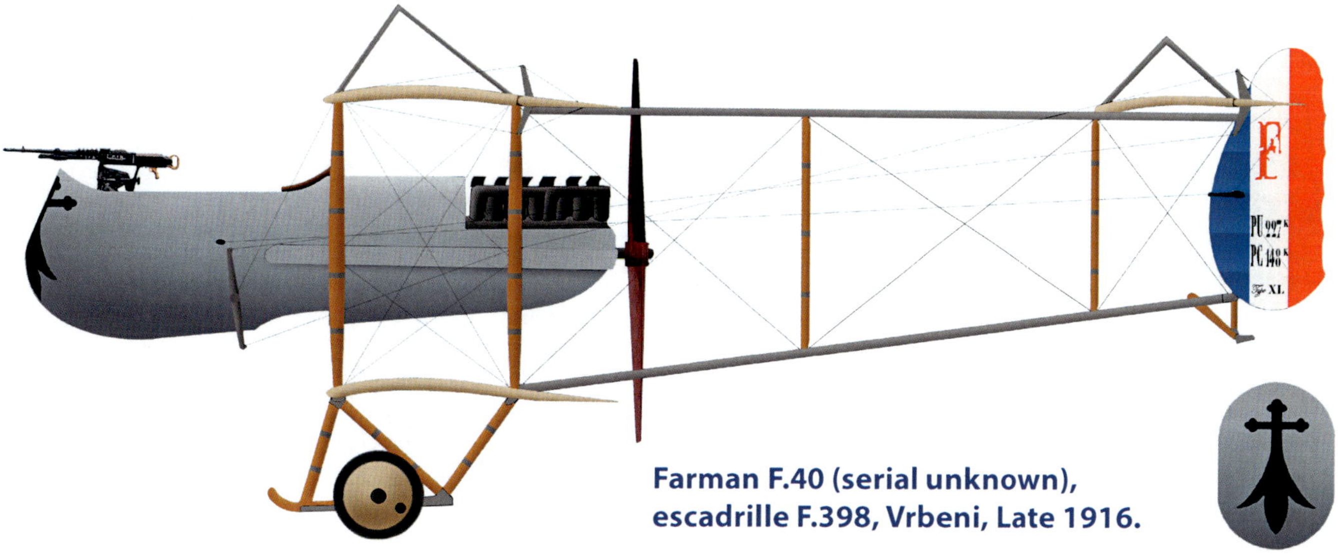

Farman F.40 (serial unknown), escadrille F.398, Vrbeni, Late 1916.

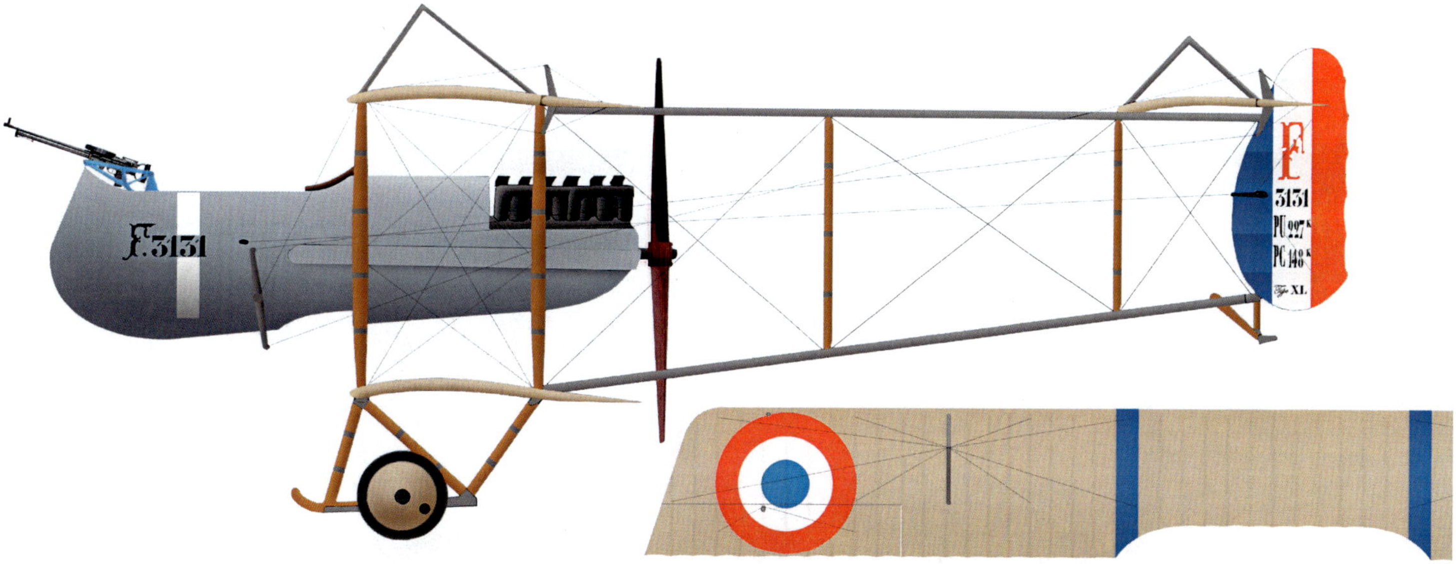

Farman F.40 n°3131, escadrille F.398, Vrbeni, Early 1917.

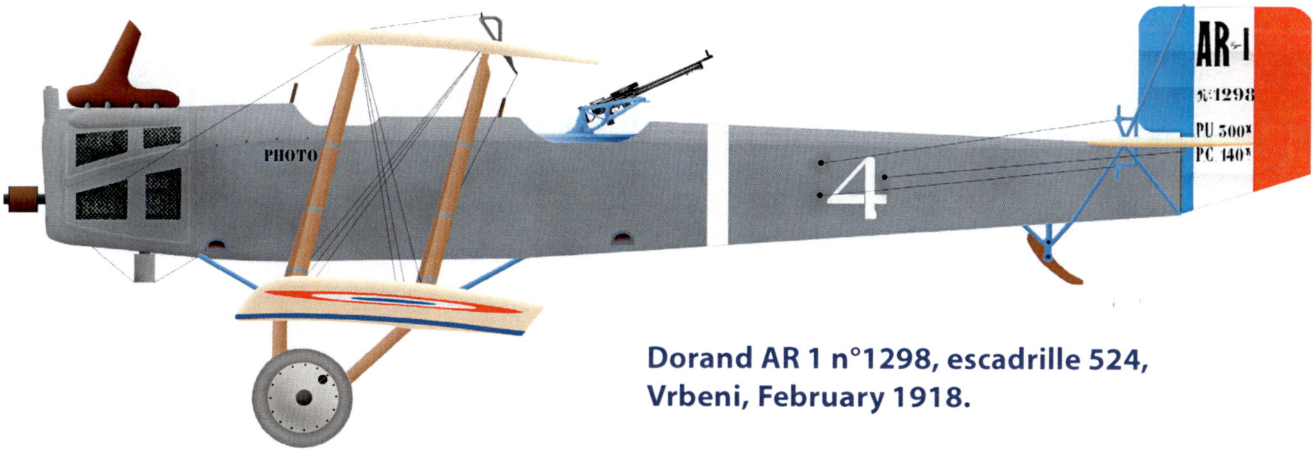

Dorand AR 1 n°1298, escadrille 524, Vrbeni, February 1918.

Escadrille 525

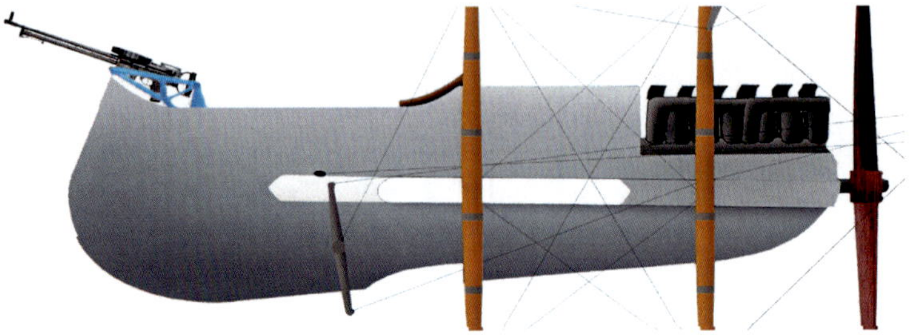

Created as MF 99 S in February 1915, renamed F 399 in August 1916 and Escadrille 525 in June 1917. On January 17, 1918, the unit is renamed as 2nd Serbian escadrille.

Locations:

- Belgrade in March 16, 1916.
- Palanka in April 6, 1915.
- Banjica in April 18, 1915.
- Detachment in Prahovo from May 8 to September 29, 1915.
- Detachment in Pozarevac from September 22 to October 16, 1915.
- Rajla in September 24, 1915.
- Detachment in Nish from October 7 to October 30, 1915.
- Detachment in Palanka from October 16 to October 30, 1915.
- Kraljevo in October 30, 1915.
- Raska in November 2, 1915.
- Mitrovica in November 9, 1915.
- Scutari (today Shköder, Albania) in December 3, 1915
- Evacuation to France via Italy on December 25, 1915.
- Escadrille MF 99 S dissolved, and recreated as MF 99 in France in June 9, 1916.
- Salonica-Sedes in June 25, 1916.
- Vertekop in July 27, 1916.
- Ostrovo, in September 8, 1916.
- Vrbeni, in November 23, 1916.
- Boresnica in December 4, 1916.
- Vrbeni in January 15, 1918.

Commanding Officers:

- Cdt Roger Vitrat in March 2, 1915.
- Cne Pierre Perrin de Brichambaut in June 9, 1916.
- Cne Georges Lecoq in August 21, 1917.
- Lt Paul Estienne in August 7, 1918.

Equipment:

- Farman MF XI in March 1915.
- Farman F.40 in June 1916.
- Dorand AR 1 during Summer 1917.
- Breguet 14 during late 1918.

The unit also had an escort section of 4 Nieuport fighters (12, 21, 23, 24, or 24bis) since June 1917.

Victories: 2 official victories.

Casualties: 1 pilot killed from enemy action.

History:

Escadrille MF 98 T, commanded by Commandant Roger Vitrat, was created in France in March 1915 in order to help Serbia left alone after the declaration of war of Austria-Hungaria in 1914. The unit joined the country via neutral Greece and the port of Salonica (at this time unoccupied by Entente troops), and established in Banjica, near Belgrade, on April 18, 1915. The detailed history of escadrille MF 99 S is told in Chapter 4. When Serbia was invaded by a general offensive of German, Austro-Hungarian, and Bulgarian troops in September 1915, Escadrille MF 99 S had to withdraw with Serbian army across the mountains of Albania, under the snow, and joined the Adriatic coast in December 1915. From here it was carried to Italy by boat, and then evacuated to France in late December 1915.

MF 99 S, which had just experienced one of the greatest odysseys of French aviation in the First World War, was then disbanded but was reconstituted on June 9, 1916 in France under the command of Captain Perrin de Brichambaut, intended to operate with the aviation of Armée d'Orient.

Above: Farman F.40 of escadrille F.399 wearing as personal insignia the monogram of Peter I, king of Serbia. (Collection Boris Ciglic)

However, it had nothing in common with the former MF 99 S except for one of its pilots, Adj Peté, and one of its observers, Captain Mavet, who were veterans of the first unit.

Very quickly renumbered F 399, the new squadron equipped with Farman F.40s landed at Mikra, south of Salonica, where the Serbian air force was in training. It was quickly put at the disposal of the Serbian army: after having been stationed in Salonica-Sedes on June 25, 1916, it reached the Vertekop airfield on July 27, 1916, thus following the Franco-Serbian troops in their offensive in the mountains of Macedonia towards the town of Monastir, which was bombed by 16 Serbian aircraft, including five from F 399, on August 16, 1916. Moving on September 8, 1916 to Ostrovo, the F 399 gains the base of Vrbeni (currently Itea, in Greece) on November 23, shortly after the capture of the city of Monastir by the Serbians. It then moved to Boresnica in December 1916, then returned to Vrbeni at an undetermined date, where it remained until the end of the war.

In January 1917, its Farman aircraft took on the distinctive signs defined for the squadrons of the Serbian army: a horizontal white band on the fuselage, and a red chevron on the wings, to which was unofficially added a wheel decoration and a personal decoration worn on the front of the nacelle. The squadron also had one Nieuport 12 two-seater and several Nieuport 21 single-seaters for escort duties. It only suffered one loss in aerial combat in 1917 – an observer died of his wounds, and a Farman was shot down by flak on July 30, 1917. The story of this Farman was published in the magazine "*La Guerre Aérienne Illustrée*": it was piloted by the squadron leader, Captain Perrin de Brichambaut, taking along as an observer Sergeant Perrin de Brichambaut, his own brother, whom he had had arrested because he had come late for the mission's take-off! The crew landed in the swamps and escaped unscathed, despite the threat of enemy fire. On August 21, the command passed to Captain Georges Lecoq and little happened in this sector of the front where Serbian and Bulgarian troops were frozen in the mountains. The squadron became the 2nd Serbian squadron on January 17, 1918, but unlike the 521 squadron, it was never 100% Serbian. Commanded by Lt Paul Estienne on August 7th 1918, it suffered its last combat loss four days later when one of its Nieuport escorts piloted by Corporal Daniel Poirier was shot down in aerial combat by the German ace Gerard Fieseler. Participating in the final offensive of September, the squadron then came under the command of the Serbian army.

Above & Below: Farman F.40 of escadrille F.399 wearing as personal insignia the monogram of Peter I, king of Serbia. (Above: Collection Boris Ciglic, Below: Collection Laurent Winer.)

Above: A Serbian general visits escadrille F.399 in early 1917 and this Farman F.40 with a cartoon-bird as personal insignia. Note the Nieuport fighters in the background. (Collection Boris Ciglic)

Below: Crashed Farman F.40 from escadrille F.399 in early 1917, with a horseshoe as personal insignia. The squadron markings, a white band on the fuselage, is clearly visible, unlike the red chevron under the wings. (Collection Boris Ciglic)

Facing Page, Above: Farman F.40 of escadrille F.399 in Boresnica during the first months of year 1917, wearing another personal insignia. (Collection Laurent Winer)

Facing Page, Below: Farman Contrary to appearances, this Farman F.40 decorated with a cross of Lorraine does not belong to the F.388 squadron (504) but to the F.399 of which it bears the regulatory markings, the white band on the sides of the fuselage. Captain Pierre Perrin de Brichambaut, squadron leader, stands just below the fuselage, with a light uniform and a riding crop under his arm. (Collection Boris Ciglic)

Right: Capitaine Pierre Perrin de Brichambaut, a pre-war aviator, had the rare privilege for a pilot from Armée d'Orient to see his name in the newsmagazine "*La guerre aérienne illustrée*" during the war.

Below & Following Page, Above: Capitaine Pierre Perrin de Brichambaut, flying with his own brother Georges as observer, crashed his Farman F.40 in a swamp on August 21, 1917. (Collection Laurent Winer)

N

Above, Right, & Facing Page, Below: Three views of Nieuport 12 n°1755 of escadrille 525 in mid-1917. The photos show the complete markings of the unit, a white band on fuselage, and a red chevron in and under the wings. (Collection Laurent Winer for top photos and Boris Ciglic for the others)

Above: Nieuport 21 n°2369 in Boresnica during mid-1917. (Collection Laurent Winer)

Below: French pilots in Boresnica in 1917, around a Nieuport 21. (Collection Boris Ciglic)

Above: Nieuport 23 n°1704 from escort section of escadrille 525 in Boresnica, 1917. (Collection Laurent Winer)

Above: Breguet 14 for escadrille 52

Maurice Farman MF 11 n°197, Escadrille MF 99S, Banjica (Serbia), Summer 1915.

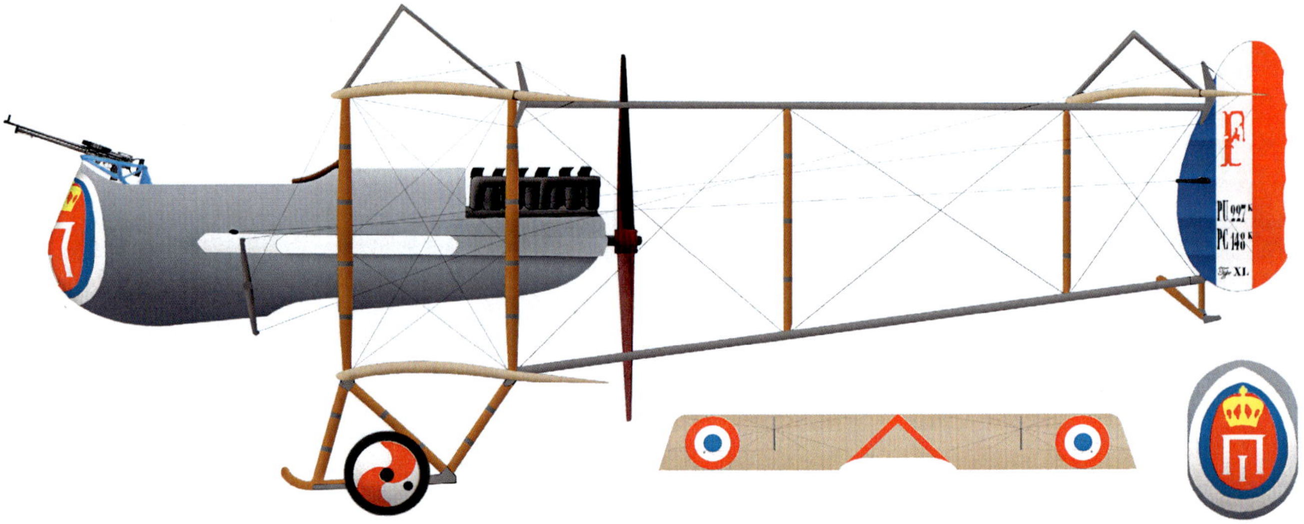

Farman F.40 (serial unknown), Escadrille F.399, Boresnica, Spring 1917.

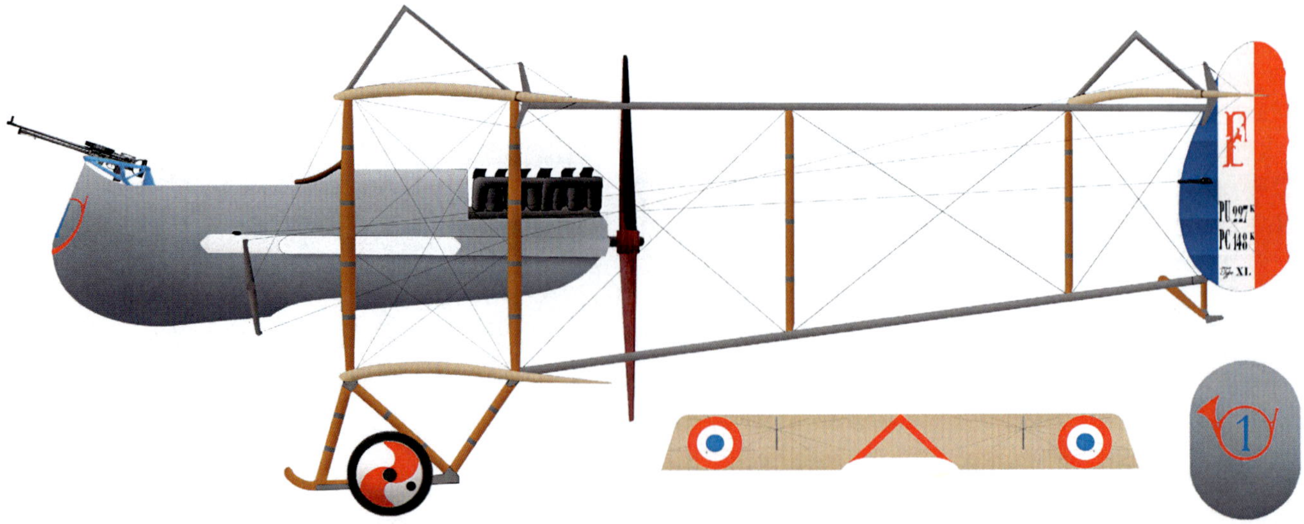

Farman F.40 (serial unknown), Escadrille F.399, Boresnica, Spring 1917.

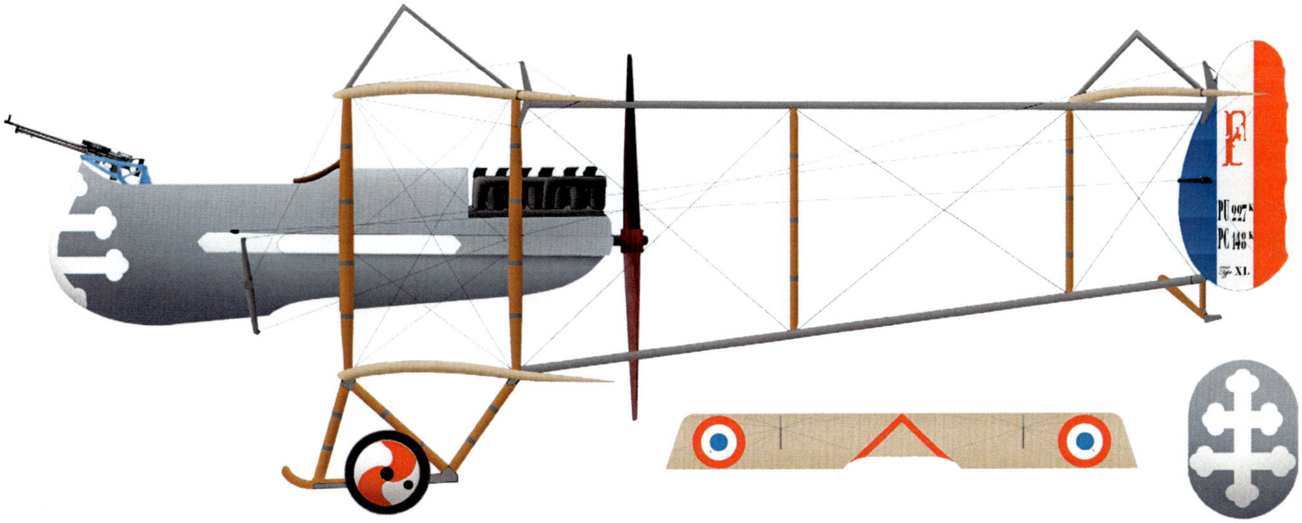

Farman F.40 (serial unknown), Escadrille F.399, Boresnica, Spring 1917.

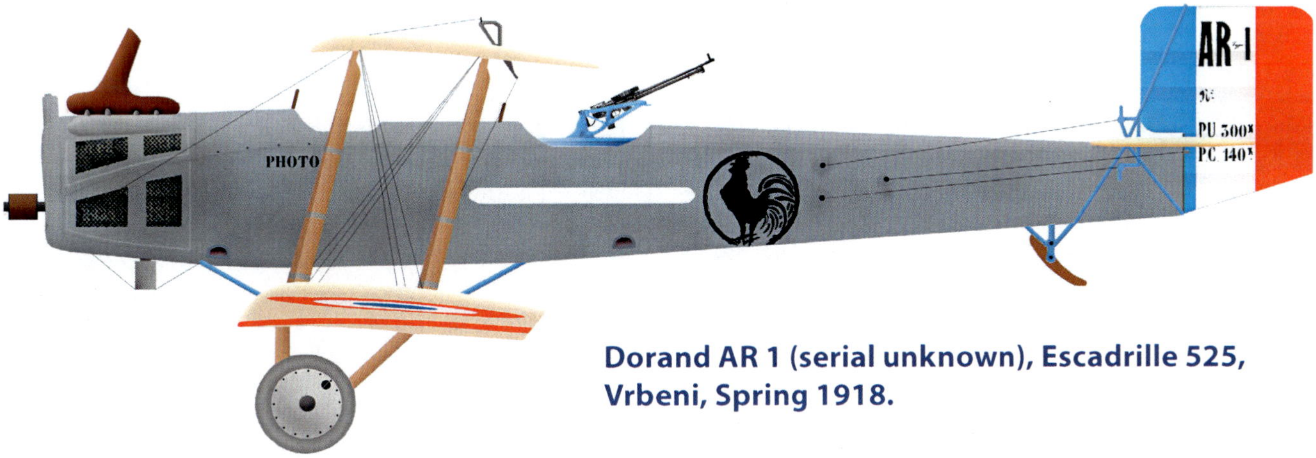

Dorand AR 1 (serial unknown), Escadrille 525, Vrbeni, Spring 1918.

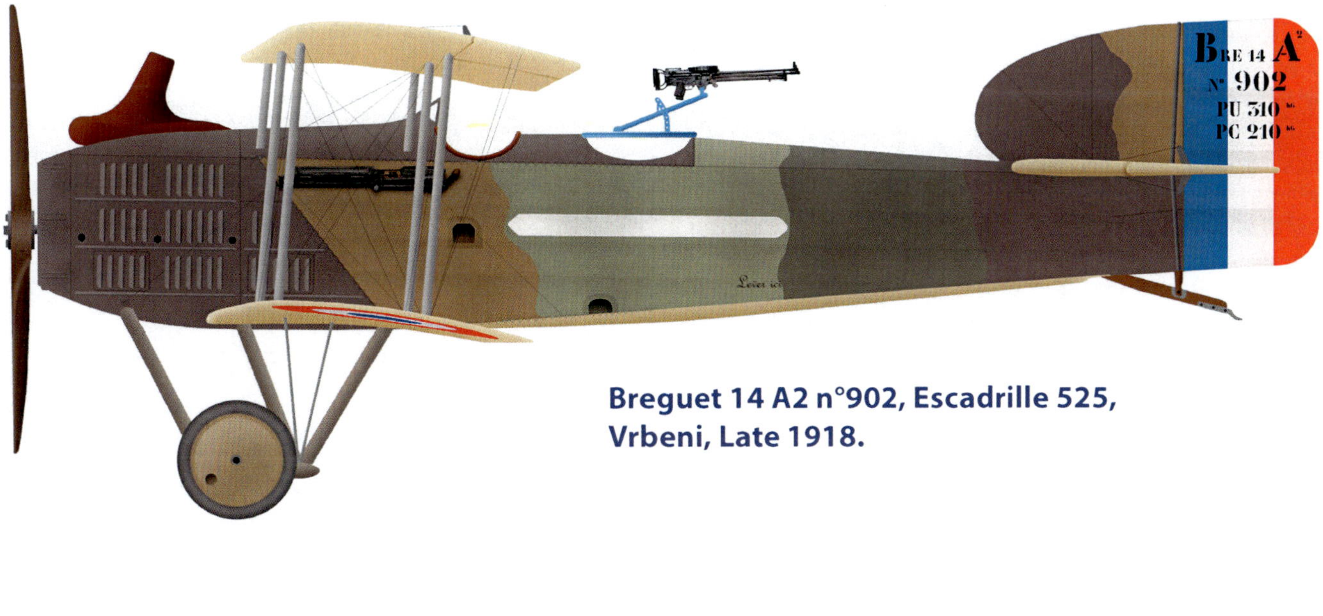

Breguet 14 A2 n°902, Escadrille 525, Vrbeni, Late 1918.

Escadrille 531

There is no evidence that Escadrille 531 ever had a unit insignia.

Locations:

- Gorgop from the creation of the unit (March 22, 1918) until the end of the war.

Commanding Officers:

- S/Lt Alexandros Zannas from the creation of the unit (March 22, 1918) until the end of the war.

Equipment:

- Nieuport 24, 24bis and 27 in March 1918.
- SPAD VII since June 1918.

Victories: 9 official victories + 3 unconfirmed

Casualties: 1 pilot killed from enemy action.

History:

The 531 squadron was created following a bargaining process between the French and Greek governments. Greece did not officially enter the conflict until 1917, at the end of an imbroglio between the King and his former Prime Minister, Eleftherios Venizelos. The latter, supported by the Entente nations who had forced the King to abdicate, was now in charge of the country and engaged the national army against the central empires by purging it of its most Germanophile officers. Modern equipment was lacking, especially aviation: the Greeks had only a handful of old Farmans, veterans of the Balkan wars of 1912–1913. Very concerned by the air force, Venizelos obtained in 1917 the creation of a corps squadron, the 1st Hellenic squadron or escadrille 532 equipped with Dorand AR 1 two-seaters and some fighters. He would have liked to have more but his demands were refused by the French army staff, which had big logistic problems to supply the Salonica front from Italian and French ports. The German and Austrian U-boats made life difficult for the supply cargo ships, and some equipment remained in the port of Brindisi. Venizelos then offered the services of the Greek merchant navy, in exchange for an additional squadron. Commander Denain, who discovered that he had several fighters in reserve in his Air Park at Sedès, accepted on condition that it was a fighter squadron. The 531 squadron, or 2nd Hellenic squadron, was created on March 22, 1918.

However, the creation of the 531 squadron corresponded to a military need: the sector of Monastir and the mountains were under the respective guard of the escadrilles 506 and 523, and the protection of the city of Salonica was ensured by escadrille 507. There was nothing in the Vardar sector where the Greek troops and the 122nd French Infantry Division were stationed, except for the fighters of the protection patrol of escadrille 502, sometimes reinforced by aircraft from Salonica. Captain Pacquin, commander of the Vardar air sector, insisted on assigning the best fighter pilots scattered in the other squadrons: he was quickly proved right since Adjudant-Chef Dieudonné Costes, considered as the best fighter of the Armée d'Orient with 4 victories, 2 of which were confirmed, was assigned there. For political reasons, the only Greek pilot with a fighter licence, SLt Alexandros Zannas, a convinced Venizelist, was assigned as unit leader.

The pilots moved to the Gorgop airfield, which they shared with the escadrilles 502 and 532. Nieuports were delivered to them and decorated with Greek cockades as attested by several official documents, but also by the testimony of one of the pilots, Sergeant Paul Andrillon, who wrote to his parents: "*I visited my old crate in Greek colors (sky blue and white like the racing club), it is in good condition and*

Above: Photo of pilots in the airfield of Gorgop, during Summer 1918. From right to left are the following fighter pilots from escadrille 531: Adjudant-Chef Dieudonné Costes (6th from right), Sergeant Paul Andrillon (7th), Sous-Lieutenant Maurice Ripoche (8th), Sous-Lieutenant Alexandros Zannas (9th), Maréchal des Logis Maurice Lashermes (12th), Adjudant Basile Sauné (14th). Costes and Sauné were the only two French aces of the Orient front. (Collection Benoit Henriet)

will be ready to fly tomorrow morning if the weather is good, which it will be, because I see the sky clearing up hour by hour."

With Gorgop just a stone's throw from the front line, skirmishes against German observation aircraft were not long in coming. As early as April 30, 1918, adjudant Basile Sauvé and Sergeant Ripoche shot down one in the morning, which fell south of Parovica. While going the same day to photograph his victim at the controls of a Sopwith 1A2 in order to obtain the homologation, he met five German fighters which are attacked by his escort made up of the pilots Costes, Andrillon, and Lashermes, who claimed one of them, without being able to confirm it. However, the trio shot down a reconnaissance plane the next day.

A third official victory was won by Adjudant Basile Sauné and Maréchal des Logis Robert Lambert on May 4, then at the end of the month the squadron supported the Greek troops who seized the Skra di Legen mountain, while the squadron received its first SPAD VII fighter. The month of June was to prove to be a great period for the 531 squadron, which won no less than five official victories, allowing Adjudant-Chef Dieudonné Costes and Adjudant Basile Sauné to become aces when they shot down their 5th official victory on June 13th 1918. Commander Denain then sent his two aces to the Monastir sector on June 17, 1918, so that they could reinforce escadrille 506, which was rather badly damaged by the enemy air force. Unfortunately, on June 20, 1918, during a patrol where Costes was unable to join due to a mechanical problem, Basile Sauné was attacked by a patrol of German fighters and was shot down by the ace Gerard Fieseler.

Costes then returned to Gorgop with his squadron and tried, unsuccessfully, to attack a Drachen on September 8, 1918, in the company of his usual teammates Maurice Lashermes and Paul Andrillon. He won the last confirmed victory of the squadron on September 10, shortly before the general offensive of the Orient Front launched on September 14, 1918, which put Bulgaria out of the war. A daring coup de main was made on the mountains in the center of the front, by French and Serbian troops who benefited from an intense artillery preparation, whose large pieces had been laboriously mounted on the mountains at arm's length. Costes and some of his comrades left Gorgop to reach Vertekop where the 523 fighter squadron was located, from where they carried out missions of air cover and ground strafing of the Bulgarian positions. A breach was quickly created, and the cavalry division of General Jouinot-Gambetta entered it, whose troops galloped on the enemy's rear until Prilep, cutting in two the routed Bulgarian troops.

Above: Panoramic view of the Skra di Legen mountain, where escadrille 531 was engaged in late May 1918 to support the attack of Greek troops.

On 28 September, the fighters of escadrille 531 landed on the Hudova field, which had been abandoned by the Germans. Two days later, the armistice was signed with Sofia, putting an end to the fighting. The French pilots being demobilized, the unit came under the control of the Greek government but was quickly disbanded in 1919, its fighters distributed among the various Greek corps squadrons.

Above: In front of a SPAD VII are standing, from right to left, Sergeant Paul Andrillon, Sous-Lieutenant Alexandros Zannas and Maréchal des Logis Maurice Lashermes. (Collection Benoit Henriet)

Above: SPAD VII of escadrille 531. The plane does not seem to wear Greek cockades, despite documentary evidence that all the first Nieuports of the unit had blue/white/blue cockades. (Collection Benoit Henriet)

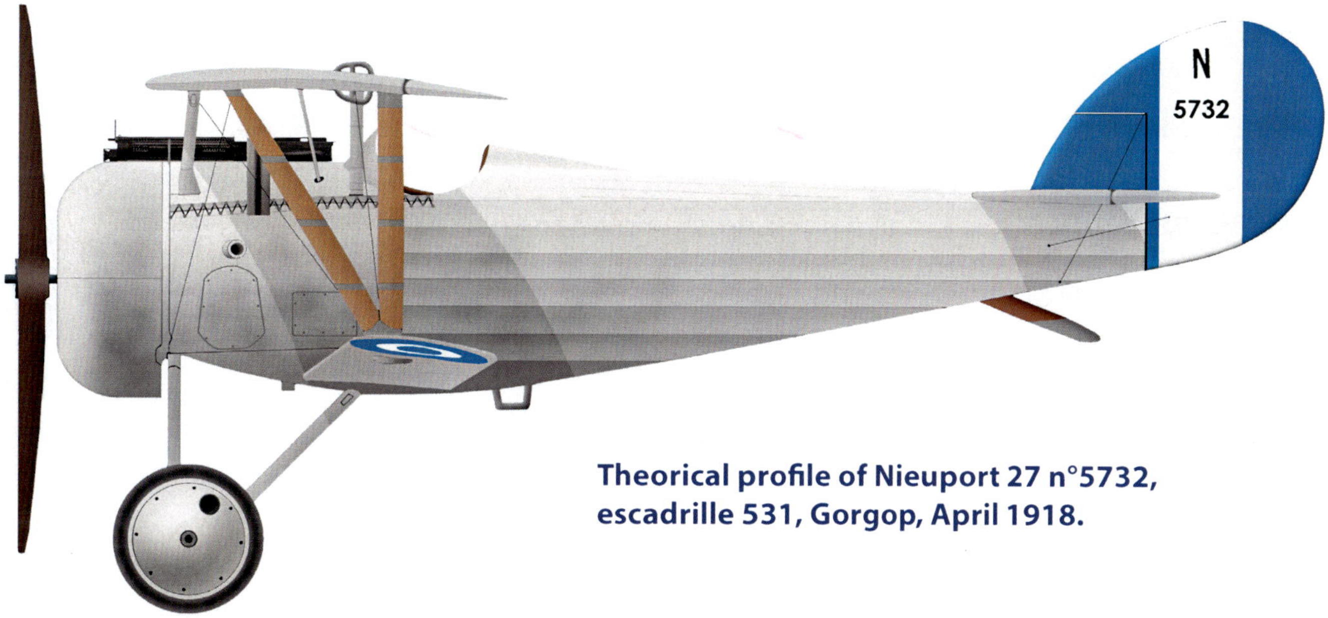

Theorical profile of Nieuport 27 n°5732, escadrille 531, Gorgop, April 1918.

Escadrille 532

There is no evidence that Escadrille 532 ever had a unit insignia.

Locations:
- Gorgop from the creation of the unit (December 20, 1917) until the end of the war.

Commanding Officers:
- Cne Henri Cottaz from the creation of the unit (December 20, 1917) until the end of the war.

Equipment:
- Dorand AR 1
- Sopwith 1A2

The unit was given 1 Breguet 14 A2 in January 1918.
The unit also had a section of Nieuport fighters (24, 24bis, or 27) for escort duties.

Victories: 2 official victories.

Casualties: 1 pilot killed from enemy action.

History:
The 532 squadron, or 1st Hellenic squadron, was created at the end of 1917 to be the first aerial component of the unified Greek army which was to go to war alongside the Entente. An army corps squadron was then organized, equipped with Dorand AR 1 and Sopwith 1A2 observation planes and including a few escort Nieuport fighters. Placed under the command of Captain Henri Cottaz, the squadron was declared operational on December 20, 1917 and settled on the airfield of Gorgop which it would not leave during the conflict. All the licensed Greek pilots it had been possible to find in the ranks of the French army were assigned to it; these being in insufficient number, it was necessary to supplement the workforce by French pilots for the Dorand and for all the fighter pilots. Among the latter, Adjudant-Chef Dieudonné Costes, who also piloted the first Breguet 14 delivered to the squadron before being transferred in March 1918. The operational beginnings were laborious: in addition to a shortage of petrol during the winter of 1917 –1918 which limits flights, the Greek observers were judged very severely by Captain Pacquin, head of aviation in the Vardar sector, who thinks that Lt de Boysson is the only observer in the squadron suitable for missions.

Performance is improving over time and the proportion of Greek pilots was gradually increasing. After the armistice, the squadron passed under the control of the Greek army of which it became the squadron (Moira) A.

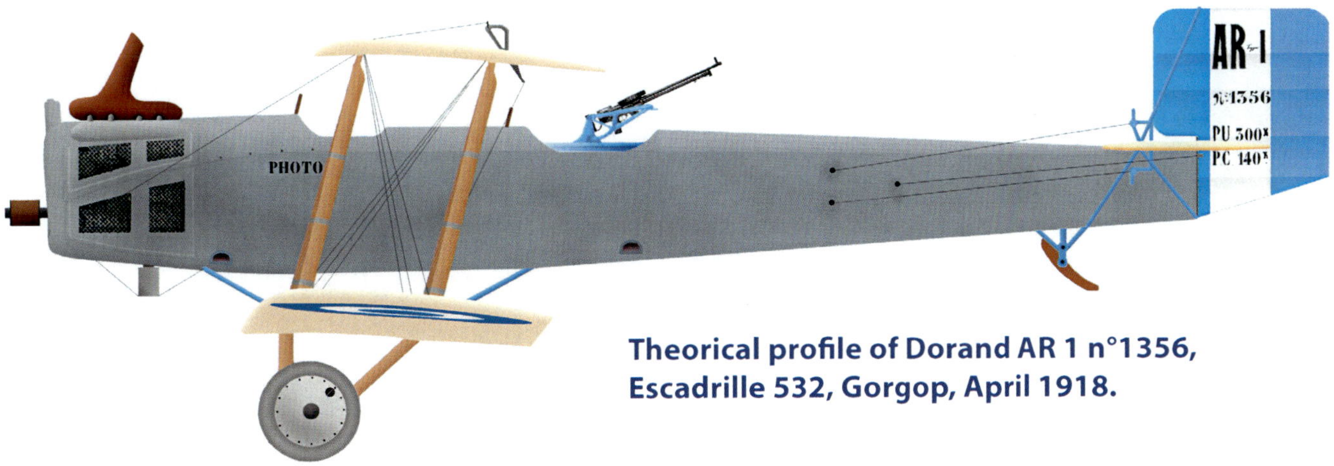

Theorical profile of Dorand AR 1 n°1356, Escadrille 532, Gorgop, April 1918.

Above: The airfield of Gorgop in 1918, shared with 3 units.

Escadrille 533

There is no evidence that Escadrille 533 ever had a unit insignia.

Locations:

- Dimitric from the creation of the unit (July 10, 1918) until the end of the war.

Commanding Officers:

- Lt François d'Anglejean-Châtillon from the creation of the unit (July 10, 1918) until the end of the war.

Equipment:

- Dorand AR 1
- Breguet 14 A2 in Late 1918.

The unit also had a section of Nieuport fighters (24, 24bis or 27) for escort duties.

Victories: No victories.

Casualties: No casualties

History:

533 Squadron, or 3rd Hellenic Squadron, was created on July 10, 1918 as an organic squadron of the 1st Greek Army Corps. Although its commander, Lieutenant d'Anglejean-Châtillon, was a French officer and a veteran of the Armée d'Orient, all the observers and most of the pilots were Greek – the few French pilots were mainly pilots of the Nieuport escort fighters. The squadron was equipped at its creation with two-seater AR 1 observation aircraft, and later with some Breguet 14s. Based at Dimitric during the whole conflict (east of the front), it carried out its missions of observation corps and artillery adjustment without suffering any loss or victory in the air, before passing under the complete control of the Greek army at the armistice, where it took the name of Moira B and was engaged in 1919 at Smyrna in the terrible Greek-Turkish war of 1919 to 1922.

Above: Dorand AR 1 with Lorraine engine from escadrille 533 in Dimitric during Summer 1918.

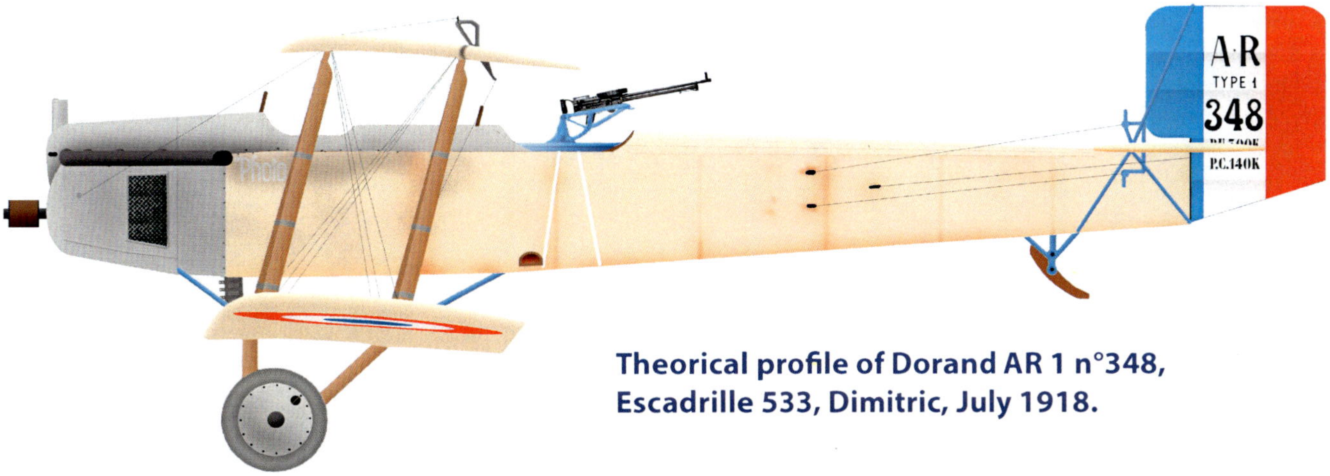

Theorical profile of Dorand AR 1 n°348, Escadrille 533, Dimitric, July 1918.

Escadrille 534

There is no evidence that Escadrille 534 ever had a unit insignia.

Locations:
- Salonica-Lembet from the creation of the unit (August 3,1918) until the end of the war.

Commanding Officers:
- Lt Economacos from the creation of the unit until the end of war.

Equipment:
- Dorand AR 1
- Breguet 14 A2 in late 1918.

The unit also had a section of Nieuport fighters (24, 24bis or 27) or SPAD VII for escort duties.

Victories: No victories.

Casualties: No casualties

History:
Escadrille 534 or 4th Hellenic squadron, has left few traces in the French archives and it is not even certain that it had an operational role during the First World War.

A document dated August 3, 1918 tells us that the formation of the squadron was currently completed but that it would only be used on the front when an operational sector is found for it. It is an army corps squadron, equipped with observation planes which are probably Dorand AR 1 accompanied by some Breguet 14 and maybe some fighter planes for the escort of the latter (Nieuport 24 or SPAD VII). The training courses for Greek personnel having had time to produce their first student pilots, the staff of the squadron was entirely composed of Greeks and the command was entrusted to Lt Economacos (also mentioned under the spelling Oikonomakos). On October 18, 1918, when the great offensive took place that broke through the front and led Bulgaria to ask for the armistice, the 534 squadron was stationed at Lembet, near Salonica, which seems to indicate that it remained in training or waiting for an assignment.

After the armistice, a document dated December 10, 1918 mentions it still in Salonica, equipped with six Breguet 14 A2 and three SPAD fighters. The Greek government of Venizelos, wishing to attract the good graces of the Franco-British governments whose support it needed for its territorial claims in Turkey, decided to send escadrille 534 to participate in the fight against the Bolsheviks in Ukraine.

The squadron landed on April 1st 1919 in Odessa and was immediately assigned a SPAD reassembled by the French aviation park in Galatz. It settled down near this one, on the same ground. Its strength was to be increased to 15 Breguet 14 B2s in order to make it a bombing squadron. It is not known if the 534 squadron had time to have any operational activity in Ukraine. Indeed, because of the situation in the area of Smyrna where the Greek troops must face the actions of the Turkish irregulars, it is envisaged to transfer there urgently the squadron. The park of Galatz receives then the order not to provide it any aircraft but to continue the training of the Greek pilots on its Breguet 14, the endowment in aircraft of the squadron having to intervene in Salonica.

The 534 squadron was then transferred to the Smyrna region at an undetermined date (probably in July 1919), and took the name of Moira C of the Greek army, where it fought until 1922 with its Breguet 14 B2s, based in northern Turkey.

A Greek Breguet 14 during the Greco-Turkish war in 1921. Escadrille 534 became Moira C during this war.

Above: Breguet 14 A2, captured by the Turks in August 26, 1922. Escadrille 534 used its material from WW1 until 1922 during the long Greco-Turkish war.

Escadrille 575

Escadrille C.575 never had a unit insignia.

Locations:
- Lyon on July 11, 1918.
- Port-Saïd (Egypt) in August 27, 1918.
- Beyrouth (Lebanon) on April 23, 1919.

Commanding Officers:
- Cne Robert de Braquillanges during all WW1.

Equipment:
- Caudron G.6

Victories: None.

Casualties: None.

History:
The C.575 squadron was a latecomer to the war, as it was created in Lyon on July 11, 1918, to be the air component of the French Detachment of Palestine-Syria (DFPS), fighting with British troops in Palestine. It did not depend from Armée d'Orient in Salonica.

The unit landed in Port Said on August 27, 1918 and was immediately directed to Ludd, a village in Palestine about ten kilometers southeast of Jaffa, not far from the front line with Turkish troops. Unfortunately, the British authorities sent the convoy back to Port Said where the aircraft were to be assembled at the beginning of September 1918, so that the battle of Meggido, which began on the 19th of that month and saw the rout of the German-Turkish troops, took place without escadrille C.575. The British armies then conquered all of Palestine and progressed to Lebanon and Syria, which were destined to be occupied by France according to the Sykes-Picot agreements.

The C.575 then moved to the Lebanese city of Beirut, where an advance party of eight Caudron G.6s left in the holds of a ship on December 10th 1918, followed by Captain de Braquilanges on January 26th 1919. The last French soldier left the Port-Saïd base on May 16, 1919 to join the unit which took part in numerous battles against the Syrian insurgents, being re-equipped with Breguet 14s in 1920.

Above: Colonel Jean Philpin de Piépape, commanding officer of D.F.P.S. (French Detachment of Palestine and Syria), on which escadrille C.575 depended. (Photo Paul Castelnau)

Above & Below: Caudron G.6 n°5472 of escadrille C.575 in the airfield of Port-Saîd in late 1918.

Above & Facing Page: One of the Caudron G.6 in Port-Saïd received a colorful shark mouth in 1918–1919.

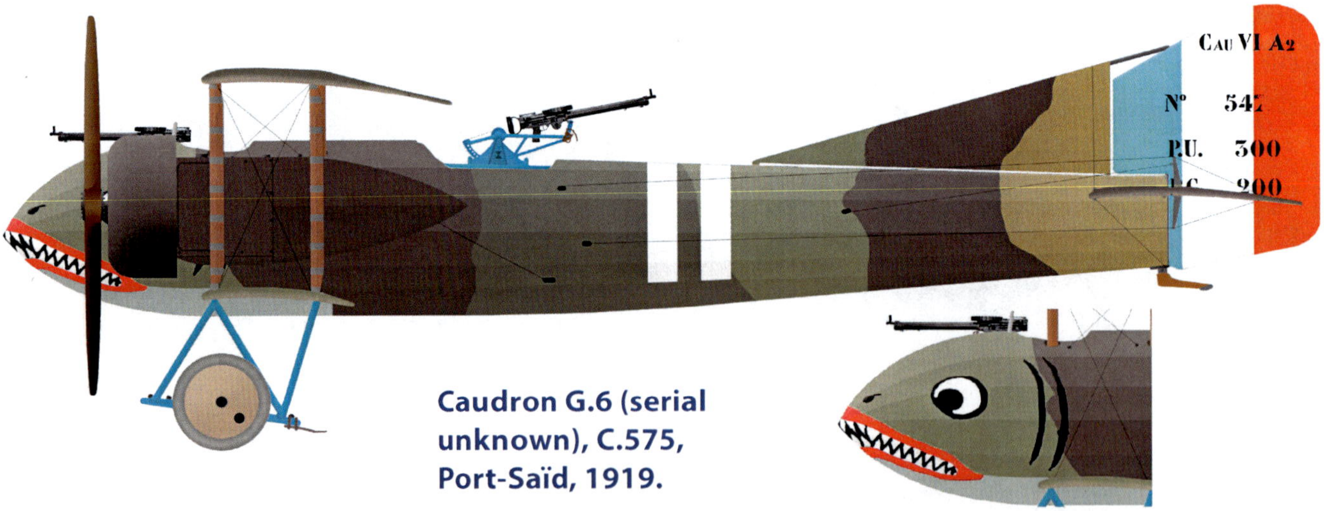

Caudron G.6 (serial unknown), C.575, Port-Saïd, 1919.

Miscellaneous Units

Short-lived units all depending from the command of Armée d'Orient.

Escadrille MF 88 bis

Insignia: No insignia

Operational Life: February 1916 – June 1916

Location: Salonica-Sedes

Commanding Officier: Cne Robert de Brun

Equipment: 8 Farman MF XI bis

Victories/Casualties: None.

History:
In order to provide an organic squadron to each of the large French units, escadrille MF 88 bis was created in February 1916 on the field located on the route de Serres, near Salonica (agricultural school) by duplicating escadrille MF 88 (future 504). The new squadron thus created was intended to carry out reconnaissance missions for the 17th colonial infantry division, even though this division, like all the French troops entrenched in Salonica, was not in contact with the enemy but was on guard around the Greek city...

Consequently, its commander, Captain de Brun, first sent his Farman MF XI bis to carry out exploratory flights on the Chalkidiki, east of Salonica. On May 13, 1916, the Farman n°1253 piloted by Adjudant Roth (a Serbian) and having Lt. Bordes as observer, was blown away by the wind and crashed on landing in a hangar, causing the loss of the aircraft and some scares for its crew who escaped unhurt. The missions were more and more distant, offensive reconnaissance where bombs are thrown on objectives as far away as the Strumica train station (April 20) and Livunovo (north of Petric, May 9). May 14, 1916 saw the last notable event of the squadron: Farman n°1251, piloted by MdL Hugon with Lt. Chambaret as observer, carried out a training mission and returned to land when it was crossed by a Greek peasant riding a donkey. The animal panicked and was hit by the aircraft, which broke one of the landing gear legs. Hugon skillfully managed to land on one leg, limiting the damage to a minimum; the Greek peasant, a certain Nicolas Koudla from the village of Karaïsi, was killed on impact.

The MF 88 bis was disbanded on June 20, 1916 and its personnel distributed to the other squadrons of the Eastern Front.

Section Italienne (Italian Section)

Insignia :

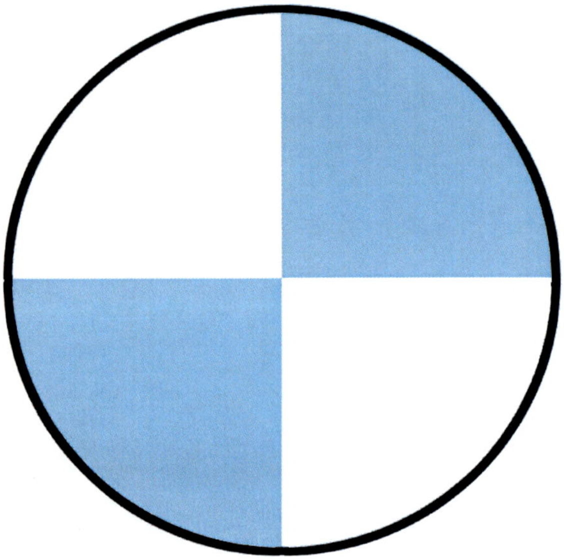

Operational Life: September 1916 – December 1916

Location: Snevce.

Commanding Officer: Cne Lucien Galloy.

Equipment: 3, then 4 Farman F.40

Victories/Casualties: None.

History:
In order to provide air cover for the 35th Italian Infantry Division stationed near Lake Doiran, a squadron detachment was created on September 3, 1916 under the command of Captain Lucien Galloy. Initially composed of three Farman F40s (increased to four in November), the unit, attached to escadrille F 388, was based in Snevce on September 9, 1916. Its aircraft were marked with a disc divided into four white and blue parts. We lose track of the detachment in December 1916, probably disbanded at the very beginning of 1917 after the arrival of the first Italian air detachments on the Salonica front.

Section de Bombardement (Bombing Section)

Insignia:

Operational life: November 1916 – February 1917

Location: Topcin.

Commanding officer: Lt André de Saint-Severin

Equipment: 6 Farman F.40 and F.60.

Victories/Casualties: No victories. One pilot lost (captured by the enemy).

History:
While the Orient Bombardment Group (GBO) was disbanded in June 1916, the need to have a squadron dedicated to bombing the enemy was felt in the Orient Aviation. The Voisin LAS of the GBO being out of breath, Major Denain decided to create a bombing section of six Farman aircraft, on November 15, 1916, under his direct orders in Command of the Allied Armies (CAA). Placed under the command of Lieutenant André de Saint-Severin, the small formation was based at Topcin and was attached to escadrille N 391 with which it shared the airfield.

Below: Farman F.40 n°3304 of Section de Bombardement, in 1916–1917.

Above: Damaged Farman F.40 seen in Boresnica in 1917 – very likely a former airplane from Section de Bombardement.

The operational activity of the section was limited by the bad winter weather. Nevertheless, some raids were carried out: on December 12, 1916, 4 Farman bombed Livunovo where a Bulgarian airfield was located. The French aircraft were attacked on their way back and one of them, piloted by SLt Richard and taking Sgt Léglise as an observer, had to land behind enemy lines where it was captured. During the night of February 5–6, 1917, only one aircraft of the bombing section (S/Lt Jean Sisson, and Sgt Blondel) attacked the Hudova airfield.

The bombing section was no longer in the news after this date, and was probably disbanded: the Farmans were slow and vulnerable to German fighters, and their action could only be ineffective, as Commander Denain wrote in a report dated February 17, 1917: "*We have to give up on major bombing raids, because there is no bombing squadron, and the Farmans are unable to undertake them now without excessive risk.*"

Despite the attempt to recreate a bombing section with the Sopwiths delivered in the spring of 1917, it is only much later, with the arrival of the Breguet 14s of escadrille 510 squadron, that the Orient Army will be able to dispose of an effective bombing force.

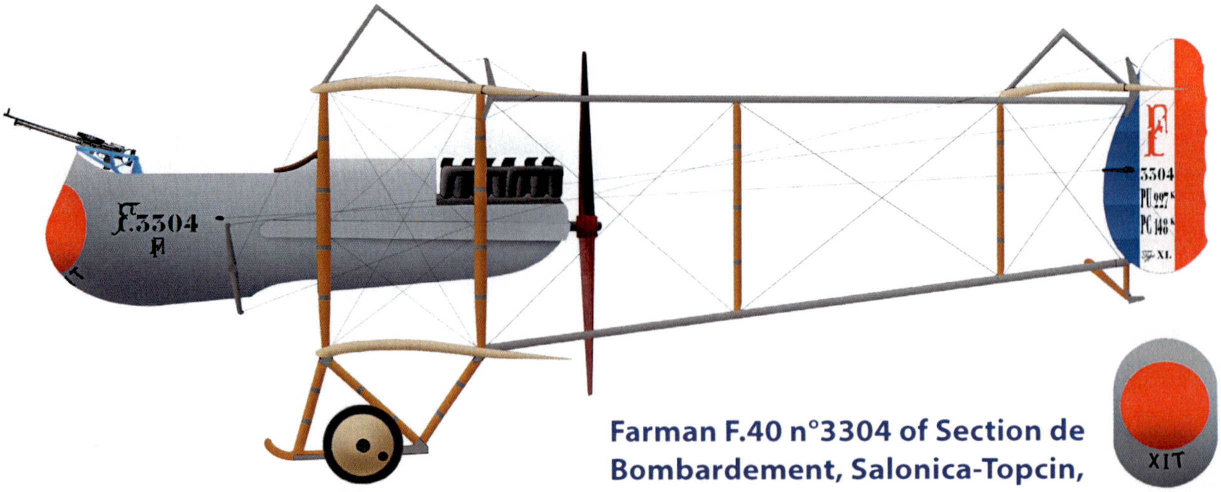

Farman F.40 n°3304 of Section de Bombardement, Salonica-Topcin, Late 1916.

Détachement de Koritza

Insignia: No insignia

Operational Life: January 1917 – September 1917

Location: Koritza (today Korçë, Albania).

Commanding Officer: Lt Georges Prevost (observer)

Equipment: 2 Farman F.40 and 1–2 Nieuport 21.

Victories/Casualties: No confirmed victory + Unconfirmed. No casualties.

History:
At the end of 1916, on the edge of the French front, in the mountains of Albania, was the region of Koritza, occupied by the Greek royal administration since the Balkan wars. Very isolated by mountains served by very poor roads, the region is at the center of all traffic and swarmed with spies in the pay of the central empires. French troops occupied it with Colonel Descoins at its head, who quickly noticed that the Albanian-speaking population of the region was very hostile to the Greek Venizelist administration that they tried to impose on them. He then decided to proclaim on December 10, 1916, with the support of General Sarrail, a "Republic of Koritza" entrusted to local notables, under the close supervision of the French army. This policy would allow until the end of the war to secure this end of the Orient front and to rally several Albanian bandits to the cause of France.

From the aviation point of view, Colonel Descoins officially received on January 7, 1917 a detachment of two planes (two Farman F.40) which settled near the city on an improvised ground. It was commanded by Lieutenant Georges Prevost (observer). This detachment was later expanded with a Nieuport 21 escort aircraft and included a total of four pilots and three observers. These aircraft carried out several observation and attack missions in the region facing the Austrian troops – an enemy air attack took place on Koritza on February 16, 1917, and probably led to the reinforcement of the local detachment, as an escort Nieuport was reported shortly afterwards. A victory was claimed by this unit on 31 March 1917, when the Farman flown by MdL Lesterpt with Lt Prevost as gunner was attacked by a single-seater Halberstadt which seemed to be hit by the Farman's return fire (140 rounds fired) as well as by the intervention of the escorting Nieuport 21 flown by Lt Marse, which emptied a magazine of its Lewis gun (47 rounds). The enemy aircraft was seen to land urgently south of Lake Ohrid by French soldiers of the 76th Infantry Division, but the victory, not mentioned in the Orient Army bulletin, was not certified. In the spring-summer of 1917, the Koritza detachment took part in the "mass" attacks carried out by the AFO air force in the lake region, by pooling the maximum number of Farman bombers and Nieuport fighters on a specific objective.

In September 1917, escadrille 501 moved to Koritza with its Dorand AR 1 and Nieuport escort aircraft. The Koritza detachment was immediately disbanded.

Escadrille 534 bis

Athens detachment in August 1917, renamed Escadrille 534 bis in August 1918.

Insignia: No insignia

Operational Life: August 1917 – October 1918

Location: Athens – Paleo Faliro.

Commanding Officer: S/lt Emile Bruninghaus in August 1917, replaced by Cne Robert (French military attaché in Athens) in July 1917.

Equipment: 3 Nieuport fighters, plus a training section of 3 Farman F.40.

Victories/Casualties: None.

History:
After a year and a half of a delicate situation, to say the least, with, in Athens, a legal Germanophile government led by King Constantine I and, in Salonica, a rebel government favorable to the Allies led by the former elected Prime Minister Elefterios Venizelos, the question of the Greek government was clarified in favor of the latter in June 1917 when King Constantine was forced into exile by the Franco-British troops.

A note dated July 22, 1917 signed by Commander Denain, head of the French air force on the Orient Front, expressed concern about the situation of the city of Athens, which was within the reach of the German air force and which should be protected by a fighter patrol of two Nieuport fighters during the day and a Farman at night. An evaluation mission was then organized from August 3 to 23, 1917, in which the pilot Adjudant-Chef Dieudonné Costes participated, to study the constitution of an airfield from which the protection planes would leave – Captain Robert, French military attaché in Athens, was in charge of the operations. The airfield of Paleo Faliro (Old Phaleros), near the port of Piraeus, was selected and a patrol of three Nieuport fighters under the command of Slt Emile Bruninghaus was sent there.

No German aircraft showed their nose over Athens for the rest of the war and the pilots of the patrol were soon idle. One of them, Adjudant Basile Sauné, relieved his boredom by performing acrobatics in the sky of the capital, which earned him the nickname of "Mr. Toumba" (barrel, in Greek) by the population, and was taken up by the press. He excelled especially during the ceremonies of the Orthodox New Year. When he was not performing as an acrobat, most of his missions as well as those of his comrades consisted of liaison flights between Athens and Salonica.

On January 14, 1918, three reformed Farman aircraft were sent to Athens with their three French pilots and an observer from the Greek army. They were put at the disposal of the Greek Army Staff to train new observers – the Athens detachment was now expanded to include a training section. On July 10, 1918, 2nd Lt Bruninghaus was transferred to take command of escadrille 531 in formation, but finally joined escadrille 508. He was replaced for the rest of the war by Captain Robert of the French military mission in Athens, although he was not an aviator. The Athens detachment was then renamed 534 bis squadron in a document dated August 3, 1918, and did not experience any significant events until the end of the war, when it was quickly disbanded and its aircraft transferred to the Greek army.

Escadrille côtière de Panomi (Panomi Coastal Squadron)

Insignia: No insignia

Operational Life: October 1917 – October 1918
Location: Epanomi.

Commanding Officier: S/lt Jean Hugon.

Equipment: 6 Farman F.40

Victories/Casualties: None

History:
In order to monitor the naval approaches to the port of Salonica, a coastal surveillance squadron was created in November 1917, at the own initiative of Major Denain who equipped it with second-hand Farman F.40s that were withdrawn from service in the Orient front units. The squadron would have been equipped with six Farmans, and is indifferently designated "Naval squadron" or "Farman detachment for coastal surveillance" then finally "Panomi squadron", from the name of its base located near the Greek village today called Epanomi.

Commanded by S/Lt Jean Hugon (a pilot), it was initially composed of four other pilots and three observers. Its six mechanics, supervised by a sergeant, came from the military aviation, but the maritime aviation was called upon to provide the rest of the ground staff: a discipline officer, an accountant, a secretary, an armorer, two cooks and 15 enlisted men.

Attached directly to the CAA (Command of the Allied Armies, in Salonica), as well as to the escadrille 502 for its accounting of the technical material, its surveillance of the Gulf of Salonica have been carried out without the slightest incident having caused a loss in the personnel. The Panomi squadron was dissolved shortly after the armistice.

Farman F.40 n°3066, Escadrille côtière de Panomi, 1918.

Right: Farman F.40 from Escadrille côtière de Panomi. The Lewis gun turret seem to have been removed, as no German planes are supposed to be met in this area of patrol, far away from the frontline. (Collection Albin Denis)

Below: Pilots and mechanics of Escadrille côtière de Panomi – a mixture of Army and Navy personnel. (Collection Albin Denis)

Navy Units

All the following air units were not under the command of the Armée d'Orient, but under the Ministry of the Navy and worked to secure the maritime lines to supply the Armée d'Orient in Salonica. Most of them were *Centres d'Aviation Maritime* (CAM – Marine Aviation Center) equipped with seaplanes, but there were also two units *from Aéronautique Militaire* equipped with landplanes and attached to the command of the naval army. The CAM also organized some remoted supply bases, called *Postes de Combat* (PC – Combat Posts)

Centre d'Aviation Maritime de Port-Saïd

Insignia: No insignia.

Location: Port-Saïd (Egypt)

Commanding officer: Lieutenant de vaisseau Henri de l'Escaille from November 1914 to May 1916.

Equipment: Nieuport monoplanes.

Victories/Casualties: 2 pilots lost by enemy action (captured), 1 lost from accident.

History:
The Nieuport monoplanes of the French Navy came in Port-Saïd on November 30, 1914 and formed escadrille de Port-Saïd. Their reconnaissance flight helped to secure the Suez canal from the Turkish army in 1914, then they flew many reconnaissance all over the Eastern Mediterranean and the Red sea, using several cargo boats as mobile bases. The history of this unit is detailed in Chapter 2.

On January 1, 1916, the escadrille administratively became a Centre d'Aviation Maritime. On May 17, 1916, it was disbanded and its personnel and aircraft left to form the CAM of Argostoli.

Above: Nieuport monoplane N 16 in Centre d'Aviation Maritime de Port-Saïd, on the shore of Suez Canal. (Collection ARDHAN)

Centre d'Aviation Maritime de Salonique

Insignia: No insignia. The center was given the letter "U" as identification letter, but due to lack of photo evidence, it was probably never painted on the flying boats.

Location: Salonica. The CAM organized several combat posts :

- Cassandra, from December 15, 1916, to the end of the war.
- Skiatho, from March 24, 1918, to the end of the war.

Commanding officer:

- Lieutenant de vaisseau Maurice Thierry on December 2, 1915.
- Lieutenant de vaisseau Maurice Larrouy on February 1, 1916.
- Lieutenant de vaisseau Alfred Cintré on March 24, 1916.
- Lieutenant de vaisseau Henri Lefranc on February 6, 1917.
- Enseigne de vaisseau 1ere classe Jean Rousselin (interim) on February 12, 1917.
- Lieutenant de vaisseau René de Prunières on April 23, 1917.
- Lieutenant de vaisseau Pierre Darchis de Lantier on November 20, 1918.

Equipment: The CAM Salonique had 8 flying boats on line (+ 4 in reserve) in January 1916. In October 1916, these number rose to 12 flying boats on line, + 12 in reserve.

The planes used were (several types used in the same time) :

- FBA 100 hp Rhome (Type C) on December 1915,
- The Caudron R 150 hp prototype from March 1916 to March 1917,
- Donnet-Denhaut 160 hp Canton-Unné on May 1916,
- FBA 150 hp Hispano on November 1916,
- FBA 200 hp Hispano on July 1918,
- Tellier 200 hp Hispano with cannon on July 1918.

Victories/Casualties: The CAM Salonique bombed eight times enemy submarines. Two of his pilots were lost by accident.

History:
CAM Salonique (Salonica) was created on November 6, 1915 to be send to Moudros island, in order to give air protection to the French ships from the Gallipoli expedition. When the unit settled in Moudros on November 24, 1915 the Gallipoli expedition was almost finished and the CAM receives the order to move to Salonica one month later. It settled just on the south of the city, in the shores of the Great Air Park.

CAM Salonique, with his two Combat Posts in Cassandra and Skiatho and the help of Escadrille de Panomi in 1918, will guard the naval approaches of the city until the end of the war, attacking eight times enemy submarines menacing the naval convoys of Armée d'Orient. It would be disbanded on May 1919.

Right: FBA 100 hp Rhone (type C) used by CAM Salonique in 1916.

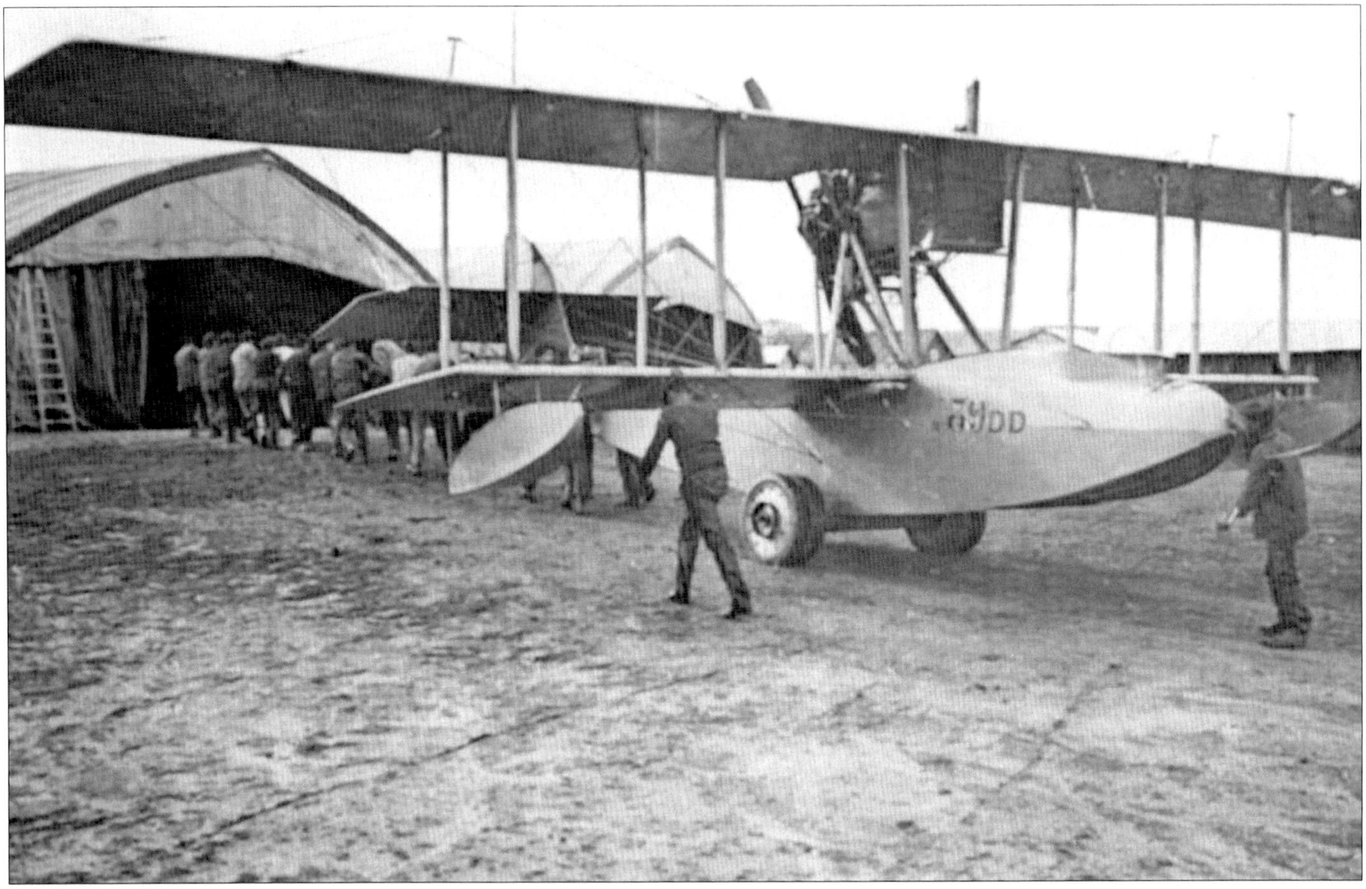

Above: Donnet Denhaut 160 hp Canton-Unné (n°39) seen in CAM Salonique in November 9, 1916.

Below: A unique bird, the Caudron R 150 hp Canton-Unné, served one entire year in CAM Salonica. It was among the very rare planes equipped with floats used by French Navy, who preferred flying boats with hulls. (ARDHAN)

Above: A FBA 150 hp Hispano (Type H) of CAM Salonique makes a stop in Escadrille de Panomi in 1918. (Collection Albin Denis)

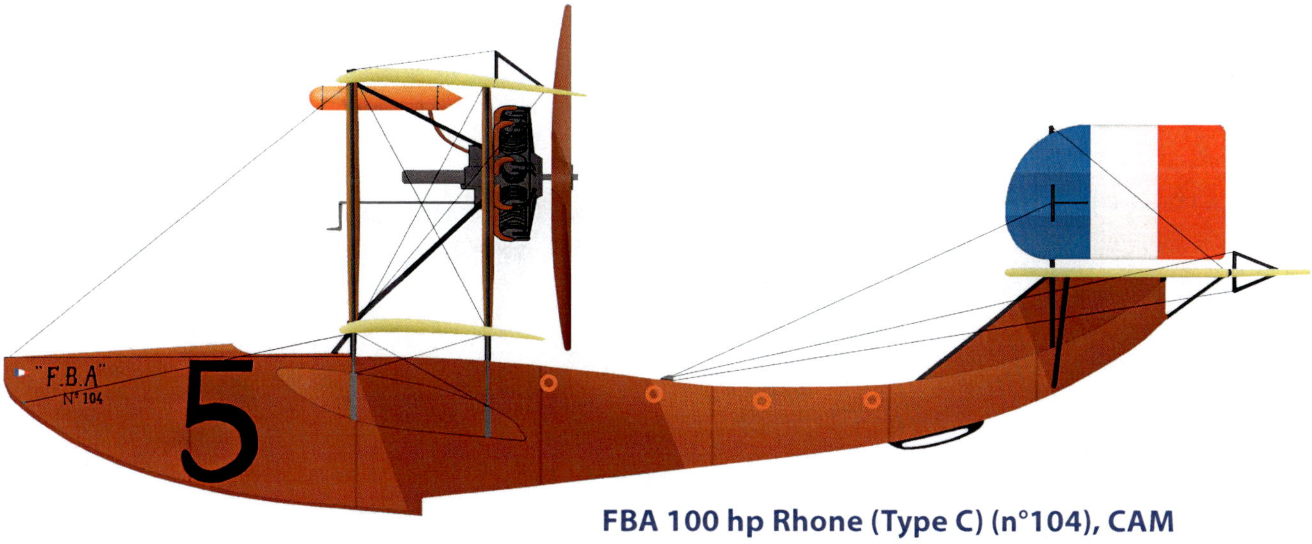

FBA 100 hp Rhone (Type C) (n°104), CAM Salonique, May 1916.

Centre d'Aviation Maritime de Corfou

Insignia: CAM Corfou never had an insignia, but an identification letter painted on its seaplanes.

Letter "C" from May 1916 to March 1917.
Letter "O" from March 1917 to November 1917.
Letter "J" from November 1917 to March 1918.
Letter "O" after March 1918.

Location: Corfou. The CAM never had any operational combat post, this despite the work organized in Mourtzo, on the coast of Greece, but still unfinished at the end of the war.

Commanding Officer:

- Lieutenant de vaisseau Maurice Larrouy on March 1, 1916.
- Enseigne de vaisseau de 1ere classe André Lorfèvre on June 14, 1916.
- Lieutenant de vaisseau Henri de l'Escaille on August 20, 1916.
- Enseigne de vaisseau de 1ere classe André Lorfèvre (interim) on December 24, 1916.
- Lieutenant de vaisseau Henri de l'Escaille on January 28, 1917.
- Lieutenant de vaisseau Jean-Baptise Hautefeuille on August 2, 1917.

Equipment: This CAM, one of the biggest of French Aviation Maritime, had 12 flying boats on line in late 1916 and ended the war with 40. The planes delivered were :

- 25 Donnet-Denhaut 160 hp Canton-Unné in May 1916.
- 20 Donnet-Denhaut 150 hp Hispano in October 1916.
- 15 Tellier 200 hp Hispano in March 1917.
- 5 Donnet-Denhaut 160 Lorraine in June 1917.
- 5 Sopwith 130 hp Clerget in August 1917.
- 15 FBA 150 hp Hispano (Type H) in February 1918.
- 10 Donnet-Denhaut 200 hp Hispano in March 1918.
- 20 FBA 200 hp Hispano (Type S) in April 1918.
- 5 Tellier-Cannon 200 hp Hispano in September 1918.

Victories/Casualties: CAM Corfou bombed 15 times enemy submarines and one time surface ships. Two of his pilots were lost from enemy action, and two others by accident.

History:
As soon as war was declared against the Central Empires, the French Navy entered the Adriatic Sea and sought a confrontation with the Austro-Hungarian Navy, which took refuge in the Cattaro Mouths (today Kotor) and remained a potential threat. The French Navy remained on guard in the southern Adriatic but faced major logistical problems because its closest home port was Malta, controlled by the British ally.

The situation improved in the summer of 1915 when the Italians entered the war. During the winter of 1915, Serbia was invaded by a combined attack of German, Austrian, and Bulgarian troops and its army retreated through the mountains of Albania. A crowd of tens of thousands of starving soldiers (and civilians) camped on the shores of Albania and the French and Italian navies took charge of evacuating them for treatment. They found a landing place, the Greek island of Corfu, which was occupied on January 11, 1916 without asking the Greek government, which was forced to bow to the fait accompli.

The island of Corfu thus became a major base for the French navy, and Commander Noël, commander of the CAM in Bizerte, suggested setting up a CAM there, which was accepted by ministerial decision on April 6, 1916. Two detachments left from St Raphaël and the center was officially created on May 1st 1916, under the command of LV Larrouy. The aircraft it received were Donnet Denhaut 160 hp with Canton-Unné engines, which turned out to be of poor quality, and moreover, were flown by inexperienced pilots, which meant that many of them wore out prematurely to the point that the CAM had only one operational aircraft left on June 17, 1916. Eight new aircraft were delivered on July 1st, but the commander of the CAM, LV Larrouy, was judged responsible for the poor state of the center and was dismissed. He was temporarily replaced by his second in command, Ensign Lorfèvre, and then by Lieutenant de l'Escaille on August 24th 1916.

With these air reinforcements, the flights could resume and the CAM could concentrate on its main mission, which was the surveillance of the Otranto dam. This was a gigantic floating net erected between the heel of the Italian boot and the island of Corfu, equipped with floating mines and along which armed trawlers and other larger vessels of the Italian, French, and British navies patrol in order to prevent any Austro-Hungarian ships and submarines from leaving the Adriatic. This barrage, which was constantly reinforced during the war, only achieved a few results, but it did not prevent the movement of enemy submarines towards the Mediterranean. Its surveillance was completed by aerial patrols carried out by the Italians from Brindisi in the west, and the French in the east from Corfu. Considering the strategic position of the island, the Minister of the Navy wished to develop to the maximum the capacities of the CAM, which reached a theoretical endowment of 30 aircraft, and would count about 50 by the end of the war.

On September 15, 1916, the first air attack took place with the participation of aircraft piloted by EV1 Albert Duval, Yves Mahéas, and SM Jean Pierre Salaün, with EV1

Above: First hydroplane used by CAM Corfou in 1916, a Donnet Denhaut 160 hp Canton-Unné, seen here in Brindisi after having crossed the Adriatic. (Collection Paolo Varriale)

Raoul Naudin, SM Edouard Drillet, and QM Jean Riolfo as observers, all of whom received a citation. Like all the submarines of the First World War, this one sailed for most of its time on the surface, and preferred not to dive during the attack to face the seaplanes with its flak guns... Both submarines and seaplanes returned safely after this attack, but it was different during the second encounter which took place on 16 October 1916, when the attacked U-Boat remained on the surface and shot down with its flak the seaplane piloted by SM Drillet and taking as observer QM Fernand Lefranc, who were killed when they were shot down at sea.

That same month, the CAM of Corfu carried out air coverage of the Italian troops landing operations in the Albanian port of Santa Quaranta (Sarandë). Winter limited the number of flights, while more reliable Donnet-Denhaut 150hp Hispano were delivered, followed by Tellier 200hp Hispano in January 1917. On February 11, 1917, the first confrontation with Austro-Hungarian maritime aviation took place when the K 183 (pilot: Konjovic, observer: Pramer), K172 (Nostiz, Schreinzer), and K177 (Nardelli, Brooser) seaplanes attacked the port of San Quaranta, where a new landing of Italian troops was taking place, dropping 160 kg of bombs. Several seaplanes of the CAM took off to intercept them but did not manage to find them: the Austro-Hungarian aircraft continued their flight for a reconnaissance on Corfu, where they were hit by a violent flak fire, and then returned to the base after having flown 720 km. A new submarine was encountered and attacked a few days later on February 16, 1917, and the bad weather of March strongly limited the flights.

The month of May 1917 was rich in action and saw the CAM fully resume its activities by making nearly 250 crossings of the Adriatic. On May 9, 1917, a submarine was attacked by a seaplane piloted by EV1 Georges Bragayrac with QM mechanic Jean Lamarque as observer. Two other submarine attacks were carried out during the month, for which several pilots were cited, although the details are not known. On 15 May 1917, the Austro-Hungarian Navy carried out a raid against the Otranto dam by several of its ships led by Admiral Horty (future leader of Hungary after the war): several seaplanes of the CAM attacked the enemy ships with bombs, without success. Finally, on 26 May 1917, three Austrian seaplanes, K154 (Hell, Engel), K152 (Klimburg, Schramek), K153 (Zelzeny, Dörfler), took off from the port of Durazzo for a reconnaissance mission over Corfu. In order to remain discreet, they calculated to arrive at noon over the objective in order to take advantage

Above: Hydroplane Tellier 200 hp Hispano, used by CAM Corfou after March 1917.

Facing Page & Following Page: Two views of license-built Sopwith 130 hp Clerget (Sopwith Baby) sent in Corfou in August 1917 to give a fighter cover to the Centre d'Aviation Maritime. The performance of those fighters revealed unsatisfactory and this job was later given to escadrille 562 in Potamos. (ARDHAN)

of the meal break... Meal or not, the alarm sounded at the CAM and four seaplanes took off to try to intercept them. But the Austrians fly at 3200 meters of altitude according to a South-East axis. The K152 and K154 remained to turn to the North of the island in waiting, while the K153 left to fly over the city of Corfu to take photographs. According to the Austrian archives, a French seaplane flying 300 meters lower attacked the K153 but visibly its machine gun jams – the Austrian defends himself by shooting with the flare gun on his assailant, which leaves to land in the port. Then the naval flak came into action and two Nieuport fighters were seen by the Austrian crew taking off from Santa Quaranta on the Albanian coast but heading in the wrong direction. The K153 then left towards the north to join its two crew members, then returned all together to Durazzo after a flight having lasted nearly 5 hours and 40 minutes.

This raid by the Austrian air force made people aware of the lack of aircraft adapted for interception. Sopwith Baby fighter aircraft were then delivered and became operational in August 1917, at the time when the command of the center was returned to LV Hautefeuille. In the meantime, patrols continued and on 29 May 1917 a submarine was sighted and attacked three times. The next day, an Austrian submarine sank the French ship *Italia* after a hard fight, and two seaplanes from CAM participated in the rescue of its crew.

No less than four submarine attacks were carried out in June 1917 by the crews of the CAM, where EV1 Bragayrac stood out. During this month, a British maritime aviation center was set up in Corfu, which led to a new distribution of patrol zones. In July, the French sailors suffered from the malaria endemic in Greece at that time. On 28 July 1917, the SM Corret was cited for having attacked and put to flight an enemy seaplane, allowing other seaplanes to attack a submarine. The Austrian archives only mention a submarine reconnaissance patrol conducted on 29 July by their seaplanes L131, L132, and L135.

Three new submarine attacks were carried out until the end of the year, on 29 and 30 September and on 25 November 1917, while FBA 160 and 150 hp seaplanes were delivered. Winter reduced the number of flights and at the beginning of 1918, as it was noted that the Otranto dam was only crossed at night by Austrian and German submarines, the missions of the French seaplanes were progressively oriented towards escorting naval convoys. However, on February 20, 1918, an enemy submarine was spotted and attacked by the seaplanes piloted by EV2 Pierre Gourvest, SM Plau Gélau, and QM Michel Le Bloa, which were all cited, as well as the observer QM Louis Tilly – the submarine appeared to be hit and was believed sunk, although the German or Austrian archives do not indicate any submarine losses that day.

From March to May 1918, the CAM of Corfu would

OB

Above: FBA 200 hp Hispano (Type S) with tactical number O-2 in CAM Corfou, 1918.

experience reduced activity due to the poor state of its air equipment, whose maintenance was complicated by the diversity of aircraft types in service. Nevertheless, according to Austrian sources, French aircraft participated at least five times in the raids conducted in April and May on the ports of Durazzo and Kumbor, along with Italian and British aircraft.

On June 6, 1918, two Austrian reconnaissance seaplanes, the K169 (Holoubek, Kohlhauser) and K226 (Scheschinger, Riedl) carried out a reconnaissance on the Otranto dam and passed by Corfu. The alarm sounded at the CAM and two Sopwith fighter planes took off to intercept them. They did not manage to reach the Austrians, who flew at an altitude of 3100 meters and returned to their base without being bothered.

In September 1918, the last attack of the center against an enemy submarine took place, while the Spanish flu epidemic hit the island and 200 HP Tellier Canon seaplanes were delivered. It was on board one of these that SM André Boucher (pilot) and the observer Roger Gérard disappeared at sea, the last operational loss of the center. At the time of the armistice, the CAM was maintained for several months as a regrouping center for other naval aviation units in the region. It was finally disbanded in 1919.

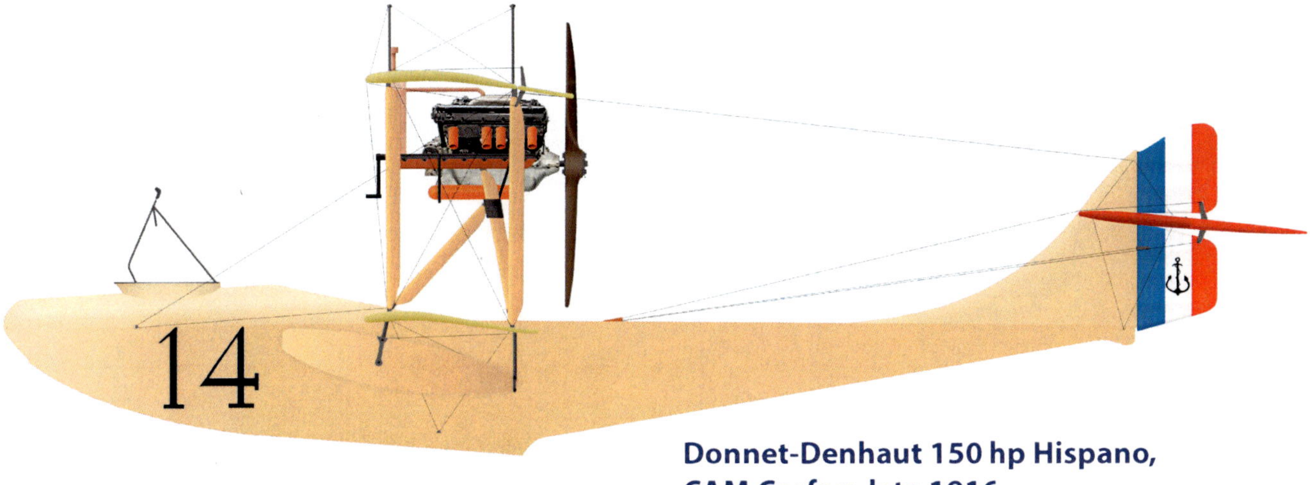

Donnet-Denhaut 150 hp Hispano, CAM Corfou, late 1916.

Tellier 200 hp Hispano n°23, CAM Corfou, 1917. The plane had not yet received a tactical number and the identification letter.

Sopwith 130 hp Clerget (Sopwith Baby as British designation), CAM Corfou, June 1917.

FBA 150 hp Hispano (Type H) n°363, CAM Corfou, Spring 1918.

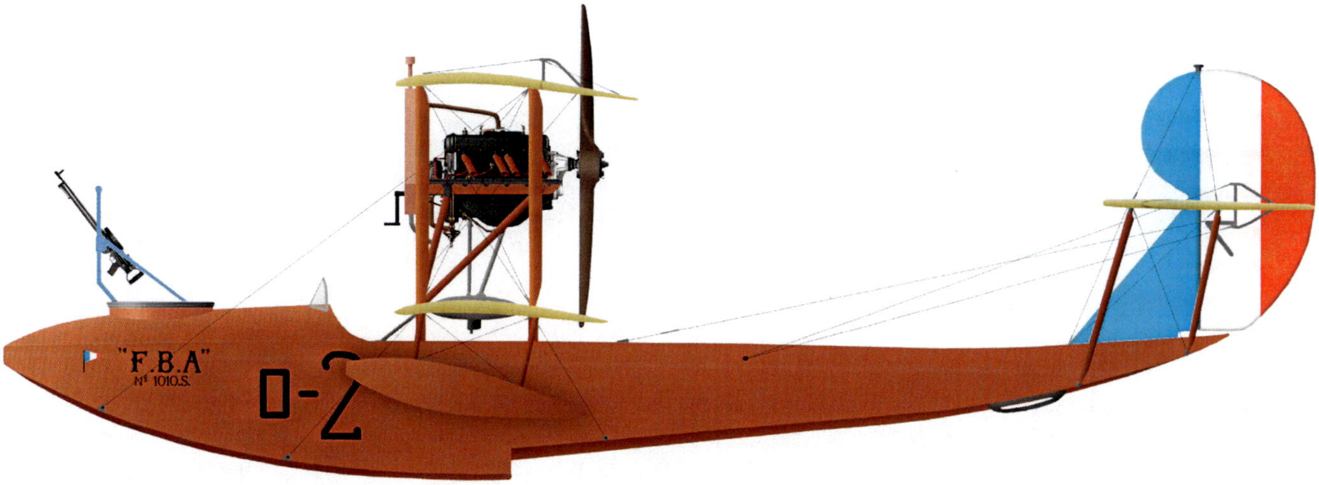

FBA 200 hp Hispano (Type S) with tactical number O-2 in CAM Corfou, 1918.

Centre d'Aviation Maritime d'Argostoli / Mytika / Platéali

Insignia: The CAM never had an insignia, but had the identification letter "S" painted on its seaplanes in 1917, then "G" when based in Platéali since November 1917.

Location: The CAM was first based in Argostoli in May 1916, then moved to Mytika in July 1917, and finally in Platéali in November 1917.

A Combat Post was organized in Vathi from August 1917 to February 1918.

Commanding Officer:
- Lieutenant de vaisseau Henri de l'Escaille on May 14, 1916.
- Lieutenant de vaisseau Marcel Destrem on July 1916.
- Enseigne de vaisseau de 1ere classe Louis Guitton on March 1917.
- Lieutenant de vaisseau Gabriel Fradin on December 1917.

Equipment: This CAM, one of the biggest of French Aviation Maritime, had 9 flying boats on line in March 1916 and ended the war with 33. The planes used were:

- Nieuport monoplanes in May 1916 until May 1917.
- FBA 100 Rhone (Type C) in October 1916.
- FBA 150 hp Hispano (Type H) in 1917.
- Donnet-Denhaut 150 hp Hispano in 1917
- Donnet-Denhaut 200 hp Hispano in 1918.
- FBA 200 hp Hispano (Type S) in 1918.
- Tellier-Cannon 200 hp Hispano in 1918.

Victories/Casualties: CAM Argostoli / Mytika / Platéali bombed 2 times enemy submarines and contributed to sink one. Four of his pilots were lost by flying accidents.

History:
The history of this CAM, which changed location twice, began with the will of the naval staff to open a new center in still neutral Greece to monitor the approaches to the Corinth Canal. The personnel and equipment, Nieuport single-seater seaplanes, were taken from the disbanded CAM Port Saïd and were transported on May 14, 1916 by the seaplane carrier *Campinas* to Argostoli, on the island of Cephalonia. Flights started immediately from the ship although the shore installations were completed on June 1, 1916. The sturdy Nieuport monoplane, whose production dates back to before the war, was replaced by 100 hp FBAs (Type C) from October 1916 but continued to be used until May 1917.

In June 1917, Greece officially joined the Entente camp and declared war on the central empires. In August 1917, the CAM was moved to Mytika on the Greek mainland, in order to ensure better surveillance of the convoys entering and leaving the Corinth Canal through the Gulf of Mytika towards Italy, where military equipment destined for the armies in Salonica was transiting. The choice of Mytika proved to be disastrous, however, because the area was infested with malaria and all the personnel of the center quickly fell ill to such an extent that it was necessary to quickly find another location, which was finally found in Platéali (20 km to the south). The new move took place in November 1917 and the activity of the CAM was slowed down by the installation works and the poor health of the staff.

At the same time as the first move to Mytika, a combat post was created on August 5, 1917 in Vathi on the island of Ithaca, housing a section of two 100 hp FBAs coming in rotation from the Mytika CAM and then from Platéali after the second move. This Vathi battle station had its moment of glory on December 14, 1917 when it participated in the destruction of the German submarine UC-38.

A naval convoy left Taranto the day before and entered the pass between the islands of Leucate and Cephalonia at dawn on 14 December 1917. It was composed of the escort vessels *Lansquenet*, *Mameluck*, the auxiliary cruiser *Rouen*, led by the fast cruiser *Chateaurenault*. Underwater, two torpedoes were fired at the smaller ships of the convoy and missed them. But the one intended for the *Chateaurenault* hit the target and exploded on the sides of the ship at 6.47 am. The French sailors were in a state of panic: the escorts rushed to the supposed origin of the wake, without finding anything, while a trawler, the *Balsamine*, and the other ships collected the crew of the *Chateaurenault*. During this time, the two 100 hp FBAs of the Vathi battle station, n° 221 coded "S 10" of second maître (SM) Robert Leclerc and matelot Emile Gazagne, as well as n° 236 coded "S 11" of SM Emile Jeanblanc and matelot Léon Trouillet, were on patrol in the Gulf of Molou when a visual signal from the ship *Aigrette* called them. Jeanblanc landed his FBA in the vicinity and learned that an SOS had just been launched by French ships in the pass south of the island of Leucate. He immediately took off again and flew to the scene accompanied by Leclerc's FBA. When they arrived north of the island of Ithaca, at Cape Agios Nikolaos, at about 8.55 am, a second torpedo had already hit the *Chateaurenault* at 8.40 am and sank it. The FBAs saw three trawlers busy collecting survivors swimming among the debris, as well as the *Lansquenet* moving to launch 7 depth charges at the origin of the torpedo, which all exploded. The two FBAs separated to explore the surroundings: the "S 11" of Jeanblanc flew over the shore of the island of Ithaca towards the west, then obliqued towards the north once the island of Cephalonia had been reached, crossed the pass to Cape Dukato on the island of Ithaca and after having obliqued towards the east returned to the south where the patrol boats were. The submarine, hit by the grenades of the Lansquenet which opened a waterway, had to surface after about twenty minutes and was attacked

Above: One of the Nieuport monoplane of the CAM Argostoli, veteran of the CAM Port Said (ARDHAN)

again with guns and then grenades by the *Mameluk*, which made it dive again. It was at this moment that the FBA of Jeanblanc arrived on the scene. The observer Trouillet saw a whitish conical shape and signaled to his pilot who flew at 500 meters of altitude. The FBA spiraled down to 250 meters where Trouillet threw a bomb which fell 20 meters from the eddy left by the submarine. It surfaced once again: its crew evacuated it. It was cannoned by the two French escorts, which probably hit its torpedo compartment because a violent explosion made it dive and sink in ten seconds. Jeanblanc's FBA, although at an altitude of 300 meters, was even shaken by the power of the explosion. Patrolling further north on the island of Ithaca, Leclerc's FBA "S 10" arrived after the battle and dropped its bombs on the oil stain left by the submarine. The two seaplanes, after having flown for 20 minutes over the scene of the tragedy, returned to Vathi after a flight which lasted about two hours.

The two FBA crews were cited in the army order and could be proud of being the only French naval airmen of the war to have contributed to the sinking of a submarine, the UC-38. The majority of its crew members, including its commander, Oblt z.S. Hans Hermann Wendlandt, survived the sinking, having been rescued by the French ships.

The Combat Post at Vathi was closed on February 23, 1918 because the local facilities could not accommodate the more powerful seaplanes that now equipped the CAM, Donnet Dennaut and FBA 150 hp as well as the 200 hp. Now installed at Platéali, the latter resumed its activity and accidentally lost an FBA 150 hp on December 29, 1917. On April 12, 1918, a section spotted a submarine at the entrance to the Gulf of Patras and bombed it. The submarine dived and the convoy it was likely to attack was diverted. A new crew disappeared on May 14, 1918 on an FBA 150 hp and another one fell victim to an accident on a Donnet Dennaut 150 hp on August 23. Malaria, endemic in Greece at that time, struck again the staff and five mechanics died on the spot or in the hospital of Corfu. The activity of the CAM was so slowed down at the end of the war that its closure was considered. It was finally maintained until the armistice and dissolved shortly afterwards, the move to Corfu of the staff and equipment ending in February 1919.

Above: View of an FBA Rhone 110 hp, whose drift has been dismantled, in Argostoli Bay at the end of 1916 (ARDHAN)

Facing Page, Above: A view of the drama of 14 December 1917: the *Chateaurenault*, hit by a torpedo, will soon sink. It will be avenged by the ships of its escort which will sink its aggressor, the UC-38, with the help of the FBA seaplanes of the CAM of Plateali.

Facing Page, Below: FBA 150 hp Hispano of the CAM of Plateali, coded G.21, during the year 1918 (ARDHAN)

Below: One of the FBA 110 hp Rhone in its hangar at Argostoli, 1916-1917 (ARDHAN)

G 21
B.A

Above & Left: Two views of FBA 200 hp Hispano, equipping the CAM of Platéali from the summer 1918. They both carry the identification letter "G" attributed to the CAM. (ARDHAN)

Nieuport monoplane with 80 hp Clerget engine, CAM Argostoli, Summer 1916.

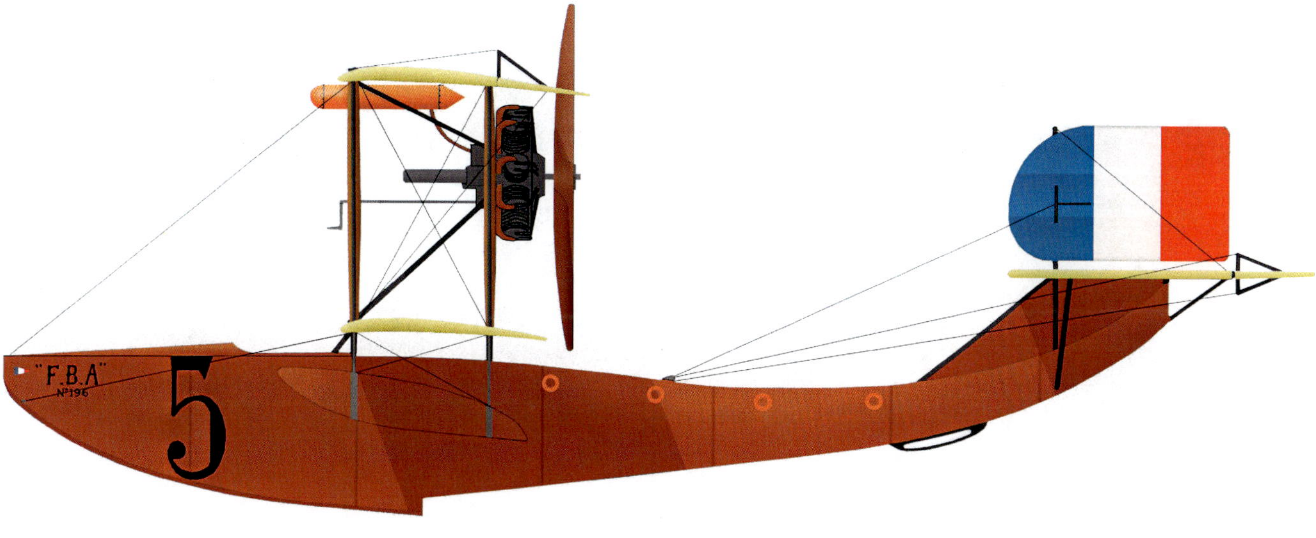

FBA 100 hp Rhone (n°196), CAM Argostoli, Late 1916.

FBA 150 hp Hispano (n°648), CAM Platéali, Summer 1917.

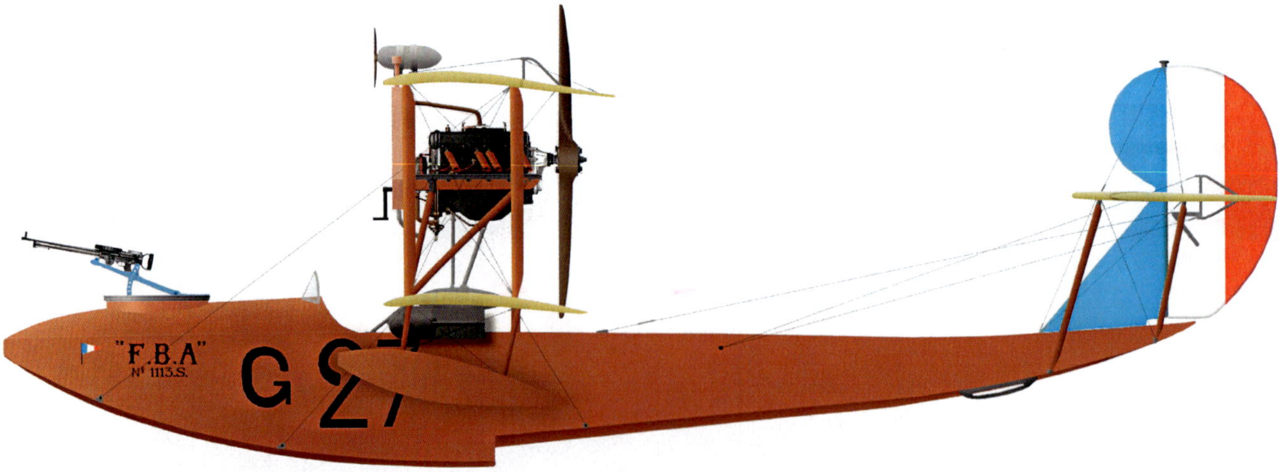

FBA 200 hp Hispano (n°1113), CAM Argostoli, Summer 1918.

Centre d'Aviation Maritime de Milo

Insignia: The CAM never had an insignia, but had the identification letter "M" painted on its seaplanes.

Location: The CAM was first based in the island of Milos, a Greek island in the Aegean Sea, just north of the Sea of Crete. It had no Combat Post attached.

A Combat Post was organized in Vathi from August 1917 to February 1918.

Commanding Officer:

- Enseigne de vaisseau de 1ere classe Guilbaud on January 1918.

Equipment: This CAM Milo, had 12 flying boats on line at his beginnings and ended the war with 18.

The only planes used were FBA 200 hp Hispano (Type S) and some Tellier-Cannon 200 hp Hispano.

Victories/Casualties: CAM Milo suffered no casualties among its pilots.

History:
The CAM of Milo (Milos), in the Cyclades archipelago, was created by ministerial decision on December 26, 1917, in order to protect the entrance to the Aegean Sea and the approaches to the port of Piraeus. The development works started immediately under the command of enseigne de vaisseau de première classe Guilbaud and were completed in May; the aircraft, FBA 200 hp HS and a handful of Tellier Canon, were delivered two months later. The seaplanes delivered in crates suffered somewhat from the trip and at least two FBAs were refused at the reception; the center was supposed to have 16 seaplanes, 12 of which were armed. The patrols were conducted over the Cyclades archipelago, going west to Cape Matapan and the island of Grabusa west of Crete. No U-boat was encountered during the four months of operational activity of the center, which only suffered one loss, a few days after the armistice, shortly before its disbandment.

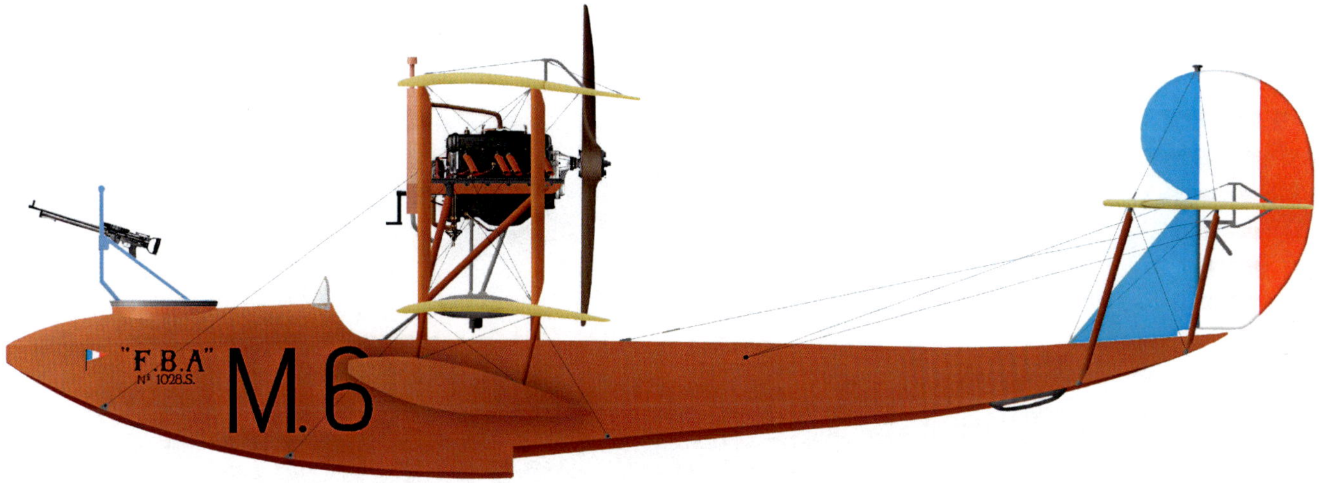

FBA 200 hp Hispano (n°1028), CAM Milo, October 1918.

Escadrille 562

Insignia: No insignia

Location: Potamos

Commanding Officer:
- Sous-lieutenant Victor Philibert on July 25, 1917.
- Lieutenant Pierre Bourgoin in July 1918.

Equipment: Nieuport 24bis and 27.

Victories/Casualties: No victories – No casualties.

History:
This squadron was created because of an incursion in the roadstead of Corfou of two Austrian aircraft on May 24, 1917, which could not be intercepted by the French seaplanes of the CAM of Corfu because of their heavy ascent. The chief of the French navy in Corfou complained about it to the under-secretary of state for aeronautics in a telegram dated May 28, 1917, and proposed that fighters settle on the ground of Potamos formerly used by the squadron of Farman MF 82 of Corfou at the beginning of 1916.

After talks with the under-secretary of state for aeronautics, the minister of the navy announced on July 25, 1917 to the commander of the first naval army the creation by the department of the war of a fighter squadron, which would take its quarters in Potamos, a coastal village located at 4 km of the CAM of Corfou, itself located at Govino (Gouvia), on the east coast of the island. Its armament consisted of 10 single-seat Nieuport 24 bis 120 hp Le Rhône, of which six were armed and four kept in reserve. Its commander, Sous-Lieutenant Victor Philbert, as well as its six non-commissioned pilots, were all from military aviation but received their orders from the commander of the CAM Corfou, who was in charge of the refueling. Their journey took place by train from Toulon to Taranto, then by sea to Corfou, the personnel arriving on August 30, 1917, the squadron being fully operational in October.

Nothing is known about the operational activity of the squadron, which was in fact a half-squadron compared to the standards of the French front, except that it carried out numerous missions to protect the Corfou naval base, undoubtedly intimidating the Austrian seaplanes which reduced their incursions. It is even probable that the Nieuports carried out offensive missions against the enemy, as the Austrian archives of the Durazzo base mention raids by French landplanes ("franz. Landflugzeug"). In July 1918, Lieutenant Bourguin took command of the squadron, which did not win any aerial victories or suffer any losses in combat, apart from three aircraft lost in accidents with no damage to the pilots.

In September 1918, the under-secretary of state for aeronautics planned to rearm the squadron with Hanriot HD.1s, probably taken from stocks destined for Italian aviation. The armistice put an end to the project and N 562 was disbanded shortly afterwards, its aircraft being transferred to the Greek government in January 1919.

Nieuport 24bis n°4550, Escadrille 562, Potamos, Summer 1918.

Right: Sgt André Christaud, pilot of escadrille 562, poses in front of his Nieuport 24 bis n° N 4550 named "*Denise*".

Escadrille G.488

Insignia: Escadrille G.488 had no insignia but the letter "R" as identification marking.

Location: Kourtesis

Commanding Officer: Lt André Bastien

Equipment: Caudron G.4 and G.6

Victories/Casualties: No victories. 1 pilot killed in an aerial accident.

History:
Escadrille G.488 was created as a coastal squadron, intended for aerial surveillance of enemy submarines, by ministerial decision of August 15, 1917. Initially planned to be based in Bizerte (Tunisia), it was actually created in Greece, in Courtesis (Kourtesis), on October 1, 1917, 50 km southwest of Patras, northwest of the Peloponnese peninsula, in order to monitor the entrance to the Gulf of Corinth. Directed by Lt André Bastien, it received its Caudron G.4 delivered in flight, from Corfu, where they were assembled on the airfield of Potamos.

Initially composed of four Caudron G.4s, the G.488 squadron started its operational activity on May 14, 1918 and had to face an implacable enemy: the endemic malaria which was rampant in Greece at that time. On September 1st 1918, with eleven Caudron G.4 and G.6 aircraft at its disposal, the health of the personnel (flu and malaria) was such that the squadron was evacuated to Patras, thus ending its operational activity.

It was disbanded there on December 31, 1918, at the same time as all the coastal squadrons, without having encountered an enemy submarine. At that time, the squadron had twelve aircraft, nine of which were armed. Its flying personnel were: three pilot officers, six non-commissioned pilot officers, and eight petty officer observers. During its short existence, it lost only one pilot, Cal Paul Beer, killed in an air crash on April 21, 1918.

Eight of its Caudron G.4s were transferred, with their spare parts, to the Greek Air Force. The other planes and the rest of the equipment, including the hangars, were evacuated to the large park in Salonica.

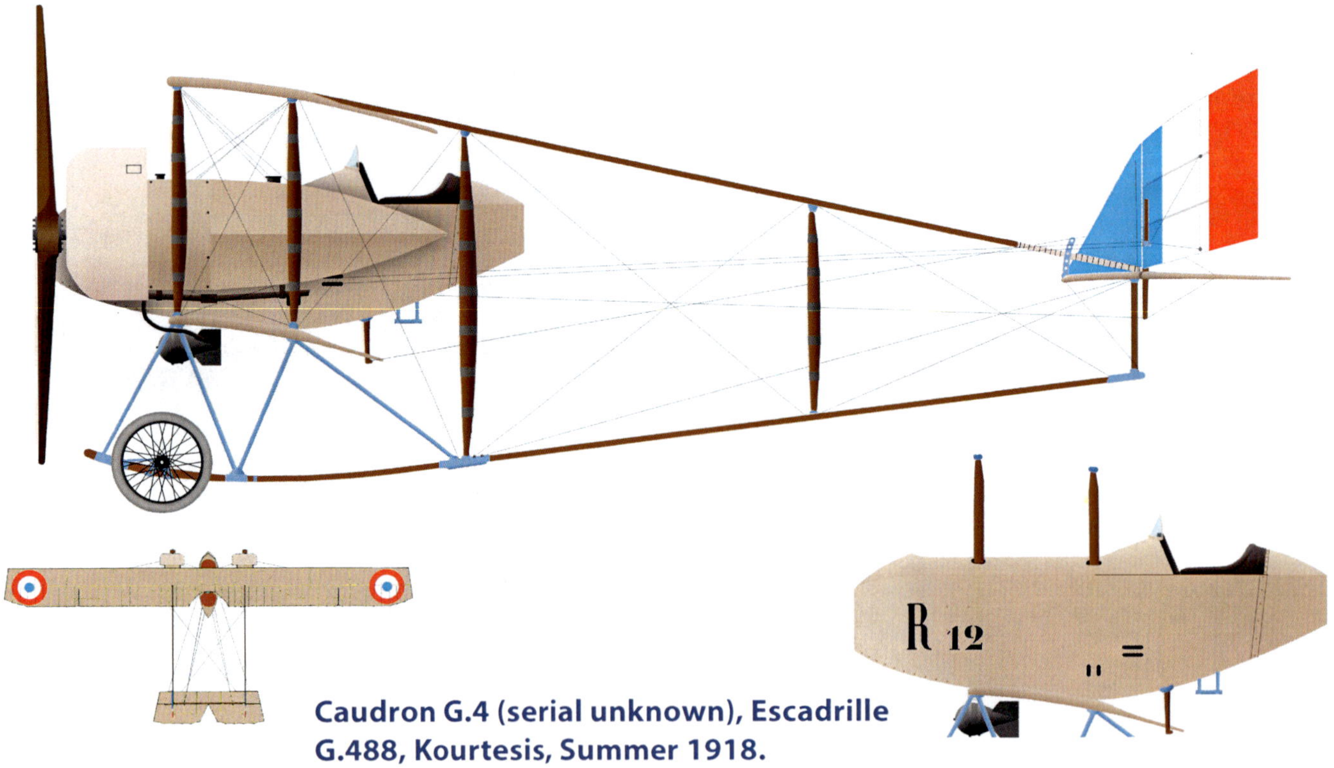

Caudron G.4 (serial unknown), Escadrille G.488, Kourtesis, Summer 1918.

Above: Caudron G.4 with tactical number R 12, is armed with English bombs. The plane does not carry his forward machine gun. (Collection Albin Denis)

Right: Bulgarian airmen in front of a Roland D.III fighter of the Bulgarian air service. This one has a Mercedes engine. At least six of these fighters were supplied to Bulgaria.

Appendix 3: Victories and Losses

Notes on the Air Victories of the Eastern Air Force

The French Aéronautique Militaire awarded its pilots (and machine gunners) a victory when an enemy plane was shot down in the French lines, or when it was considered as destroyed in the enemy lines (which excluded a plane simply forced to land in the enemy lines). For the second case, it was necessary to receive several written testimonies from combatants who witnessed the fight, whether they themselves were pilots, observers in a plane, in a balloon, or even on the ground. The reports must be communicated to the commander of the aeronautics of the local French army, who makes the decision of confirmation (homologation). As far as we are concerned, these decisions were made by Major Victor Denain, head of the aviation d'Orient.

The problem is that there is no document in the archives of the Orient Army (just like the other French armies, by the way) listing the victories awarded to the pilots. For the year 1917, there is a list of the enemy planes that fell and, if applicable, the name of the victorious allied pilot. For the year 1918, there is an even more detailed document, "Etat des avions ennemis abattus", describing in detail each air battle for the French, British, Italian, Serbian, and Greek air forces. But it also describes fights involving German planes that are documented as not having been confirmed to the pilots, as is the case with the claim made on April 30, 1918 by the pilots of the Franco-Greek 531 squadron Dieudonné Costes, Maurice Lashermes, and Paul Andrillon – Paul Andrillon's correspondence is unequivocal on the fact that the victory

Victory Claims of French Aéronautique

#	Date	Pilot or gunner	Unit	Plane piloted	Plane claimed
					Year
1	9-Jun-15	Adj-ch Thirouin Marcel (Pilot)	525 - MF 99 S	MF XI n°194	EA
		Sgt-Major Magnoni (Gunner)			
2	23-Jul-15	Cne Paulhan Louis (Pilot)	525 - MF 99 S	MF XI n°193	EA
		Cprl Lestradet André (Gunner)			
					Year
3	13-Jan-16	Sgt Denti Charles (Pilot)	524 - MF 98 T	Morane L	EA
		Cprl Weisshaupt André (Gunner)			
4	1-Feb-16	Sgt Terme Alain (Pilot)	507 - N 91 S	Nieuport 10	Albatros C
		MdL Astor Henri (Gunner)			
5	17-Feb-16	Sgt Fétu Adrien (Pilot)	507 - N 91 S	Nieuport 10	Albatros C
		S/Lt de Minteguiaga Paul (Gunner)			
6	23-Mar-16	Lt Mezergues Albert (Pilot)	506 - V 90	Voisin Canon	EA
		Matelot Pacatte Georges (Gunner)			
7	24-Mar-16	Adj Bonnard Clément (Pilot)	507 - N 91 S	Nieuport 10	EA
		Cprl Genevois Joseph (Gunner)			

was not confirmed to them.

The only tangible document to confirm the homologation of the victories remains the citations to the order of the army awarded to the victorious pilots, and published in the Journal Officiel de la République Française, which are the proof of the homologation because this journal is by definition official... It is, however, not the absolute proof because it can happen that a victory has been forgotten in the administrative process. Nevertheless, in the table on the following pages, there is a column listing the documents of the Army of the Orient documenting the claims, as well as a column specifying the citation in the Journal Officiel proving the homologation of the victory. The truth is somewhere between the two columns... The ace of the aces of Salonica, Dieudonné Costes, claimed throughout his life to have obtained 9 aerial victories, whereas the Journal Officiel only confirmed six.

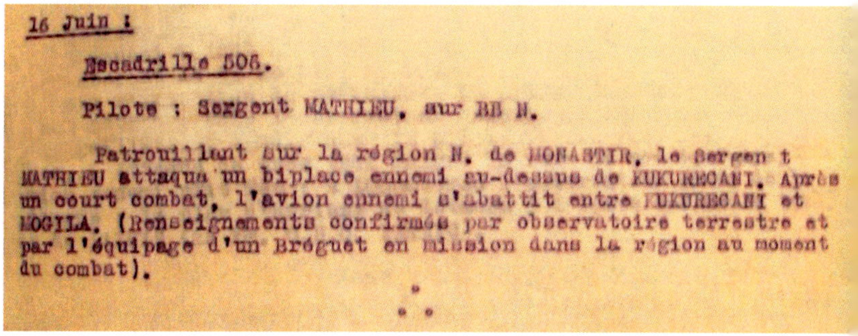
16 Juin :

Escadrille 506.

Pilote : Sergent MATHIEU, sur BB N.

Patrouillant sur la région N. de MONASTIR, le Sergent MATHIEU attaqua un biplace ennemi au-dessus de KUKURECANI. Après un court combat, l'avion ennemi s'abattit entre KUKURECANI et MOGILA. (Renseignements confirmés par observatoire terrestre et par l'équipage d'un Bréguet en mission dans la région au moment du combat).

Above: Combat description of the victory claim of Sgt Mathieu, escadrille 506, on June 16, 1918. As the enemy plane fell in his own lines but as the fall was seen by a terrestrial observer and the crew of a Breguet 14, the victory was confirmed to Mathieu who later received a citation in Journal Officiel.

Militaire on Front d'Orient

Place	Source	Confirmation	Notes
1915			
Bavaniste (Serbia)	Report Escadrille 525	Citation OA(MM) 12-06-1915 - JORF 14-06-1915 p.3882	
		Citation OA(MM) 12-06-1915 - JORF 14-06-1915 p.3882	
Temess, N. Belgrade (Serbia)	Report Escadrille 525	Citation OA 27-07-1915	EA bombed after being forced to land
1916			
Dardanelles Straits	Oral testimony Charles Denti	Citation OCA 13-08-1916	
		Citation OCA 13-08-1916	
S. Palihor (12:45)	Journal de marche 122e	Citation OA(MM) 02-02-1916 - JORF 18-02-1916 p.1383	FA 66 - Hptm Guido Knörzer & Vfw Bollmann, both POW. Plane exposed in Salonica.
	DI _ 01-01-1916	Citation OA(MM) 02-02-1916 - JORF 18-02-1916 p.1383	
Karasuli	Report Escadrille 507	Citation OA(MM) 18-02-1916 - JORF 18-05-1916 p.4367	Crew captured, plane exposed in Salonica.
		Citation OA(MM) 18-02-1916 - JORF 18-05-1916 p.4367	
Lake Doiran		Citation OA(LH) 12-04-1916 - JORF 04-05-1916 p.3825	
		Citation OA 12-04-1916 - JORF 12-3-1917 p.1965	
Kilindir / Doiran, around 16:00		Citation OA(MM) 12-04-1916 - JORF 04-05-1916 p.3825	
Falled bt. Doiran and Rabrovo.		Citation OA 12-04-1916 - JORF 12-03-1917 p.1965	

Victory Claims of French Aéronautique					
#	**Date**	**Pilot or gunner**	**Unit**	**Plane piloted**	**Plane claimed**
8	27-Mar-16	Sgt Terme Alain (Pilot)	507 - N 91 S	Nieuport 10	Albatros C
		MdL Astor Henri (Gunner)			
		Lt Mezergues Albert (Pilot)	506 - V 90 S	Voisin Canon	
		Aspirant Naud Jean (Gunner)			
9	25-Apr-16	S/lt Terme Alain (Pilot)	507 - N 91	Nieuport 10	Albatros C
		S/lt Astor Henri (Gunner)			
-	5-May-16	23rd section d'auto-canons	-	-	Zeppelin LZ 85
	5-May-16	Sgt Deloupy Guy (Pilot)	506 - V 90 S	Voisin Canon	
		Matelot Lelong Robert (Gunner)			
10	26-Jul-16	Sgt Leblond Germain (Pilot)	505 - V 389	Caudron G 4 n°1495	EA
		OA 2e cl Vaisse Jules (Gunner)			
-	17-Aug-16	Lt Thirouin Marcel	523 - N 387	Nieuport 21	EA (FTL)
11	19-Aug-16	S/Lt Renaud Clovis (Pilot)	521 - F 82	Farman	EA
		S/Lt du Mesnildoz Hervé (Gunner)			
12	25-Aug-16	Lt Thirouin Marcel (2)	523 - N 387	Nieuport 21	EA
-		Lt Delrieu Louis (Pilot)	523 - N 387	Nieuport 10	EA (FTL)
		Cne Mavet Richard (Gunner)			
-		Lt Delrieu Louis (Pilot)	523 - N 387	Nieuport 10	EA (FTL)
		Cne Mavet Richard (Gunner)			
-		Lt Delrieu Louis (Pilot)	523 - N 387	Nieuport 10	EA (FTL)
		Cne Mavet Richard (Gunner)			
13	13-Sep-16	Sgt Vermonet Abel	506 - N 390	Nieuport 21	EA
14	23-Oct-16	Sgt-Major Sicot Emile (Pilot)	507 - N 391	Nieuport 10	Albatros
		Lt Leune Jean (Gunner)			

Militaire on Front d'Orient

Place	Source	Confirmation	Notes
Lake Amatovo	Report Escadrille 506	See 25-04-1916	FA 69 - OfStv. Max Koestle & Ltn Helmut Camin
		See 25-04-1916	
	Report Escadrille 506	Citation OA(LH) - JORF 04-05-1916 p.3825	
		Citation OA(MM) - JORF 04-05-1916 p.3825	
S. Kar, E. Lake Amatovo - cote 585	Report Escadrille 507	Citation OA - JORF 16-10-1916 p.9009 "3rd victory"	Plane captured and exposed in Salonica, crew escaped to German lines
		Citation OA - JORF 16-10-1916 p.9009 "3rd victory"	
Vardar river		Citation OA - JORF 12-03-1917 p.1964	Wreck and crew captured
	Report Escadrille 506	Citation OA - JORF 12-03-1917 p.1965 "attacked a Zeppelin"	
		Citation OA - JORF 12-03-1917 p.1965 "attacked a Zeppelin"	
Doret-Hissar	Report Escadrille 505	Citation OA - JORF 07-03-1918 p.2148	French crew killed after his victory
		Citation OA - JORF 07-03-1918 p.2148	
Vertekop	JMO Serbian aviation	-	Thirouin had his own propeller damaged by enemy bullet
Dobropolje (FTL)	JMO Serbian aviation	Citation OA - JORF 20-03-1917 p.2222	Renaud and du Mesnildoz had their own plane damaged with 7 bullets
		Citation OA - JORF 20-03-1917 p.2222	
Trésina (near Vertekop)	JMO Serbian aviation	Citation OA(LH) - JORF 24-10-1916 p.9268	Enemy crew captured - Thirouin's plane damaged with 14 bullets.
Enemy lines	JMO Serbian aviation	-	During the same mission
Enemy lines	JMO Serbian aviation		
Enemy lines	JMO Serbian aviation		
N. Bogdanci	CROA September 1-15	Citation OA 26-01-1918 - JORF 07-03-1918 p.2149	German lines - Wreck confirmed by a reconnaissance done by Dieudonné Costes
Hudova (16:30)	Report Escadrille 507	Citation OA - JORF 29-07-1917 p.5903	
		Citation OA - JORF 21-04-1917 p.3154	

Victory Claims of French Aéronautique					
#	**Date**	**Pilot or gunner**	**Unit**	**Plane piloted**	**Plane claimed**
15	29-Oct-16	Sgt-Major Sicot Emile (Pilot)	507 - N 391	Nieuport 10	Fighter
		S/Lt Astor Henri (Gunner)			
16	6-Nov-16	MdL Mortagne Jacques	506 - N 390	Nieuport 21	EA (FTL)
		MdL du Lac Henry	506 - N 390	Nieuport 21	
17	19-Nov-16	Lt Mahieu Jules (Pilot)	Air Park	Farman F 40	Aviatik C
		Sol Maréchal Emile (Gunner)			
18	19-Nov-16	Lt Mahieu Jules (Pilot)	Air Park	Farman F 40	Aviatik
		Sol Maréchal Emile (Gunner)			
					Year
19	5-Feb-17	MdL Viallet Georges	523 - N 387	Nieuport 21	EA
20	11-Feb-17	Sgt Pellerin Fernand (Pilot)	502 - F 385	Farman F60	Halberstadt D
		S/Lt Puisieux André (Gunner)			
21	18-Feb-17	MdL Lemut Alfred	502 - F 385	Nieuport 21	Fokker D
-	20-Feb-17	MdL Domicent Lucien	505 - C 389	Nieuport 21	Albatros
	27-Feb-17	French artillery			Halberstadt D
22	2-Apr-17	S/Lt Miletic Miodrag (Pilot)	523 - Dét. Nieuport	Nieuport 12	DFW C
		Cne Vukosavljevic Branko (Gunner)	523 - Dét. Nieuport		
-	3-Apr-17	Lt Miletic Miodrag	523 - N 387	Nieuport 23	Two-engined bomber
-	24-Apr-17	S/Lt Cousinery Marcel	507 - N 391	Nieuport 23	Two-engined bomber
23	29-Apr-17	Adj Costes Dieudonné	507 - N 391	Nieuport 23	Two-engined bomber
	2-May-17	French artillery			Halberstadt
24	20-May-17	Lt Thirouin Marcel	523 - N 387	Nieuport 23	Albatros D (FTL)
		MdL Martin Charles	523 - N 387	Nieuport 23	

Militaire on Front d'Orient			
Place	**Source**	**Confirmation**	**Notes**
SE. Lake Doiran, NE. Akinjali	Report Escadrille 507	Citation OA - JORF 08-04-1917 p.2769	
		Citation OA - JORF 08-04-1917 p.2769	Astor's 4th victory
Kravari	Report Escadrille 506	Citation OA - JORF 03-04-1917 p.2647	Enemy plane later bombed on the ground by 2 Farmans from Escadrille F.386
		Citation OA - JORF 03-04-1917 p.2647	Unclear if Mortagne and Du Lac, both pilots, were flying in a two seater or two different planes
Drama (enemy lines)		Citation OA - JORF 08-04-1917 p.2770	During a ferry flight Salonica - Romania
		Citation OA - JORF 08-04-1917 p.2770	
Drama (enemy lines)		Citation OA - JORF 08-04-1917 p.2770	During a ferry flight Salonica - Romania
		Citation OA - JORF 08-04-1917 p.2770	
1917			
Vasak (SE Prilep)	JMO Serbian aviation	Citation OA 16-02-1917 - JORF 08-06-1917 p.4461	
Dzéovo	Report Escadrille 502	Citation OA 01-03-1917 - JORF 08-06-1917 p.4461	
		Citation OA 01-03-1917 - JORF 08-06-1917 p.4461	
Dzéovo	List of downed EA - Armée d'orient	Citation OA 01-03-1917 - JORF 08-06-1917 p.4461	Jasta 25 - Lt Bodo Von Lyncker KIA from air collision
Kaljani (SE Prilep)	List of downed EA - Armée d'orient	-	
Gumendje	List of downed EA - Armée d'orient	-	Kaghol I - Ltn Franz Kaestner captured
N. Tusin / W. Norste	JMO Serbian aviation	Citation OA 13-04-1917 - JORF 23-07-1917 p.5692	Enemy plane seen crashed by soldiers from 58e DI.
		Citation OA 13-04-1917 - JORF 23 -07-1917 p.5692	
N. Preslap, 100m N. Dobropolje	List of downed EA - Armée d'orient	-	
E. Prilep	List of downed EA - Armée d'orient	-	Dives vertically and disappears in a cloud.
	List of downed EA - Armée d'orient	Citation OA 01-06-1917 - JORF 15-08-1917 p.6409	Costes is armed with a vertical machine gun. Enemy seen falled by gound observers.
Kalinovo	List of downed EA - Armée d'orient		
Cerna river, S. Drenovo	List of downed EA - Armée d'orient	Citation OA 11-06-1917 - JORF 22-07-1918 p.6325 "3rd victory"	
		Citation OA 13-06-1917 - JORF 22-07-1918 p.6325	

Victory Claims of French Aéronautique					
#	**Date**	**Pilot or gunner**	**Unit**	**Plane piloted**	**Plane claimed**
25	24-May-17	S/Lt Emile Bruninghaus (Pilot)	508 - SAL 1	Sopwith 1A2	EA
		S/Lt Guy Picard-Destelan (Gunner)	508 - SAL 1		
-	6-Jun-17	Cprl Laffon Lucien	506	Nieuport	Two-seater (FTL)
26	11-Jun-17	Sgt Voituron Léon	506	Nieuport	Two-seater
27	13-Jun-17	S/Lt Nast Robert	506	Nieuport 21 n°1649	Two-seater (FTL)
		Mdl Laumonier Fernand	506	Nieuport	
28	2-Jul-17	MdL Laumonier Fernand	506	Nieuport	Fighter
		MdL Domicent Lucien	506	Nieuport	
29	9-Jul-17	MdL Meignen Fernand	523	Nieuport	Two-seater
30	21-Jul-17	S/Lt Tomic Miodrag	523	Nieuport	Albatros C 200 Hp
		MdL Martin Charles	523	SPAD VII	
		MdL Mollien Robert	523	Nieuport	
-	20-Aug-17	MdL de Christmas Valdemar	507	Nieuport	Fighter
31	21-Aug-17	Lt Lebouc Jacques	506	SPAD VII	LVG C
		MdL Hébert René	506	SPAD VII	
	21-Aug-17	French artillery		-	DFW C
32	16-Oct-17	S/Lt Nast Robert	506	SPAD VII	Fighter
-	14-Nov-17	S/Lt Bonnetête Louis	502	Nieuport	Fighter
-	15-Nov-17	MdL Drozière Gaston	506	SPAD VII	Rolland C
	29-Nov-17	mitrailleuses d'infanterie françaises	30e DIC	-	EA
	15-Dec-17	French artillery		-	LVG C

Militaire on Front d'Orient			
Place	**Source**	**Confirmation**	**Notes**
N. Cote 1050, Cerna river loop	List of downed EA - Armée d'orient	Citation OA 21-06-1917 - JORF 12-09-1917 p.7184	Burns inside german lines
		Citation OA 21-06-1917 - JORF 12-09-1917 p.7184	
Capari (19:00)	List of downed EA - Armée d'orient	-	
Piton rocheux (07:00)	List of downed EA - Armée d'orient	Citation OA 21-06-1917 - JORF 12-09-1917 p.7184	
Baranci (09:00)	List of downed EA - Armée d'orient	Citation OA 17-07-1917 - JORF 22-07-1918 p.6325	FTL and machine-gunner on the ground
		Citation OA 17-07-1917 - JORF 22-07-1918 p.6325	
Dedebal / Budakovo (NNE. Novak)	List of downed EA - Armée d'orient	Citation OA 17-07-1917 - JORF 22-07-1918 p.6325 "2nd victory"	
		Citation OA 17-07-1917 - JORF 22-07-1918 p.6325	
S. Skra Di Legen, N Livadi	List of downed EA - Armée d'orient	Citation OA(MM) 27-07-1917 - JORF 15-11-1917 p.9137	Plane captured and exposed in Salonica
Toroche / 10 km SE. Kapinjani	List of downed EA - Armée d'orient	Citation OA 29-07-1917 - JORF 28-10-1917 p.8575	FA 246 - Pilot unknown POW, observer Ltn Georg Engelmann (WIA-POW-DOW)
2,5 km E. Subotsko		Citation OA 29-07-1917 - JORF 28-10-1917 p.8575	
		Citation OA 29-07-1917 - JORF 28-10-1917 p.8575	
Kanatlarci	List of downed EA - Armée d'orient	-	
SW. Florina	List of downed EA - Armée d'orient	Citation OA(LH) 26-07-1918 - JORF 05-09-1918 p.7787	FA 246 - Pilot unknown POW, observer Obltn Herbert Esche (WIA-POW-DOW)
		Citation OA 24-08-1917 - JORF 15-11-19117 p.9137	
Monastir, French lines	List of downed EA - Armée d'orient		Fwbl Johann Krause and S/Lt Reiselt POW
NE Prilep, enemy lines	List of downed EA - Armée d'orient	Citation OA 19-10-1918 - JORF 24-01-1919 p.926	
Kizil Doganli	List of downed EA - Armée d'orient	-	
N. Kanatlarci	List of downed EA - Armée d'orient	-	
E. Magarevo, near Monastir	List of downed EA - Armée d'orient		
Mala-Rupa / Vardar	List of downed EA - Armée d'orient		FA 34 - Vfw Karl Twittmann & Ltn d R Ernst Loose KIAs

Victory Claims of French Aéronautique					
#	Date	Pilot or gunner	Unit	Plane piloted	Plane claimed
					Year
33	6-Jan-18	Sgt Ripoche Maurice	502	Nieuport	Aviatik
		Sgt Le Guyader André	502	Nieuport 24 n°5361	
-	24-Jan-18	Sgt Perrier Claudius	506	Nieuport	Monoplace
		Cprl Lefevre Charles, Désiré	506	Nieuport	
34	26-Jan-18	A/C Costes Dieudonné	532	Nieuport 27 n°5523 ou 5545	Roland
		Brig Lashermes Maurice	532	Nieuport 27 n°5527 ou 5546	
35	27-Jan-18	S/Lt Nast Robert	507		
-	11-Mar-18	Sgt Lecuir Marcel (Pilot)	510	Bréguet 14	Single-seater
		Lt Apostolou (Gunner)	510		
-	13-Apr-18	S/Lt Fortuny	506	Nieuport	Single-seater
		Adj Perrier Claudius	506	Nieuport	
		Cprl Aumaitre Joseph	506	Nieuport	
	23-Apr-18	French artillery	32e cie ballons	-	Albatros D III
	25-Apr-18	French artillery	32e cie ballons	-	Albatros D V
36	30-Apr-18	Adj Sauné Basile	531	Nieuport	Fighter
		Sgt Ripoche Maurice	531	Nieuport	
-	30-Apr-18	A/C Costes Dieudonné	531	Nieuport	Two-seater
		Sgt Andrillon Paul	531	Nieuport	
		MdL Lashermes Maurice	531	Nieuport	
37	1-May-18	A/C Costes Dieudonné	531	Nieuport	Two-seater
		Sgt Andrillon Paul	531	Nieuport	
		MdL Lashermes Maurice	531	Nieuport	
38	4-May-18	Adj Sauné Basile	531	Nieuport	Two-seater
		MdL Lambert Robert	531	Nieuport	

Militaire on Front d'Orient

Place	Source	Confirmation	Notes
1918			
Dzeovo	List of downed EA - Armée d'orient	Citation OA 27-03-1918 - JORF 07-05-1918 p.3977	
		Citation OA 27-03-1918 - JORF 07-05-1918 p.3977	
Budakovo	List of downed EA - Armée d'orient	-	Seen diving in his lines
Mrzenci / Negorci	Etat des avions ennemis abattus	Citation OA(MM) 14-02-1918 - JORF 24-03-1918 p.2644	Costes used a vertical machine gun
			Enemy seen spinning on the ground
Golfe d'Orfano	Etat des avions ennemis abattus	Citation OA 19-10-1918 - JORF 24-01-1919 p.926	MdL René Hébert took a major part in the victory and was not credited for it
Mileteovo-Hudova	Etat des avions ennemis abattus	-	Seen falling in his lines, damaged
Côte 1248	Etat des avions ennemis abattus	-	Damaged, retreats in his lines
		-	
		-	
	Etat des avions ennemis abattus		
13 km SE. Monastir, Vacuf-Egri	Etat des avions ennemis abattus		Jasta 25 - Vfw Heinrich Schott KIA
Negorci / Pardovica (07:50)	Etat des avions ennemis abattus	Citation OA(MM) 02-05-1918 - JORF 29-06-1918 p.5607	Photo of the wreck
		Citation OA 21-05-1918 - JORF 05-07-1918 p.5822	
NW. Negorci (09:30)	Etat des avions ennemis abattus	-	
NW. Mrezenci, 3 km N. Guevgueli	Etat des avions ennemis abattus	Citation OA 21-05-1918 - JORF 05-07-1918 p.5822	Photo of the wreck
		Citation OA 21-05-1918 - JORF 05-07-1918 p.5822	
		Citation OA 21-05-1918 - JORF 05-07-1918 p.5822	
3 km W. Doiran, Spanish hill / Dolina	Etat des avions ennemis abattus	Citation OA 21-05-1918 - JORF 05-07-1918 p.5822	Photo of the wreck
		Citation OA 21-05-1918 - JORF 05-07-1918 p.5822	

Victory Claims of French Aéronautique

#	Date	Pilot or gunner	Unit	Plane piloted	Plane claimed
39	1-Jun-18	Adj Sauné Basile	531	Nieuport	EA
		MdL Delaye Pierre	531	Nieuport	
40	4-Jun-18	Adj Sauné Basile	531	Nieuport	EA
41	8-Jun-18	A/C Costes Dieudonné	531	Nieuport	EA
42	13-Jun-18	A/C Costes Dieudonné	531	SPAD VII	Two-seater
		Adj Sauné Basile	531	SPAD VII	
43	15-Jun-18	MdL Lambert Robert	531	Nieuport	Fighter
44	16-Jun-18	Sgt Mathieu Paul	506	Nieuport	Two-seater
-	24-Jun-18	A/C Costes Dieudonné	531	Nieuport	LVG C (FTL)
		MdL Hébert René	531	Nieuport	
		Sgt Leman Henri	506	Nieuport	
		Cprl Aumaitre Joseph	506	Nieuport	
	3-Jul-18	French artillery		-	EA
45	9-Jul-18	Lt Bonnetête Louis	532	Nieuport	Two-seater
-	25-Jul-18	Mdl Laffont Louis	524	Nieuport	Albatros D III
46	28-Jul-18	Sgt Giron Alexis (Pilot)	506	Breguet 14	Albatros D V
		Sgt Garat Bernard (Gunner)			
-	5-Aug-18	S/Lt Cochet-Forthuny Frédéric	506	Nieuport	Fighter
		Adj Perreau	506	Nieuport	
		Sgt Leman Henri	506	Nieuport	
47	7-Aug-18	Lt Miletic Miodrag	523	SPAD VII	Biplace (FTL)
-	16-Aug-18	MdL Duthoit Georges	504	Nieuport	Fighter
48	25-Aug-18	Sgt-Maj Hodiera Mateya	523	Nieuport	Two-seater

Militaire on Front d'Orient

Place	Source	Confirmation	Notes
N. Mojina	Etat des avions ennemis abattus	See 13-06-1918	
Kovanec (08:30)	Etat des avions ennemis abattus	See 13-06-1918	
Borlovo (16:30)	Etat des avions ennemis abattus	Citation OA (1st) 05-08-1918 - JORF 27-08-1918 p.7549 "4th victory"	Photo of the wreck
S. Hudovo / Kovanec (06:45)	Etat des avions ennemis abattus	Citation OA (2nd) 05-08-1918 - JORF 27-08-1918 p.7549 "5th victory"	Crashed on ground
		Citation OA 05-08-1918 - JORF 27-08-1918 p.7549 "5th victory"	
N. Gurincet	Etat des avions ennemis abattus	Citation OA 05-08-1918 - JORF 27-08-1918 p.7549 "2nd victory"	Crashed on ground
Kukurcecani / Mogila	Etat des avions ennemis abattus	Citation OA 30-06-1918 - JORF 02-08-1918 p.6702	Observer DOW
E. Gradesnica (E Monastir, E Cerna)	Etat des avions ennemis abattus	-	
6 Km NW. Dobropolje		-	
		-	
		-	
16,8 km ENE. Monastir, "Piton brûlé"	Etat des avions ennemis abattus		
Borgorodica (11:00)	Etat des avions ennemis abattus	Citation OA(LH) 12-07-1918 - JORF 05-09-1918 p.7787	Fallen in no man's land. Photographic confirmation.
Zovik	Etat des avions ennemis abattus	-	Seen spinning
Road Podmol - Dobrusovo	Etat des avions ennemis abattus	Citation OA 09-08-1918 - JORF 28-09-1918 p.8474	
		Citation OA 09-08-1918 - JORF 28-09-1918 p.8474	
Sud Podmol	Etat des avions ennemis abattus	-	Seen spinning
		-	
		-	
Kozjak (Dobropolje)	Etat des avions ennemis abattus	Citation OA 15-09-1918 - JORF 08-12-1918 p. 10568 "2nd victory"	
S. Kanatlarci	Etat des avions ennemis abattus	-	
Chlem (9 km ENE Dobropolje) (10:45)	Etat des avions ennemis abattus	Citation OA 15-09-1918 - JORF 08-12-1918 p.10568	

Victory Claims of French Aéronautique					
#	**Date**	**Pilot or gunner**	**Unit**	**Plane piloted**	**Plane claimed**
49	2-Sep-18	Sgt Verdier André (Pilot)	510	Breguet 14	Fighter
		Cne Gruel (Gunner)	2e bureau		
	10-Sep-18	Cpt Perrin de Brichambaut Pierre	531	SPAD VII	Fighter
50		S/Lt Costes Dieudonné	531	SPAD VII	
		Sgt Andrillon Paul	531	SPAD VII	
-	10-Sep-18	Cpt Perrin de Brichambaut Pierre	531	SPAD VII	Fighter
		S/Lt Costes Dieudonné	531	SPAD VII	
		Sgt Andrillon Paul	531	SPAD VII	
-	11-Sep-18	Lt Bonnetête Louis	506	?	Two-seater
		Sgt Mathieu	506		
-	14-Sep-18	Cprl Castaings	508	Nieuport	Fighter
51	19-Sep-18	Lt Jacquemart Georges (Pilot)	510	Breguet 14	Fighter
		MdL Hérault (Gunner)	510		
		Mdl Chappaz Vincent (Pilot)	510	Breguet 14	
		S/Lt Mateitch (Gunner)	510		
		Sgt Vilon (Pilot)	510	Breguet 14	
		Cprl Reydel (Gunner)	510		
-		Lt Jacquemart Georges (Pilot)	510	Breguet 14	Fighter
		MdL Hérault (Gunner)	510		
		Mdl Chappaz Vincent (Pilot)	510	Breguet 14	
		S/Lt Mateitch (Gunner)	510		
		Sgt Vilon (Pilot)	510	Breguet 14	
		Cprl Reydel (Gunner)	510		

Victory Summary Leading Pilots & Gunners					
Rank	**Name**	**First Name**	**Claims**	**Confirmed**	**Notes**
S/Lt	Costes	Dieudonné	9	6	
S/Lt	Sauné	Basile	5	5	
S/Lt	Astor	Henri	4	4	As gunner
Lt	Thirouin	Marcel	4	3	
Sgt	Andrillon	Paul	4	1	
S/Lt	Terme	Alain	3	3	

Militaire on Front d'Orient

Place	Source	Confirmation	Notes
Hudova airfield	Etat des avions ennemis abattus	Citation OA(MM) 27-09-1918 - JORF 07-01-1919 p.252	Photo of the wreck
N.E. Guevgueli	Etat des avions ennemis abattus	-	Seen spinning. According to the combat report, MdL Maurice Lashermes was here instead of Cne Perrin.
		Citation OA(LH) 14-11-1918 - JORF 04-04-1919 p.3480 "6th victory"	
		-	
N.E. Guevgueli	Etat des avions ennemis abattus	-	Seen spinning. According to the combat report, MdL Maurice Lashermes was here instead of Cne Perrin.
		-	
		-	
Resna	Etat des avions ennemis abattus	-	Wreck found by French troops
		-	
	Etat des avions ennemis abattus	-	Seen spinning
Trojaci	Etat des avions ennemis abattus	Citation OA 20-11-1918 - JORF 27-04-1919 p.4361	Seen spinning
		-	
		-	
		-	
		-	
		-	
Trojaci	Etat des avions ennemis abattus	-	Seen spinning
		-	
		-	
		-	
		-	
		-	

Victory Summary Leading Pilots & Gunners (continued)

Rank	Name	First Name	Claims	Confirmed	Notes
S/Lt	Nast	Robert	3	3	
S/Lt	Miletic	Miodrag	3	2	
MdL	Lashermes	Maurice	3	2	
MdL	Hébert	René	3	1	+2 confirmed victories in France
Lt	Bonnetête	Louis	3	1	
Lt	Delrieu	Louis	3	0	

French Losses

Date	Pilot (P) or Observer (O)	Unit	Plane Piloted	Cause
1914				
23-Oct-14	LV Destrem Marcel (P), OK	Escadrille Monténégro	Nieuport hydroplane N6	Engine problems
31-Dec-14	QM Grall Hervé (P), WIA	CAM Port-Saïd	Nieuport hydroplane N13	Engine problems
	Cne Sterling Walter Francis (O - British), OK			
1915				
27-Jan-15	QM Le Gall Jean-Marie (P), KIA	CAM Port-Saïd	Nieuport hydroplane N14	Engine problems
	Lt Partridge (O - British), KIA			
18-May-15	S/Lt Beausire de Seyssel (-de) Pierre (P), OK	524 - MF 98 T	MF XI	Engine problems
	EV Delaye (O), OK			
13-Jun-15	S/Lt Saint-André Marcel (P), OK	524 - MF 98 T	MF XI n°461	Engine problems
	S/Lt Pasquet (O), OK			
6-Aug-15	Sgt Dumas Henri (P), OK	524 - MF 98 T	MF XI n°459	Flak
	Cne Aubert (O), OK			
25-Aug-15	Sgt Dumas Henri (P), KIAcc	524 - MF 98 T	MF XI	Accident
	S/Lt Saint-André Marcel (O), KIAcc			
31-Aug-15	Sgt Garsonnin Louis (P), WIA	524 - MF 98 T	MF XI	Out of gas
	S/Lt Allier (O), OK			
1-Oct-15	Adj Brézillon André (P), OK	524 - MF 98 T	MF XI	Engine problems
	Lt Masnou François (O), OK			
11-Oct-15	QM Trouillet (P), POW	CAM Port-Saïd	Nieuport hydroplane N14	Engine problems
	Lt Paul (O - British), POW			
22-Dec-15	LV Barthélémy de Saizieu (P), POW	CAM Port-Saïd	Nieuport hydroplane N17	Engine problems
	2nd Lt Ledger H.M. (O - British), KIA			
27-Dec-15	S/Lt Grosourdy de Saint-Pierre (-de) Guy (P), OK	524 - MF 98 T	MF XI	Engine problems
	Lt Lecompte Jules (O), OK			

Front d'Orient

Place	Notes
1914	
Scutari Lake (Montenegro)	Crashed on landing – Plane destroyed
Ghadian (Palestine)	Crashed in desert, both man returned to British ship *Minerva* in Aquaba gulf
Today Yotvéta, Israël	
1915	
Bir El Abd (Sinaï)	Crew killed by mistake by British troops while gaining Allied lines in Suez canal. Plane abandonned and recovered by British boat.
On sea, Cape Hellès	Crew saved by Allied tug.
On sea, 7 km cape Hellès	Crew and plane saved by allied destroyer.
On sea, Tenedos / Asian coast	Crew saved by fishing boat.
Mudros	Crash after takeoff.
Sulva	Force landed, plane destroyed on ground by enemy artillery.
Rabbit islands	Plane crash-landed, crew unhurt.
Beer-Sheba (Palestine)	Force-landed in desert, plane destroyed and crew captured.
Beer-Sheba (Palestine)	Crash landed in desert. British observer killed on ground by turkish troops.
On sea, Sarros Gulf	Saved by British tug.

French Losses

Date	Pilot (P) or Observer (O)	Unit	Plane Piloted	Cause
1916				
6-Jan-16	Lt Lecompte Jules (P), KIA	524 - MF 98 T	MF XI n°942	Aerial combat
28-Feb-16	Cpt Césari Antoine (P), OK	524 - MF 98 S	MF XI bis	Engine problems
6-Mar-16	MdL Marseille Robert (P), WIAcc	503 - MF 86	MF XI n°939	Accident
	Cprl Guéranne Georges (O), WIAcc	501 - MF 83		
23-Mar-16	Sgt Girard Marcel (P), KIA	506 - V 90 GBO	Voisin canon	Flak
	Soldat Misrachi Humbert (O-Greek), KIA			
23-Mar-16	Adj Costantini Pierre (P), OK	506 - V 90 GBO	Nieuport 10	Flak
	S/Lt Ferret Préjelan René (O), OK			
24-Mar-16	Cne Serre (-de) Gaston (P), WIAcc	501 - V 83 GBO	Voisin LAS	Engine problems
	S/Lt de Montvel (O), OK	Nieuport detachment		
24-Mar-16	S/Lt Gomberg Alexandre (P), OK	506 - V 90 GBO	Voisin LAS	Enemy fighter
	Lt Richepin Jacques (O), WIA			
23-May-16	Adj Dieval Léon (P), WIA - DOW	504 - MF 88		?
29-May-16	Lt Delauney Georges (P), OK	503 - MF 86	Farman	Accident
	S/Lt Boyer (O), KIAcc			
21-Jun-16	MdL Bernarjuzan Jean (P), WIA	505 - C 89	Caudron G 4	Air Combat
	OA 2e cl Vaisse Jules (O), OK			
26-Jul-16	Sgt Leblond Germain (P), WIA-DOW	505 - C 89	Caudron G 4 n°1495	Flak
	OA 2e cl Vaisse Jules (O), KIA	505 - C 89		
2-Aug-16	Cprl Guéranne Georges (P), KIAcc	501 - MF 83	Farman F 40 n°3048	Accident
	Sol Roodt Henri (O), KIAcc			
14-Aug-16	S/Lt Chambaret Emile (O), KIAcc	505 - C 89	Caudron G.4	Accident
27-Aug-16	MdL Pécrus Marcel (P), KIAcc	503 - MF 86	Farman MF 11 bis n°1228	Accident during test flight
	Sgt Desmoulin (O), KIAcc	235e Infantery Regiment		
16-Sep-16	MdL Bruninghaus Emile (P), OK	501 - F 383	Farman	Air combat
	Lt Rouget (O), WIA			
16-Sep-16	Lt Sarda Jean (O), KIAcc	505 - C 89	Caudron G.4	Accident
27-Sep-16	Adj Negrié Louis (P), OK	525 - MF 399	Farman	Air combat
	S/Lt Brusley Robert (O), KIA			
1-Oct-16	Cal Rouable Maurice (P),WIA-POW	522 - MF 384	Farman	Air combat (Bulgarian Fokker E)
	S/Lt Maoumovitch (O-Serbian), WIA-POW			
8-Oct-16	SM Drillet Edouard (P), KIA	CAM Corfou	Donnet-Denhaut 150 hp n°67 "C.12"	Flak
	QM Le Franc Casimir (O), KIA			

Front d'Orient

Place	Notes
1916	
On sea, Gallipoli	
On sea Mythilène / Mudros	Pilot saved by allied boat, plane lost.
Salonica Air Park	
Doiran lake	During a bombing raid on Volovec.
S. Doiran lake	Crew force-landed and walked to french lines after buring their plane.
Vatiluk	
Bulgarian lines	
Kavalla	During mission - Observer unknown, and unhurt.
Kasimli (auxiliary landing field)	Strong wind made plane crash during landing
Janès	
Doret-Hissar	Pilot WIA, died while landing his plane, observer dying in crash.
Samli	
Acret Hissar	Unknown circumstances.
Snevic	
Kenali / Florina	Plane damaged, landed with engine out of order
Around Salonica	
Ostrovo	Pilot brings back the plane home with his observer killed.
Near Sofia	
Otranto barrage	Shot down by Flak from U-35.

French Losses

Date	Pilot (P) or Observer (O)	Unit	Plane Piloted	Cause
28-Oct-16	S/Lt Lhuilier Julien (P), KIAcc	505 - C 89	Caudron G 4	Accident
30-Oct-16	EV Regnard (O), WIA	32e cie aérostiers	Balloon	Thunderstorm
30-Oct-16	Cne Petraskovic Radoslav (P-Serbian), WIAcc	522 - MF 384	Farman	Drunk...
2-Nov-16	Sgt Vermonet Abel (P), KIAcc	506 - N 390	Nieuport 21	Accident
4-Nov-16	Cprl Arcaute Jean (P), WIAcc	505 - C 389	Caudron G4 n°9106	Accident
	Lt Boutet de Montvel (O), WIAcc	505 - C 389		
19-Nov-16	Cprl Champy Louis (P), KIAcc	524- MF 398	Farman	Accident
19-Nov-16	Lt Théodoresco Jean (P), KIAcc	Salonica Air Park	Farman F 60	Accident on landing
	Sol Manche Jules (O), KIAcc			
19-Nov-16	MdL Lambert Robert (P), POW	507 - N 391	Nieuport 10	Accident on landing behind enemy lines
27-Nov-16	Cprl Ristić Rista (O-Serbian), KIA	524 - MF 398	Farman	Air combat
10-Dec-16	Sgt Guillot Pierre (P), KIA	507 - N 391	?	Air combat
	S/Lt Soulier (O), WIA-POW			
12-Dec-16	S/Lt Richard Jean (P), POW	Section de bombardement	Farman	Unknown causes
	MdL Léglise (O), POW			
13-Dec-16	Sgt Gernez Louis (O), OK	32e cie aérostiers	Balloon	Fighter attack
16-Dec-16	Cpt Gavrilović Mihajlo (P-Serbian), KIAcc	524 - MF 398	Farman	Accident
	Cpt Ristić Luka (O-Serbian), KIAcc			
17-Dec-16	MdL Du Lac Emile (P), WIA	506 - N 390	Nieuport 21	Air combat
22-Dec-16	S/Lt Perrody (O), OK	35e cie aérostiers	Balloon	Air combat
23-Dec-16	Cprl Battini Lucien (P), OK	524 - MF 398	Farman	Air combat
	S/Lt Dersoir (O), KIA			
23-Dec-16	Adj Stefanović (P-Serbian), OK	524 - MF 398	Farman	Air combat
	Cne Hadjera (O-Serbian), WIA			
26-Dec-16	Cal Dupon Pierre, (P) MIA-POW	522 - MF 384	Farman	Accident
30-Dec-16	Adj Freudenberg André (P), KIAcc	522 - MF 384	Farman	Accident
	Asp Gros Jean (O), KIAcc			

Front d'Orient	
Place	**Notes**
Rosna (E. Florina)	
	Observer broke his leg after his parachute jump - Time 08:15.
Kajmakcalan	Flies from Vrebeni to Sorovich to buy wine and food. Drunk, crashes on the return flight.
Florina	
Fort Karakorum	Crash during a ferry flight.
Near Vertekop	Mechanic expelled from the cabin.
Sedès	
S. Gradsko	Captured during a special mission, evaded in 1918.
Vertekop	
Hudova	Lt Bodo von Lyncker, Jasta 25 (1)
Bulgarian lines	Returning from a bombing mission in Livunovo
	Time 16:00.
Čuka	
SE. Monastir	Attack from 2 enemy fighters.
Elbasan (Albania)	Lost returning from a liaison flight in Salonica.
Vrbeni	

French Losses

Date	Pilot (P) or Observer (O)	Unit	Plane Piloted	Cause
1917				
4-Jan-17	Unknown observer	35e cie aérostiers	Balloon	Fighter attack
9-Jan-17	Sgt Jefremovic Miodrag (P-Serbian), POW	524 - MF 398	Farman F.40 n°3131	Became lost
	Lt Mitrovic Borivoje (O-Serbian), POW			
9-Jan-17	MdL Figuier Louis (P), WIA	505 - C 389	Caudron G.4	Air combat
	Cpt Bizé (O), OK			
4-Feb-17	Unknown observer	35e Cie	Balloon	Fighter attack
5-Feb-17	Cprl Henri Escoffier (P) KIA	503 - F 386	Farman	Air combat
	Lt Mayer Léon (O), KIA			
7-Feb-17	S/Lt Sisson Jean Charles KIAcc	503 - F 386	Farman	Accident
	Sol Renault Maurice KIAcc			
11-Feb-17	Lt Perrin de Brichambault Pierre (P), WIA	525 - F 399	Farman	Flak
11-Feb-17	Lt Tomic Miodrag (P-Serbian), OK	521 - F 382	Nieuport	Accident
12-Feb-17	Cprl Gally Marcel (P), KIAcc	505 - F 389	Farman F 60	Engine problems
	Cprl Lemaitre (O), WIAcc			
12-Feb-17	LV Lefranc Henri (P), WIAcc	CAM Salonica	Donnet-Denhaut 150 hp n°40	Crash on sea-landing
17-Feb-17	S/Lt Spiridions Lucien (P), MIA	503 - F 386	Farman	Flak
18-Feb-17	Mdl Lemut Alfred (P), KIA	502 - F 385	Nieuport	Collision in combat
19-Feb-17	Sgt-major Pavlović Ljubiša (P-Serbian), KIAcc	524 - F 398	Farman	Accident
	Sgt Michaud Henri (O), KIAcc			
25-Feb-17	S/Lt Gaudu Charles (P), KIA	502 - 385	Nieuport	Air combat
26-Feb-17	Sgt Rotival Louis (P), WIAcc-DOW	523 - N 387	Nieuport	Accident
26-Feb-17	Lt du Barry (O), WIA sv	502 - 385	-	Enemy bombing
	Sgt Sauné Basile (P), WIA lt			
27-Feb-17	Lt Tomic Miodrag (P-Serbian), OK	523 - N 387	Nieuport 10	Accident
13-Mar-17	MdL Verdier Henri (P), WIA	503 - F 386	Farman	Air combat
	Lt Joly (O), OK			
18-Mar-17	Cprl Bignon René (P), KIA	503 - F 386	Farman	Air combat
	S/Lt Ledré Marcel (O), KIA			
19-Mar-17	Lt Lebouc Jacques (P), WIA sv	506 - N 390	Nieuport	Air combat
26-Mar-17	Adj Murkiévitch Janko (P-Serbian), WIAcc	524 - MF 398	Farman	Accident
	Lt Giraudot Maurice (O), WIAcc-DOW			
31-Mar-17	QM Maurice Julien (P), KIAcc	CAM Corfou	FBA 150 hp (Type H) n°369	Crash-landing
	EV1 Lanusse Félix (O), KIAcc			

Front d'Orient	
Place	**Notes**
1917	
	Lt Otto Brauneck, FFA 69, flamed the balloon.
Around Drama	
	Lt Otto Brauneck, FFA 69 (5) - 2 fighters were attacking
Negocani	16h45, Obs escaped in parachute.
N.O. Mogila	Lt Otto Konnecke, Jasta 25 (1).
	Lightly wounded during a raid over Besista and the monastery of Melinica.
	Plane destroyed on landing.
Near Dragomanci	
Salonica	
Bulgarian lines	
Dzeovo, near railway	Collision during a dogfight with 5 german planes - killed Lt Von Lucke.
Vrbeni	Propeller broke in flight, plane crashed.
Guevgueli	Vfwb Trotzki, FA 34
Near Lerin	DOW March 1
Gorgop	21 german planes (among them 8 fighters) bombed Gorgop
	They destroyed 8 French planes and damaged 5 others, wounding 2 aviators
	Plane damaged on landing
Monastir	
Monastir / S. Orizari	Shot down by 3 enemy planes despite the escort of 2 Nieuports
Sakulevo	Plane crashed on landing
Corfou	

French Losses

Date	Pilot (P) or Observer (O)	Unit	Plane Piloted	Cause
4-Apr-17	Brig Camp Georges (P), WIAcc Lt	504 - F 388	Farman	Accident
	S/Lt Valzer (O), WIAcc Lt			
5-Apr-17	Cne Voukossavlievic Branko (P-Serbian), WIA Lt	523 - N 387 Vrbeni det.	Nieuport	Air combat
9-Apr-17	Cpt Martinet Robert (P), KIAcc	Air Park 302	Unknown Two-seater	Accident
15-Apr-17	Adj Géalot Louis (P), KIAcc	Air Park 302	Farman F 60	Accident
29-Apr-17	S/Lt Cousinéry Marcel (P), KIA	507 - N 391	Nieuport	Air combat
30-Apr-17	Lt Tomic Miodrag (P), OK	523 - N 387	Nieuport	Accident
9-May-17	Sgt Gernez Louis (O), KIA	32e cie aérostiers	Balloon	Fighter attack
10-May-17	S/Lt Miletic Miodrag (P-Serbian), OK	523 - N 387	Nieuport	Flak
16-May-17	Cne Voukossavlievic Branko (P-Serbian), WIA Sv	523 - N 387 Vertekop det.	Nieuport	Air combat
21-May-17	Adj Roth Fridrich (P, Serbian), KIAcc	523 - N 387	Nieuport	Accident
29-May-17	EV1 Olgati Pierre (P), MIA	CAM Salonica	FBA 150 hp (Type H) n°352	Unknown causes
31-May-17	Lt Allain-Launay Yves (P), KIA	524 - F 398	Farman F.40	Air combat
	Sgt Marc Raoul (O), KIA			
1-Jun-17	MdL Prokić Aleksandar (P-Serbian), KIAcc	524	Nieuport	Accident
20-Jun-17	Cprl Bonnet Joseph (P), KIAcc	503	Nieuport	Accident
20-Jun-17	Cprl Marturé Camille (P), POW	508	Nieuport	Air combat
25-Jun-17	Cprl Dalzon Eugène (P), OK	508	Farman	Air combat
	Sol Moreau Antoine (O), WIA-DOW			
30-Jun-17	MdL Landrin Jean (P), KIA	506	Nieuport	Air combat
2-Jul-17	Sgt Thamin Albert (P), KIA	501	Farman	Air combat
	Asp Gelin Félix (O), KIA			
2-Jul-17	Sgt Ducher Philibert (P), WIA	501	Farman	Air combat
	Sol Catherine (O), WIA			
5-Jul-17	Cprl Pellé Fernand (P), KIA	503	Farman	Air combat
	Sol Montbrun Joseph (O), KIA			
16-Jul-17	Adj Chan Henri (P), WIAcc	508	Farman	Accident
30-Jul-17	Cne Perrin de Brichambault Pierre (P), OK	525	Farman	Flak
	Sgt Perrin de Brichambault Georges (O), OK			
2-Aug-17	Sgt Bonsergent Gabriel (P), WIAcc	525	Farman	Accident
	Sol Salles Marius (O), KIAcc			
2-Aug-17	Sgt Grillet Joseph (P), OK	525	Dorand AR	Accident

Front d'Orient

Place	Notes
	Lightly wounded in cheek, several bullets in airplane
Mikra	Killed with aBritish passenger
Sedès	
Skočivar	Attacked by 4 enemy fighters
	Plane damaged on landing
Kravari	Vfw Treptow, Jasta 25 (1)
Bach	Forced to land, plane destroyed
	Seriously wounded in the shoulder, lands east of Brod
Vertekop	
Lost on sea	
Monastir, Cerna river, near Caniste	Time 8:20 - Obltn Friedrich-Karl Burckhardt, Jasta 25 (2)
Vertekop (Kamenita Kosa)	Killed while performing acrobatics
Kénali, near Bach	
Road to Prilep, Bulgarian lines	
N.W. Orle	
E. Kravari	
Puturos	Raid over Vasareica - Ltn Rose, Jasta 25
Beranci	Raid over Vasareica
Bulgarian lines	Attacked by 3-4 fighters, fell behind enemy lines
S. Bach	Landed in the night after a bombing raid, plane destroyed
Allied lines	Plane destroyed, crew safe
Mikra	Accident in the Air Park
Mikra	Plane destroyed, pilot safe

French Losses

Date	Pilot (P) or Observer (O)	Unit	Plane Piloted	Cause
7-Aug-17	QM Le Coz Alex (P), KIAcc	CAM Salonica	FBA 150 hp (Type H)	Crash during sea-landing
	SM Guyardeau Charles (O), KIAcc			
21-Aug-17	Cne Perrin de Brichambault Pierre (P), WIA	525	Farman	Accident after air combat
	Sgt Perrin de Brichambault Georges (O), WIA			
10-Sep-17	S/Lt Rajicevic Stevan (P- Serbian), OK	521	?	Flak
	French observer (O), OK			
12-Sep-17	Lt Marinkovitch Michel (P), WIA sv	523 (CO)	Nieuport	Air combat
11-Oct-17	Sgt Grillet Joseph (P), KIA	522	Farman or Dorand AR	Flak
	Cpt Antić Aleksandar (O-Serbian), KIA			
14-Oct-17	Sgt Passini André (P), KIAcc	521	Nieuport	Accident
16-Oct-17	Sgt Laffon Lucien (P), MIA-POW	506	Nieuport	Air combat
16-Oct-17	Sgt Parailloux Joseph (P), OK	505	Dorand AR	Air combat
	Soldat Charles Desprez (O), WIA Sv			
16-Oct-17	MdL Lhermitte (-de) Anatole (P), WIA	505	Dorand AR	Air combat
	Sgt Flahaut (O), WIA			
27-Oct-17	Sgt Rouyé René (P), WIA Lt	504	Nieuport	Air combat
28-Oct-17	Sgt Stefanović Siniša (P-Serbian), POW	523	Nieuport	Air combat
11-Nov-17	SM Allemand Auguste (O), KIAcc	CAM Salonica	FBA 150 hp (Type H)	Accident
14-Nov-17	Lt Vignoud Marcel (P), WIA Sv	502	Dorand AR	Air combat
	Lt Campredon Elie (O), WIA-DOW			
14-Nov-17	S/Lt Lapeyre Paul (P), WIA	522	Dorand AR	Air combat
	Lt Lebrun (O), OK			
15-Nov-17	MdL Le Pitancier de Montigny René (P), KIA	524	Dorand AR	Air combat
	MdL Cochennec Jean (O), KIA			
15-Nov-17	Veysset René (P), KIA	505	Dorand AR	Flak
	Asp de la Bastide Martin (O), KIA			
20-Nov-17	Sgt Vukotić Savo (P-Montenegrin), MIA	521	Dorand AR	Air combat
	S/Lt Caudy Pierre (O), KIA			
24-Nov-17	S/Lt Perrody, Pierre (O), WIA-DOW	35e Cie aérostiers	Balloon	Fighter attack
25-Nov-17	MdL Tampier Joseph (P), WIAcc-DOW	-	Dorand AR	Accident
28-Nov-17	MdL Lefresne Pierre (P), KIA	507	Nieuport	Air combat
3-Dec-17	Cprl Cardi Albert Titus (P), POW	Air Park	Nieuport	Became lost
29-Dec-17	Matelot Boriès Louis (P), KIAcc	CAM Platéali	FBA 150 hp (Type H) n°497	Accident
	Matelot Fauvet Jean (O), KIAcc			

Front d'Orient	
Place	**Notes**
Salonica	
	Their plane was damaged after air combat, and crashed on landing
	Plane broken on landing
Serbie	Severely wounded during a crash on landing, returning from a fighter mission.
Gradsko	
Vertekop	During a test flight
Enemy lines	
Mission over Prilep	Attacked by 4 enemy aircraft
Around Starovina	
	Wounded on left arm while monitoring french artillery
	Killed in 1918
Salonica	Pilot unknown
French lines	Force to land in French lines
NW. Kojuh	Wounded on right foot, comes back with 42 bullets in plane
Zovica	Attacked by 2 enemy fighters
Dedebal (enemy lines)	Raid over Kanatlarci
NW. Mogina, Negorci / Gjevgueli	Lt Otto Spielberger, Jasta 25
secteur du Vardar	Lt Otto Spieltberger, Jasta 38 (5). Burned by his balloon. DOW in Gumendzé
Otranto (Italy)	During a ferry flight from the Otranto suppluy base to Greece
Stoyakovo (W. Doiran lake)	VfW Lage, Jasta 38 (1)
Near Xanthi	
Platéali	

French Losses

Date	Pilot (P) or Observer (O)	Unit	Plane Piloted	Cause
1918				
21-Jan-18	Sgt Maître d'Hôtel Jules (P), WIAcc-DOW	508	Nieuport	Accident
30-Jan-18	Lt Beaurepaire Charles (P), KIA	509	Nieuport	Air combat
21-Mar-18	Lt Rose Yvon (P), KIAcc	510	Breguet 14	Accident (?)
	S/Lt Photiades (O-Greek), WIAcc			
5-Apr-18	Brig Montoya Léopold (P), KIA	504	Nieuport 24 bis n°4047	Air combat
12-Apr-18	Lt Defferre Henri (P), KIAcc	505	Nieuport	Accident
21-Apr-18	Cal Beer Paul (P), KIAcc	488	Caudron G.4	Accident
	Sgt Lavauzelle Edmond (O), WIAcc			
24-Apr-18	Lt Pivain Fernand (P), KIA	506	Nieuport	Air combat
25-Apr-18	Cprl Lefevre Charles (P), KIA	506	Nieuport	Air combat
25-Apr-18	Unknown observer (OK)	?	Balloon	Fighter attack
4-May-18	Sgt Rolland René (P), KIA	508	Dorand AR	Air combat
	Lt Noyer Robert (O), KIA			
7-May-18	Sgt Courant Charles (P), KIA	525	Nieuport	Air combat
14-May-18	Matelot Leloup Marcel (P), KIAcc	CAM Platéali	FBA 150 hp n°573 "G.4"	Accident
	EV1 Verdelhan des Molles (-de) Louis (O), KIAcc			
22-May-18	Brig Lagache Victorien (P), WIA	501	Dorand AR	Air combat
	S/Lt Guinard (O), WIA			
29-May-18	Sgt Ralli Dimitri (P-Greek), KIA	532	Dorand AR	Flak (friendly fire)
	Asp Laillet René (O), KIA			
30-May-18	Sgt Gollion Joseph (P), WIA	507	Nieuport	Flak (friendly fire)
1-Jun-18	Brig Olibe Marcel (P), POW	501	Dorand AR	Unknown causes
	Lt Noir (O), POW			
20-Jun-18	S/Lt Sauné Basile (P), KIA	531	SPAD VII	Air combat
27-Jun-18	MdL Rouzier Roger (P), KIAcc	Air Park	?	Accident
28-Jun-18	Cprl Boucaut Joassy (P), KIA	505	Dorand AR	Flak
	Lt Grenier de la Tour Franck (O), KIA			

Front d'Orient

Place	Notes
Sakulevo	DOW in Florina
Moglia / Monastir	Vfw Gerhard Fieseler, Jasta 25 (2)
N. Negocani near Florina	Rose was the commanding officer of escadrille 510
4 km WSW Canista, enemy lines	Vfw Gerhard Fieseler, Jasta 25 (3)
Gorgop	Accident during landing
Kourtesis	
Presba Lake	
Gramsi	
	Vfw Heinrich Schott, Jasta 25 (1)
Suhodol (16 km E. Monastir)	Vfw A. Bauhofer, Jasta 25 (4)
W. Gradesnica (enemy lines)	Vfw E Meyer, Jasta 25 (2)
Platéali	
W. Presba lake	Ltn F Köhler, Jasta 25 (1). Time 9:00. Forced landing in Stenja, near Presba lake.
Skra di Legen	
Skra di Legen	Foot cut by a direct hit from Flak.
Koryza sector	Missing in action - Germans drop a message telling they are alive
N.E. Voprecani	Vfw Gerhard Fieseler, Js 25 (-) - Victory officialy credited to Ltn Rose
Sedes	
Isvor (10 km SW Guevgueli)	

Right: Jasta 25 lineup of Roland D.II fighters in Macedonia.

Facing Page: Kaghol 1 was composed of units equipped with a number of twin-engine bombers from AEG, Friedrichshaften, Gotha, and Rumpler. Here are three AEG G.III shown on an airfield in Macedonia.

French Losses

Date	Pilot (P) or Observer (O)	Unit	Plane Piloted	Cause
4-Jul-18	MdL Maud'hui Jean (P), KIA	508	Nieuport	Air combat
5-Jul-18	Mdl Marie Eugène (P), POW	503	Nieuport 24 n°4487	Air combat
17-Jul-18	Lt Bordes Michel (P), KIAcc	507	?	Accident
28-Jul-18	S/Lt Lebouc Jacques (P), MIA	506	Bréguet 14	Air combat
	Cprl Baville Roger (O), MIA			
31-Jul-18	MdL Jean Boyer (P), WIA Lt	509	Nieuport	Air combat
?-jul-18	Lt Kakašvil Isidor (P, Russian), KIAcc	521	Dorand AR	Accident
	Sgt Božić Božidar (O, Serbian) KIAcc			
4-Aug-18	Lt Houtmann Adrien (P), KIA	503	Nieuport	Air combat
11-Aug-18	Cprl Poirier Daniel (P), KIA	524	Nieuport	Air combat
16-Aug-18	Cprl Alban Georges (P), WIA Lt	504	AR	Air combat
	MdL Basle (O), OK			
19-Aug-18	Unknown pilot, OK	504	Bréguet 14	Air combat
	Lt Tourné Jean Marc (O), WIA-DOW			
23-Aug-18	Sgt Coucurou Romain (P), KIA	524	Nieuport	Air combat
23-Aug-18	QM Potiron Eugène (P), KIAcc	CAM Platéali	Donnet-Denhaut 150 hp n°681 "G.16"	Accident
	EV2 Guelard Louis (O), KIAcc			
4-Sep-18	MdL Sprauel Louis (P), KIA	524	Dorand AR	Air combat
	Lt Boutiny (-de), Ernest (O), KIA			
10-Sep-18	Sgt Andrillon Paul (P), OK	531	SPAD VII	Air combat
11-Sep-18	Sgt Baudin Georges (P), KIAcc	502	SPAD VII	Accident on takeoff
13-Sep-18	Sgt Bonin Raymond (P), KIA	508	Nieuport	Air combat
14-Sep-18	Cprl Pinay Raymond (P), KIAcc	505	Dorand AR	Accident on landing
	S/Lt Mauberque Henri (O), KIAcc			
15-Sep-18	MdL Meignen Fernand (P), OK	523	Breguet 14	Flak
	Lt Bremond (O), WIA			
15-Sep-18	Adj Raingo Pelouse Ernest (P), OK	524	Dorand AR	Flak
	Lt Nicot Albert (O), OK			
15-Sep-18	Adj Leblond René (P), WIA-DOW	524	Nieuport	Air combat
16-Sep-18	Adj Leman Henri (P), WIA-DOW	506	?	Air combat
16-Sep-18	Brig Nugues René (P), MIA	524	Nieuport	Air combat
18-Sep-18	Adj Perreau Jean (P), MIA	506	Nieuport	Air combat
18-Sep-18	Adj Rasting Pierre (P), WIA Lt	524	Breguet 14	Flak
	Cne Michaud (O), WIA Lt			
19-Sep-18	Unknown pilot, OK	508	Breguet 14	Air combat
	Lt Fabron Georges (P), WIA			

Front d'Orient

Place	Notes
W. Canista (5km W. Razimbrej bridge)	Vfw Reinhard Treptow, Jasta 25 (6)
Cerna river loop	Vfw Gerhard Fieseler, Jasta 25 (5) - Germans drop message in french lines
?	
Kanatlarci, 13 km SW. Prilep	Vfw Gerhard Fieseler, Jasta 25 (8)
	Wounded by a bullet in the leg during an aerial combat.
Vertekop	
Orle (23 km ENE Monastir)	Vfw Gerhard Fieseler, Jasta 25 (9) - Attacked by 2 german fighters
Gradenisca	Vfw Gerhard Fieseler, Jasta 25 (11)
N. Negocani, near Pozdès	Vfw Gerhard Fieseler, Jasta 25 (12) - French plane forced to land in allied lines
Prilep / Kanatlarci	Plane undamaged
E. Monastir - 2,5 km E. Gradenisca	Vfw Gerhard Fieseler, Jasta 25 (13) - Falled in a place called "Murailles grises"
Patras gulf	
Confluent Belavoda / Lyechnitza	Vfw Gerhard Fieseler, Jasta 25 (14)
10 km WSW Doiran	Forced to land in no-man's-land, returns to French lines
Vertekop	Detached to escadrille 523
Staravina ravine	Vfw Gerhard Fieseler, Jasta 25 (15)
Vertekop	Strong wind caused crash
Front 2e Serbian army	
Grunitza (E. Budimirci)	9:00 - Plane destroyed on the ground by enemy artillery
Razim Bey bridge	Forced to land S. Budimirci, where he dies from his wounds
Sakulejvo	DOW on September 23.
Razim Bey bridge	Killed in Gradelitza
Cerna river loop	Vfw Gerhard Fieseler, Jasta 25 (18)
Pandeli Potok	Plane destroyed after forced landing
Cerna river loop	Observer right leg amputated following his injury

French Losses

Date	Pilot (P) or Observer (O)	Unit	Plane Piloted	Cause
20-Sep-18	Sgt Mariau Edouard (P), OK	508	Nieuport	Flak
20-Sep-18	Sgt Michel Léon (P), MIA	510	Breguet 14	Air combat
	Sol Annereau Edmond (O), MIA			
20-Sep-18	Adj Mamalarkis Clearchos (P-Greek), WIA	532	Dorand AR	Flak
22-Sep-18	Cprl Portal Roger (P), OK	532	Dorand AR	Flak
	Lt Valentin Pierre (O), KIA			
27-Sep-18	MdL Meniole de Gizaucourt Guy (P), POW	508	Breguet 14	Engine problems
	S/Lt Million-Rousseau Lucien (O), POW	504		
13-Oct-18	QM Aubaud Jean-Marie (P), KIAcc	CAM Platéali	FBA 200 hp (Type S) n°1040	Accident
16-Oct-18	QM Michelet Yves (P), WIAcc	CAM Corfou	FBA 200 hp (Type S) n°1093	Accident
	SM Morin Georges (O), KIAcc			
24-Oct-18	SM Boucher André (P), MIA	CAM Corfou	Tellier cannon 200 hp n°63 "O.23"	Unknown causes
	Matelot Gérard Roger (O), MIA			
6-Nov-18	Sgt-Maj Vavasseur Marcel (P), WIAcc Lt	Air Park	Breguet 14	Accident
7-Nov-18	Adj Souleguieu Jean-Louis (P), KIAcc	504	?	Accident
9-Nov-18	Sgt Piveteau Louis (P), KIAcc	524	Breguet 14	Accident
	Sol de Ghynst (O), KIAcc			
1919				
31-Mar-19	Brig Fauveau René KIAcc	523	SPAD VII	Accident
25-Jun-19	Lt Cochet-Forthuny Frédéric (P), KIAcc	510	SPAD VII	Accident
9-Aug-19	Lt Gaborit Robert (P), KIAcc	523	Breguet 14	Accident
	S/Lt Marguerite Pierre (O), KIAcc			
5-Sep-19	S/Lt Terreaux Marcel (P), KIAcc	505	Breguet 14	Accident
	Brig Pichon Roger (O), KIAcc			

Note: 3 pilots died fromSpanish flu: Sol Bonnin Raoul (Escadrille 505, October 10, 1918), Lt Michaud Léon (Commanding

Left: Roland D.II fighter of the Bulgarian air service.

Front d'Orient

Place	Notes
Razim Bey bridge	Forced landing in Bechichte, plane broken. Walks back to his unit.
Trojarci	Vfw Gerhard Fieseler, Jasta 25 (19)
Dzena	Thight injury - Landed in Livadi
Kojnsko / Sermenli	Plane comes back to Gorgop, damaged
	Crew captured by bulgarian troops
Platéali	MIA, body found on October 20.
Corfou	
Adriatic sea	Crew disappeared on sea
Sedes	Crash on landing
Negocani	Accident on landing
Nish (Serbia)	Accident during test flight
Radotchet mountain (Serbia)	
Galatz (Romania)	
Szegedin (Hungary)	
San Stéfano (Istanbul)	

officer of escadrille 505, November 1, 1918,) and Cne Maire Louis (Commanding officer of Armée Française d'Orient, November 26, 1918).

Above: Roland D.III fighters on Levunovo airfield in 1918.

Right: Bulgarian Roland D.III fighter in a hangar.

Appendix 4: Order of Battle

Entente Air Units in Macedonian Front - August 1918

Commanding officer : Commandant Victor Denain

Commandement des Armées Alliées (C.A.A.)

Commanding officer : Capitaine Maurice Hamoir

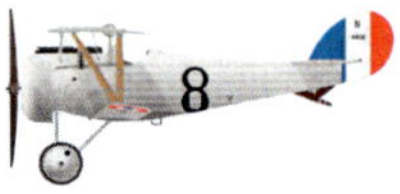
507

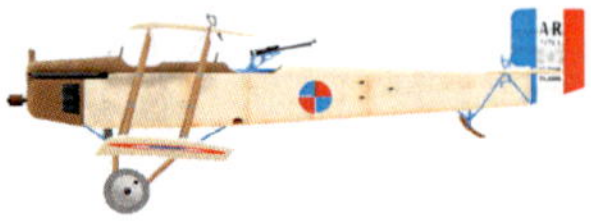
502

505

510

Armée Française d'Orient (A.F.O.) + Corpo Aeronautico Militare

Commanding officer : Capitaine Louis Maire — **CO : Capitano Mario Ajmone Cat**

506

501

503

504

73a Sq.

111a Sq.

508

509

Armée Serbe (A.S.)

Commanding officer : Commandant Geoffroy du Périer de Larsan

523

521 (1st Serbian)

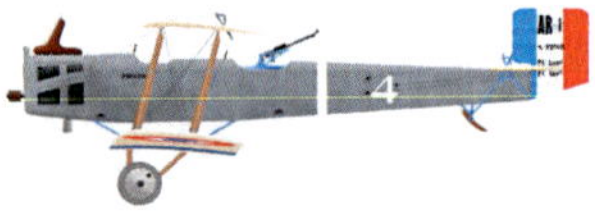
524

525

Armée Hellénique (A.H.)

Commanding officer : Commandant Marie de Thézy

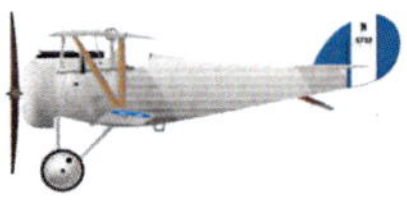
531

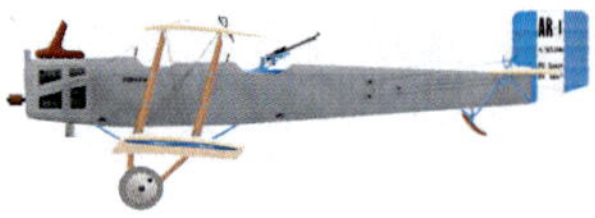
532

533

534

Royal Air Force - 16th Wing

Commanding officer : Lt Colonel Georges William Patrick Dawes

N°150 Sq.

N°17 Sq.

N°47 Sq.

Central powers units in Macedonia (1918)

Jasta 25

FA 20

FA 30

1st Bulgarian

Jasta 38

FA 34

FA 246

2nd Bulgarian

The comparison of the two charts shows the obvious numerical superiority of the air force of the Entente nations over that of the Central Empires on the Salonica front, which reached a proportion of 3 to 1 at the end of the war.

The chart also shows the overwhelming weight of the French air force, which also controls the Serbian and Greek air forces, providing them with all the equipment and most of the pilots. Major Victor Denain commanded all the Entente aviation on this front throughout the war, even having authority over the Italian aviation (two squadrons, one fighter and one reconnaissance) and the British aviation.

Not counting the seaplanes pilots in Greece, 666 French pilots were sent to the Orient front, representing 4% of the French pilots of WW1. 57 were killed in action and 44 by accident (15% of the total), a loss proportion equal to the French front.

Appendix 5: Overview of British and Italian Air Units on the Orient Front

Royal Flying Corps, Royal Navy Air Service then Royal Air Force

It is beyond the scope of this book to detail the British air force on the Salonica front, but it must be mentioned in order to put it into context.

The British troops were present in Salonica as soon as the front was formed in October 1915 and initially had no air cover apart from a few RNAS aircraft based on the island of Thasos which took part in a few engagements. The first air unit, N°17 Squadron, arrived from Egypt in July 1916, equipped with BE2c two-seaters and a few Airco DH2 and Bristol Scouts fighters.

When the Entente troops left the entrenched camp of Salonica in September 1916, the British army was assigned the eastern sector, facing the Struma River. N°17 Squadron was joined by N°47 Squadron at that time, with a heterogeneous fleet of aircraft made up mainly of Armstrong Whitworth F.K.3 and FK.8 two-seaters for observation, as well as BE 12s fighters, which also equipped N°17 Squadron. The BE 12 became the main British interceptor on the Orient Front throughout 1917. It was a heavy and unwieldy aircraft, withdrawn from the French front, but equipped with a powerful 150 hp engine, more powerful than the 80 hp of the French Nieuport 21. Also equipped with three machine guns, several pilots were successful on board, most notably Captain Ware Murliss-Green who claimed 5 of the 7 aircraft he destroyed in 1917.

In January 1918, N°17 and N°47 Squadrons received SE 5a fighters for their fighter element, which were grouped together in April 1918 to form N°150 Squadron, specialized in fighter aircraft, and also including a few Sopwith Camels and Nieuport 27s. On board this rather efficient fighter, comparable to the SPAD VII which was far from being used by all French fighter units, the British fighters multiplied their victories and claimed 40 in 1918, out of the 56 claimed by the British air force during the entire war on the Eastern Front as "Destroyed", plus some 40 extremely dubious "Out of control". This was a higher score than that of the numerically stronger French air force, but one must take into account the very lax nature of the certification of victories for the British, which was based solely on the testimony of the pilot and did not require confirmation by witnesses from outside the combat.

Several aces emerged in 1918, such as Major Gerald Ernest Gibbs (5 victories claimed against destroyed or captured aircraft, plus 5 "out of control"), Captain Acheson Gosford Goulding (Canadian, 5 victories plus 4 "out of control") and Lt Franklin Geoffrey Saunders (6 victories and 2 "out of control"), who reached the rank of ace by capturing a DFW C at Porna on January 22, 1918, which was the first victory to the credit of the SE5a's in Greece. The ace of the Entente's aces at Salonica was Canadian Captain Gerald Gordon Bell, with, according to French archives, 8 victories plus 3 "out of control", all obtained on SE5a, in addition to his 3 "out of control" victories obtained in France as a machine gunner.

It should also be noted that 26 British pilots from Squadrons 17, 47 and 150 were killed or captured in combat, very high losses which were significantly higher than those suffered proportionally by the French squadrons.

Corpo Aeronautico Militare

Finally, a few words to point out that Italy, which sent its 35th infantry division to Salonica in 1916. First without aviation (the French organized a Section Italienne on Farman to give them a minimal air cover), the local Italian forces received 47a Squadriglia in January 1917, flying Farman MF 11 and F.40 two-seaters, reenforced in June 1917 by a fighter section of 4 Nieuport 11s. After carrying out 422 missions, claming 1 unconfirmed victory and losing one aircraft in combat, the 47a Squdriglia returned to Italy in September 1917.

It was replaced by the 111a Squadriglia, flying on SAML two-seaters. This unit stayed in Macedonia until the end of the war, loosing 2 planes in combat and 2 by accident, but also claiming 2 victories.

Flighting alongside the SAML, the Nieuport 27 and SPAD VII of 73a Squadriglia came in Salonica on November 1917 and were sent in the sector of Monastir until the end of the war, claiming just one victory, without casualty.

Above: A BE 12 in Salonica.

Left: Captain Gilbert Ware Murliss-Green, seating on the Albatros C.VII he captured on January 14, 1917 – the 3rd of the 6 victories he scored in the Orient Front.

Bottom: View of the SAML of 111a Saquadriglia. (Paolo Varriale).

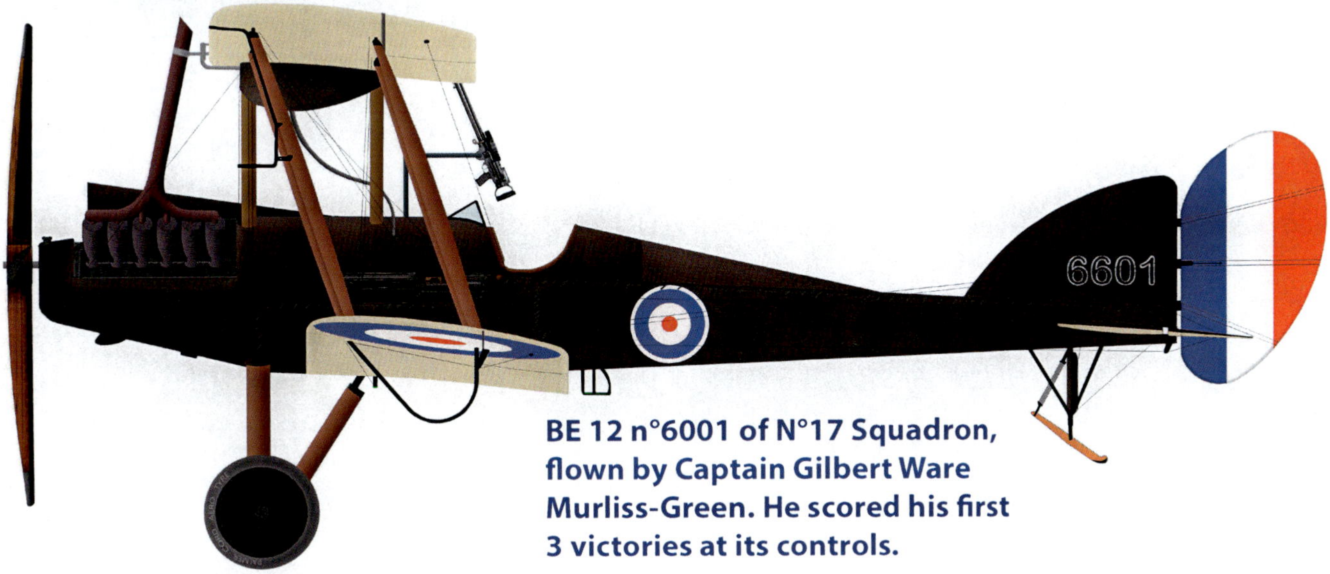

BE 12 n°6001 of N°17 Squadron, flown by Captain Gilbert Ware Murliss-Green. He scored his first 3 victories at its controls.

Appendix 6: Maps

BULGARIA
OTTOMAN EMPIRE
N
O
E
S
tric
Fort Rupel
Drama
Xanthi
Struma
ahana
Thasos
Orfano
nica
CAM Salonica (1916)
nomi (Escadrille cotière)
17)
ATHENS
Escadrille 534 bis
488
avarin
ject)
CAM Milo
(1918)
m

Above: View taken in September 1918 from the Turkish cemetery, showing the harbor of Salonika with allied warships. The triangular white tents on the left belong to one of the many military camps around the city. (Photo Fernand Cuville)

Below: Other view of the harbor by Fernand Cuville in 1918.

Above: A panorama of the city taken from the highest point, in the city wall. (Photo Léon Busy)

Below: Promenade along the narrow streets of Salonika, by photographer Léon Busy in 1918.

This Page & Facing Page: Promenade along the narrow streets of Salonika, by photographer Léon Busy in 1918.

Above & Below: Two more views of Salonica harbor.

Made in United States
North Haven, CT
04 November 2026

21725126R00217